Micro-foundations for Innovation Policy

The series 'Verkenningen' comprises studies commissioned by the WRR that are deemed to be of such quality and importance that their publication is desirable. Responsibility for the contents and views expressed therein remains that of the authors.

Scientific Council for Government Policy (WRR)
Lange Vijverberg 4-5
P.O. Box 20004
2500 EA The Haque
Tel. + 31 70 356 46 00
Fax + 31 70 356 46 85
E-mail: info@wrr.nl
Internet: http://www.wrr.nl

WRR

SCIENTIFIC COUNCIL FOR GOVERNMENT POLICY

Micro-foundations for Innovation Policy

B. Nooteboom and E. Stam (eds.)

Amsterdam University Press, Amsterdam 2008

Front cover illustration: Wassily Kandinsky, *Offenes Grün* (1923),
Roethel t. 11 n°. 704, p. 658

Cover design: Studio Daniëls, Den Haag
Layout: Het Steen Typografie, Maarssen

ISBN 978 90 5356 582 7
NUR 741 / 754

© WRR / Amsterdam University Press, The Hague / Amsterdam 2008

All rights reserved. Without limiting the rights under copyright reserved above, no part of this book may be reproduced, stored in or introduced into a retrieval system, or transmitted, in any form or by any means (electronic, mechanical, photocopying, recording or otherwise) without the written permission of both the copyright owner and the author of the book.

CONTENTS

Preface 11

Executive summary 13

Contributors 15

1 **Innovation, the economy, and policy** 17
Bart Nooteboom and Erik Stam
1.1 Introduction 17
1.2 Novelty, creativity, and innovation 18
1.3 Dimensions and measures of innovation 21
1.4 Innovation and the economy 25
1.5 What markets? 28
1.6 Innovation policy 37
1.7 Outline of the book 42

2 **Innovation and macroeconomics** 53
Gerard de Vries
2.1 Economic growth, productivity, and innovation 53
2.1.1 Introduction: The importance of economic growth 53
2.1.2 The contribution of innovation to economic growth 53
2.1.3 The contribution of labour to economic growth 56
2.1.4 Some figures 57
2.1.5 Conclusions 58
2.2 Productivity in the EU, US, and the Netherlands 58
2.2.1 International comparison 58
2.2.2 Explaining the productivity gap between the EU and US 62
2.2.3 Dutch innovative potential from an international perspective 64
2.3 Innovation in the market economy and innovation policy in the Netherlands 65
2.3.1 Innovation in the Dutch market economy 65
2.3.2 Innovation policy in the Netherlands 66
2.4 Conclusions 68
2.4.1 The urgency of reinforcing the growth potential of the Dutch economy 68
2.4.2 Macroeconomic analysis and innovation policy 69

3 **Learning, discovery, and collaboration** 75
Bart Nooteboom
3.1 Introduction 75
3.2 Evolution and learning 76
3.3 Cognition 80
3.4 Cycle of discovery 82

3.5 Market failures (with a vengeance) 89
3.6 Open innovation 92
3.7 Conclusions 96

4 Research, higher education, and innovation 103
Gerrit Kronjee and Bart Nooteboom
4.1 Introduction 103
4.2 The relationship between knowledge and innovation 103
4.2.1 Effects of knowledge generation 103
4.2.2 Science in an innovation system 105
4.2.3 Social relevance as a measure for acquisition and transfer of knowledge 108
4.3 Sources of finance 110
4.4 Research in the higher education system and innovation 111
4.4.1 Financing university research 111
4.4.2 Effect of the structure of education 113
4.4.3 Quality measurement of research and research policy 115
4.4.4 Programming and funding research in higher education 115
4.5 Research in the business sector and innovation 118
4.5.1 Fundamental research, the business community, and innovation 118
4.5.2 Sources of knowledge work 120
4.6 Specific innovative attempts in knowledge policy 121
4.6.1 European policy 121
4.6.2 Creation of technologically leading institutes (TTIs) 122
4.6.3 Stimulating the immigration of knowledge workers 123
4.7 A policy recommendation: Promoting the generation and utilisation of knowledge in 'third spaces' 125
4.7.1 Basic assumptions 125
4.7.2 A recommendation of 'third spaces' 126
4.8 Conclusions 129

5 Entrepreneurship and innovation 135
Erik Stam
5.1 Entrepreneurship and innovation 135
5.1.1 Entrepreneurship defined 135
5.1.2 Entrepreneurial opportunities 137
5.1.3 Entrepreneurship as an organisational product 139
5.2 Entrepreneurship, innovation, and economic growth 140
5.2.1 Knowledge spillovers 141
5.2.2 Experimentation 143
5.2.3 Decentralisation 143
5.2.4 Competition 144
5.2.5 Entrepreneurship and economic growth: Empirical evidence 145
5.3 Entrepreneurship in the Netherlands 147
5.3.1 International comparison 147
5.3.2 Entrepreneurship in the Netherlands over time 150

	5.3.3	Entrepreneurship and innovation in the Netherlands: Summary	154
5.4	Entrepreneurship and innovation policy		154
	5.4.1	New firm formation	156
	5.4.2	Technology start-ups	156
	5.4.3	Spin-offs and corporate venturing	158
	5.4.4	High growth start-ups	160
5.5	Summary		161
6	**Barriers to innovation**		173
	Leo van der Geest and Lars Heuts		
6.1	Introduction		173
	6.1.1	Economic and political losers	174
	6.1.2	Market failures and coordination problems	175
	6.1.3	Counterproductive policy	176
	6.1.4	Case studies	176
6.2	Blowing in the wind		177
	6.2.1	Netherlands must become a front runner	177
	6.2.2	Obstacles to wind energy	178
	6.2.3	Political windiness	180
	6.2.4	Conclusion	181
6.3	Waiting for the bus		182
	6.3.1	Advantages of the Whisper	182
	6.3.2	Bus stop	182
	6.3.3	Obstacles for the Whisper	183
	6.3.4	Conclusion	184
6.4	Virtual patient		185
	6.4.1	Advantages of the EPD	185
	6.4.2	Delayed progress	185
	6.4.3	Obstacles for the EPD	186
	6.4.4	Conclusion	190
6.5	Energy neutral houses		190
	6.5.1	From recycling house to energy neutral house	191
	6.5.2	Advantages of an energy neutral house	191
	6.5.3	Obstacles to energy neutral houses	191
	6.5.4	Energetic authorities?	192
	6.5.5	Conclusion	194
6.6	General conclusion		194
7	**Collaboration, trust, and the structure of relationships**		199
	Bart Nooteboom		
7.1	Introduction		199
7.2	Trust in whom and what?		200
7.3	Trust and control		201
7.4	Conditions and the role of government		204
7.5	Transparency and trust		207
7.6	Trust and network structure		210

7.7	Conclusions	213
8	**Innovation and organisation** *Bart Nooteboom and Robert Went*	219
8.1	Introduction	219
8.2	Deployment of people and the organisation of work	221
8.3	Ambidextrous organisation	225
8.4	Platforms	231
8.5	Innovation by lead users	233
8.6	Open source communities	235
8.7	Innovation in government services	239
8.8	Conclusions	242
9	**Innovation and creativity in organisations: individual and work team research findings and implications for government policy** *Neil R. Anderson and Rosina M. Gasteiger*	249
9.1	Introductory comments	249
9.2	Defining innovation at work	250
9.3	Antecedents and facilitators of innovation	251
	9.3.1 Individual-level antecedent factors: Individual and job characteristics	253
	9.3.2 Work group level antecedent factors	257
9.4	The process of innovation	258
9.5	Conclusions	262
10	**Inter-organisational networks and innovation** *Marius T.H. Meeus, Leon A.G. Oerlemans, Patrick Kenis*	273
10.1	Introduction	273
10.2	The importance of cooperation for innovation: Some empirical underpinnings	273
10.3	A framework linking innovation policy to effects of networks on innovation	275
10.4	The literature review	280
	10.4.1 Method of literature review	280
	10.4.2 Findings: Networks of inter-organisational relations and innovation	282
	10.4.3 Summary: Main findings in the literature	291
10.5	Innovation policy and networks: The policy section	297
10.6	Some concluding remarks	304
11	**Regional innovation policy** *Ron Boschma*	315
11.1	Introduction	315
11.2	Variety and regional development	316
11.3	System failures as basis for regional innovation policy	325
11.4	How to design regional innovation policy?	327

11.5	Policy options	331
11.6	Conclusions	333
12	**Conclusions for innovation policy: Opening in fours**	343
	Bart Nooteboom and Erik Stam	
12.1	Introduction	343
12.2	Perspectives	344
12.3	Targeted industrial policies	347
12.4	Failures of governments, markets, and systems	350
	12.4.1 Government failures	350
	12.4.2 Market failures	352
	12.4.3 System failures	354
12.5	Generation and utilisation of knowledge	356
	12.5.1 Valorisation	356
	12.5.2 Funding university research	358
12.6	Entrepreneurship	359
12.7	Organisation	361
12.8	Networks and regions	362

PREFACE

This edited volume serves as a background study for the WRR (Netherlands Scientific Council for Government Policy) advice on innovation policy in the Netherlands. This WRR-advice, which is released separate, emphasizes the importance of effective functioning of innovation as a key determinant of wealth creation and the role and responsability of the government for the functioning of the innovation system.

The Netherlands seems to lose momentum with respect to innovation. Before we can say anything about policy directions for innovation, we need to have an adequate understanding of how innovation works. To reach this understanding the WRR has asked several experts to conduct indepth studies on the different facets of innovation. These studies resulted in the notion of innovation as a system which needs further development of insight in what goes on in the interactions between individual and organizational actors, in the creation of ideas their application in innovations, and their diffusion. A central element of innovation is that it thrives on cooperation between actors that are at a cognitive distance that is sufficiently large to lead to a novelty, but sufficiently small in order to make collaboration possible between the actors involved.
This background-study also builds bridges in order to come to a better understanding of innovation and the foundation of the possible role of government policy. This book not only functions as a background study for the advice on innovation policy in the Netherlands. It aims to make a contribution to the international debate on innovation and innovation policy, especially in the European Union.

The volume brings together insights from WRR staff (Gerrit Kronjee, Gerard de Vries and Robert Went) and external staff from a range of disciplines such as economics, sociology, geography and psychology. We (the WRR) are indebted to Neil Anderson, Ron Boschma, Rosina Gasteiger, Leo van der Geest, Lars Heuts, Patrick Kenis, Marius Meeus and Leon Oerlemans for the extensive scope of their research, productive questions and comments during the preparation and execution of this project and the meetings they were willing to attend.

Prof.dr. W.B.H.J. van de Donk
WRR Chairman

EXECUTIVE SUMMARY

This volume moves beyond macroeconomic conditions for innovation to micro-level processes involving individual and organisational agents and their interactions in the generation, utilisation, and diffusion of ideas, products, and processes. We use the concept of 'innovation systems', to take into account interactions between a variety of agents, various dimensions of innovation – going far beyond science and technology– and a variety of economic and institutional conditions. The analysis applies to innovation systems in general and gives an application to the Netherlands.

The main line in our policy approach is a trade-off between arguments for and against policy interventions. One reason for restraint in policy derives from the insights of Austrian economics (Hayek) that knowledge is diverse and is distributed throughout society, and that government should mobilise that knowledge rather than impose its own. A second argument against policy interventions, derived from an evolutionary position, is that innovation is highly unpredictable and full of surprises, which government should not try to second guess, and is a system phenomenon, with complex interactions that produce partly unintended consequences. On the other hand, there are market and system failures in innovation that government should address. The question then is how to do this while respecting unpredictability and the variety and distribution of knowledge. This yields a policy perspective that keeps innovation as open as possible.

We call for an innovation policy that is open in four dimensions: open with respect to collaboration with others, via open communication; open for surprises and changes of direction during innovation projects and research and development (R&D); open to new players ('challengers'), in particular entrepreneurs; and open to the world beyond the own region or country concerned and beyond established industries, to prevent the confinement of innovation to individual regions, countries, and existing industries. The different chapters of this volume develop insights as to how these different forms of openness may be organised.

Radical, path-breaking innovation is rare. Most innovative activity lies in incremental innovation, diffusion, and imitation. A key question that is asked in policy is where we should choose to lead, creating radical innovation, and where we should simply follow. Should we 'pick winners' or 'back winners', or are both choices problematic?

This volume starts with an introductory chapter (chapter 1) on the meaning of innovation, its economic significance, the roots and development of innovation policy, and the perspectives on these topics presented in this book. Chapter 2 supplies a macroeconomic analysis of the economic importance of innovation, as a source of productivity growth. Among other things, it confirms the idea that we need to delve further into the micro-level. For that, we emphasise issues of knowledge and learning, with a conceptual elaboration in chapter 3. This is needed as a basis for the analysis, and connects the book with the basic theme of the 'Lisbon agenda' for an EU knowledge economy. The volume proceeds with an analysis of the main features of innovation systems: the relation between research, higher education, and innovation (chapter 4), entrepreneurship (chapter 5) and the obstacles it meets (chapter 6), institutional conditions for trust to support learning by interaction (chapter 7), forms of organisation for innovation and innovation of organisation (chapter 8), conditions for innovation within organisations (chapter 9), features of network structure between organisations (chapter 10), and properties of regional systems of innovation (chapter 11).

Each chapter ends with implications and recommendations for policy. Here, recommendations are directed primarily at public policy, but secondarily also at policies of firms and other organisations, in the private and public sectors. The latter recommendations, oriented towards industry, are still relevant for public policy with a view to agenda setting, coordination, and stimulation. The final chapter, chapter 12, compiles and integrates the implications for policy.

In particular, the following major policy issues are addressed:
- How can we manage the combination and tension between exploitation and exploration and between stability and change, in research and in the relations between university and industry, in the organisation of firms and government, and in inter-organisational collaboration and networks?
- Can we determine where, in innovation, a country should be a leader and where a follower? What are the merits and drawbacks of the current policy of 'backing winners'? Can such choices be made while maintaining openness to outsiders and challengers?
- How can we further improve openness for entrepreneurs, and deal with obstacles, in market and system failures? How can this be done while taking into account government failures?
- How can we provide openness to surprises in innovation projects and processes? In government-funded projects and programmes, can we switch from *ex ante* safeguards that lock up innovation to *ex post* accountability that allows for changes of direction?
- How can we organise openness for collaboration and communication, in alliances and networks? What is the role and what are the limits of trust? How to combine trust and transparency? What are the effects of network structure?
- How can we combine the strengths of the local with the need for openness to the non-local and the global, in organisations, networks, regions, and countries?

CONTRIBUTORS

Neil R. Anderson, University of Amsterdam

Ron Boschma, Utrecht University

Gerard de Vries, Netherlands Scientific Council for Government Policy (WRR)

Rosina M. Gasteiger, University of Amsterdam

Lars Heuts, Nijenrode Forum for Economic Research

Patrick Kenis, Tilburg University

Gerrit Kronjee, Netherlands Scientific Council for Government Policy (WRR)

Marius T.H. Meeus, Tilburg University

Bart Nooteboom, Netherlands Scientific Council for Government Policy (WRR) & Tilburg University

Leon A.G. Oerlemans, Tilburg University & University of Pretoria

Erik Stam, Netherlands Scientific Council for Government Policy (WRR) & University of Cambridge & Utrecht University

Leo van der Geest, Nijenrode Forum for Economic Research

Robert Went, Netherlands Scientific Council for Government Policy (WRR) & University of Amsterdam

1 INNOVATION, THE ECONOMY, AND POLICY

Bart Nooteboom and Erik Stam

1.1 INTRODUCTION

Innovation has been essential for economic growth in the last centuries (Mokyr 1990; Baumol 2004). In western economies, the importance of innovation for economic growth has become more pressing given ageing populations, global competition, increasing product variety, and shortening product life cycles. Innovation is also needed to solve problems in society such as those concerning climate change, health, and congestion. Innovation also has intrinsic value as a manifestation of creativity. In this volume the focus is on the economic significance of innovation, and on its conditions, in innovation systems. The aim is to provide an analytical basis for innovation policy, primarily that of governments.

While macroeconomic analyses show the importance and the effect of innovation for productivity and growth, as will be investigated in chapter 2, and show effects of macro-level factors such as labour markets, they give little insight into the underlying processes of innovation. As a result, innovation policy is still very much a matter of improvisation and trial and error. In this book we aim to improve insight into the micro level of individual agents and corporate agents (organisations) and their interactions, in the generation, utilisation, and diffusion of ideas, products, and processes. As noted by Lundvall and Borras (2005: 614): "Innovation policy calls for 'opening the black box' of the innovation process, understanding it as a social and complex process". The aim of this volume is to contribute to that. Another reason for a focus on the micro level of the economy is that this volume is oriented towards the national government, and increasingly macro-economic policy has shifted to the supra-national level of the EU.

Hence the title: micro-foundations for innovation policy. This does not mean that we discount the value of macro policies (for example concerning taxes, labour, and social security): these indeed present vital conditions for innovation and economic growth. However, they have already been analysed to a large extent in other studies, are commonly recognised as important elements of economic policy, and we have little to add on that score. We take a complementary, relatively new perspective of 'innovation systems', that takes into account interactions between a variety of agents, various dimensions of innovation (going far beyond science and technology), and a variety of economic and institutional conditions. The analysis applies to innovation systems in general, with special attention to the Netherlands.

The main line in our policy approach is a trade-off between arguments against and in favour of policy interventions. One reason for restraint in making policy interventions derives from the insight gained from Austrian economics that knowledge is diverse and is distributed throughout society, and that government should mobilise that knowledge rather than impose its own. A second reason, derived from an evolutionary perspective, is that innovation is highly unpredictable and full of surprises that government should not aim to outguess, and is a system phenomenon, with complex interactions that produce partly unintended consequences. On the other hand, there are market and system failures in innovation that government should address. The question then is how to do this while respecting unpredictability and the variety and distribution of knowledge, and taking into account government failures. This yields a policy perspective that keeps the innovation process as open as possible. Open in four ways. Openness to uncertainty in the innovation process, as opposed to determining the innovation trajectory *ex ante*. Openness for collaboration at a sufficient 'cognitive distance', as opposed to cooperation with equals in knowledge. Openness for new entrants, as opposed to coalitions of incumbents. Openness to the world outside the own region or country and outside established industries, as opposed to inward-looking, parochial innovation policy. Such a policy would strengthen the position of the Netherlands as an open knowledge economy, in multiple respects.

In this introductory chapter we discuss the meaning of innovation, its economic significance, and the roots and development of innovation policy. After a brief introductory section, the second section concerns the key notions of novelty and creativity, and definitions of innovation. In the next section we give an inventory of dimensions and measures of innovation. The fourth section concerns the role of innovation in the economy, in the context of globalisation. This includes the necessary openness of innovation to global markets. In the fifth section we consider the position that innovation cannot be left to markets, and what perspective of markets we can then take. We contrast 'mainstream' economic views with views derived from Austrian and evolutionary economics, which have led to the 'innovation systems' view that is now current in many policy debates. We specify what we make of these perspectives for our own approach. In the sixth section we provide a historical view on innovation policies. A final section outlines the further contents of this book.

1.2 NOVELTY, CREATIVITY, AND INNOVATION

Innovation denotes both an activity and its outcome. In its outcome, it is a novel function or a novel way of performing an existing function. This is wider than new goods, services, and processes in market sectors, and includes innovation in the public sector. In its process, innovation denotes invention, the creation of innovations, and their diffusion. A key element of innovation is novelty; literally innovation means to introduce a novelty. In the current debates around innovation, its meaning is largely reduced to the economic domain, and the dominant interpretation is innovation as the successful exploitation of new ideas. In this view, the creation of new ideas precedes innovation. Before we zoom in on the economic aspects of innovation, we first explore the general meaning of innovation.

It is debatable whether sheer novelty is possible at all: paraphrasing the biblical Preacher (Ecclesiastes 1:9) and the economist Schumpeter "there is nothing new under the sun, only new combinations". Of course, new ideas do not arise from a cognitive vacuum, but are somehow constructed , from ideas and experience adopted, constructed, and accumulated along a path of life. This is further discussed in chapter 3. The term 'combination' carries a connotation that appears to underestimate the cognitive transformation or emergence involved in idea generation. That is creation. Creativity is making a connection between or combining two elements that have not previously been connected or combined (Feinstein 2006: 31). This connects with Schumpeter's view of innovation as novel combinations. Cognitive psychologist Sternberg (1996: 375) defines creativity as "the process of producing something that is both original and worthwhile". Two central elements are thus novelty (originality) and value (worthwhile). Creativity includes the process of finding a novelty and then transforming findings into observable products (Schweizer 2004). Does the condition, universally recognised to apply to innovation, that the novelty has direct value for, and is appreciated by others than the creator, also apply to creativity? We don't think so. It is widely recognised and documented (see e.g., Mokyr 1990) in science, technology, and art that new ideas often find a useful application and acceptance only later, sometimes much later.

Examples abound. In mathematics, consider non-Euclidean geometry. It was later taken up in physics, but prior to that it was appreciated as an intriguing and consistent system of new axioms and theorems. In technology there is the classic example of Marie Curie's discovery of radioactivity, later to be applied to X-ray photography, for example. In art the French impressionists were at first ridiculed and scandalized, and were barred from museums and galleries. Yet the impressionists did value each other's work and later met with huge popularity.

Would it make sense to say that those ideas become creative only then, when they were more widely applied or appreciated, long after they were produced? If value is a criterion for creativity, value to whom, and to how many? There is a double subjectivity here. First, whether someone identifies something as a novelty depends to a large extent on his prior knowledge. Second, when this novelty is turned into an innovation, its performance and value depends on the subjective judgment of its selectors. Some people may value an idea because they like it for itself, or because it inspires the further development of ideas, without any application beyond that. Creativity may have mostly or even exclusively intrinsic value to the creator. A scientist or artist who produces '*l'art pour l'art*' is not thereby barred from creativity. Both the recognition one may get from innovative accomplishment and the intrinsic value underlying creation contribute to the flourishing of people. So, we propose to allow for creativity to include novelty even when it has no identifiable value beyond intrinsic value to the creator.

This issue is more than an idle, innocuous semantic squabble. By looking only at value in application or use, in innovation policy, one runs the risk of neglecting invention and discovery. While innovation may usefully be defined as applied and accepted new ideas, there is no innovation without the underlying idea generation, prior to application. Therefore, innovation policy should be oriented not only toward innovation but also toward underlying creativity (as defined here). The relation between idea creation (exploration) and application (exploitation) is analysed in chapter 3.

According to the definition of innovation given above, innovative performance depends on social judgement of others. In the case of technological creativity, innovative performance depends on 'peer reviewed' journals, patent bureaus, but also specific prizes (e.g., of national science foundations and Nobel Prize committees). Innovativeness in the arts is assessed for example by publishers and arts foundations that provide grants and prizes. Economic creativity – entrepreneurship – is rewarded by the market, i.e., customers that pay more for the products than the costs of the sum of its inputs: from a company perspective this involves value creation (sales) and capture (profits).

"The current discussion on the so-called 'knowledge paradox' can be translated as a lack of economic creativity following on technological creativity: 'In the end the translation and transformation of knowledge into concrete products and services leads to productivity and welfare in a broad sense. In essence it is about the utilization, in a competitive environment, of opportunities that markets and newly developed knowledge offer. Innovation and entrepreneurship'" (Innovation Platform 2006: 17).

Innovation is mostly seen as primarily of economic value and purpose. However, it is widely recognised that regardless of economic growth innovation is also of value in solving societal problems, and has cultural value and intrinsic value for the flourishing of people, in activities of creation and self-realisation. From a more utopian perspective, innovation is seen as a means for world peace and ecological preservation. These dimensions of value are not necessarily separate. Solving societal problems may well go together with the furthering of economic growth. As argued in chapter 4, university research may contribute to both culture and economy. In chapter 9 it will be argued that creative teams in organisations, which further innovation, are conditioned at minimum by certain dimensions of the quality of labour, such as intrinsic motivation and autonomy.

Until now we have discussed innovation, creativity, and novelty in general terms. The current public debate on innovation is however much narrower in focus, in that it largely focuses on the economic aspects of innovation, as a 'solution' for economic 'problems' such as increasing global competition and ageing populations. The macroeconomic significance of innovation is further analysed in chapter 2. Here we discuss the position of innovation in economic structure and processes. First we specify the dimensions, the further 'ins and outs' of innovation.

1.3 DIMENSIONS AND MEASURES OF INNOVATION

The foundational definition of innovation as new combinations was provided by Joseph Schumpeter (1934: 74). The carrying out of new combinations can take several forms: (1) the introduction of a new good or quality thereof, (2) the introduction of a new method of production, (3) the opening of a new market, (4) the conquest of a new source of supply of new materials or parts, or (5) the carrying out of the new organisation of any industry. In contrast to the often-used narrow definition of innovation as technological change, Schumpeter also recognised innovation as organisational change.

Economically, innovations can be 'smaller or larger', in terms of the volume of sales involved, number of producers and users, the degree to which it is creatively destructive of existing assets and competencies, with all the implications for the destruction of capital, employment, and the restructuring of markets, industries, and institutions (such as technical, administrative and educational standards, legal rules, etc.). A well-known distinction is that between 'radical' or path-breaking, or 'paradigm-switching' innovation and 'incremental' innovation along a 'technological trajectory' (Dosi 1982, 1984). Radical innovation tends to have wider repercussions, yielding more creative destruction, involving larger volumes of sales and numbers of consumers and producers, and more markets and industries.

The transition from analogue to digital photography, for example, yielded a way to fulfil an existing function (making pictures), but added new features and functions, such as miniaturization of cameras, direct availability of pictures, digital modification and mixing with other media (image, text, sound), and the possibility to transmit pictures by phone. It was radical in using a novel (digital) technology, and had a large volume of use and production of cameras.

The so-called shifts in techno-economic paradigms (Dosi 1984; Freeman and Louca 2002) reflect the emergence and diffusion of new general-purpose technologies (GPTs) like the steam engine, the combustion engine, nuclear power, electricity, and information and communication technologies. The emergence, diffusion, and adoption of these GPTs is said to initially lower productivity levels and economic growth (due to problems of integration and restructuring), but in the longer term to improve productivity levels and ultimately economic growth (see chapter 2). Technological innovation is often seen as the major motor of economic growth (Mokyr 1990). Some authors however are quite critical of the assumed relationship between the invention of GPTs and national economic growth. According to Edgerton (2007) most economic growth does not take place in countries that heavily invest in the invention of new technologies, but takes place in countries that adopt, adapt, and copy new technologies. The fact that prosperous countries spend relatively much money on research and development (R&D) is in this light seen as an effect of wealth, not a cause. Other GPTs, such as nuclear power, are far from driving economic growth, in contrast perhaps to their geopolitical influence. Some very important innovations are not based on spectacular technological inventions. A striking example in this respect is the introduction of the shipping container, which has led to major cost-reductions in transportation, and enabled for example the rise of China as a low-cost producer country (Levinson 2006).

The distinction between incremental and radical innovation is related to the distinction between "first and second order change" (Bateson 1972), "single and double loop" learning (Argyris and Schön 1978), and "exploitation and exploration" (March 1991), which denote change and learning within established basic principles or logics, and changes of such principles and logics. A big difference is that these concepts are relative to the level of analysis, while the notion of radical innovation is more absolute, related to the world economy. A new way of performing a small task within a firm may deviate from established logic or design principles and would thus be exploratory; on the level of an entire industry its impact may be small and it would hardly be called a radical innovation.

Generally, a distinction is made between invention/discovery, innovation, and adoption/diffusion. Innovations can be new to a person, an organisation, a market or industry, a country or the world. Such distinctions are made, for example, in the EU Community Innovation Survey (CIS). When the definition includes novelty only for a given individual or organisation it overlaps with the notion of adoption. There may be arguments for such a broad definition, in the analysis of intra-organisational conditions and consequences of adoption, which may entail organisational innovation. This happens in chapter 9 of this volume, for example. However, in general we use the term innovation to denote something new for others than only the producer or adopter of an innovation.

There are stages in innovation processes, from idea creation and knowledge sourcing via development and testing, in exploration; to application (with new products or processes); improvement; diffusion; and differentiation (exploitation). As one progresses along these stages, uncertainty mostly decreases. This used to be seen as a linear process, in one direction. Later, it was recognised that there are feedback loops. In particular, experience with testing and application yield adaptation, differentiation, extension, and inspiration for new ideas, so that the process is circular rather than linear (Kline and Rosenberg 1986). How that process works is of course of crucial importance and will be analysed in chapter 3.

Empirical measures of innovation relate to different aspects of innovation, different stages of the innovation process, and different conditions for innovation. The development of innovation indicators reflects the development of policy (to be discussed later). First efforts, in science policy (S), concentrated on indicators of science related to R&D. The fundamental shortcoming of this is that it concerns invention rather than its use in innovation. While R&D may provide an input, innovation also requires prototype development, testing, design, organisation, marketing, and distribution. Those, in turn, require collaboration with customers, suppliers, distributors, and others. Later, attention included some output indicators concerning technology and its production and use, in a shift to science and technology policy (ST). Most recently, in the development of innovation policy – including science, technology, and innovation (STI) – the arsenal of indicators has been widened considerably, with plans to include not only product and process innovation, but also marketing innovation (including design) and organisational innovation (including knowledge management, the organisation of collaboration, and even measures of social capital), as reported in the third version of the *Oslo Manual* (OECD 2005, 2007). In other words, there has been a growing recognition of the non-technological aspects of innovation. Also, in addition to R&D another input is adoption, and we require measures of diffusion. Gault (2007: 16) reports that 41 percent of innovative firms in the Community Innovation Survey (CIS) innovate by adopting technology from other organisations. The present repertoire of indicators can be ordered as concerning inputs, process (e.g., linkages), outputs, and impacts. They are summarised in table 1.1

The categorisation in terms of inputs, process, outputs, and impacts is somewhat problematic in the non-linear, recursive view of innovation where inputs may also be outputs and vice versa. Skill, for example, is both an input and an output.

The indicators hardly correlate (cf. Kleinknecht et al. 2002), which suggests their complementary nature. Some of the improved impact indicators still carry problems. An innovation that is new to a firm can be due to internal development but also due to simple purchase of new technology from a supplier, in the form of machinery or instruments (Arundel 2007). An innovation may be new to the market of a firm not because the firm is very innovative but because that market is backward. Some indicators are still lacking. Impacts from innovation take time, can be indirect, and, in part, are difficult to trace (such as cultural effects or the wide repercussions of creative destruction). We still lack indicators of institutional effects that may enhance or obstruct innovation. Ultimately, of course, the question is not what the merit is of individual indicators, but how they can be used to measure the impact of policy. That requires an often difficult matching of different databases. Generally, to evaluate the impact of a policy measure, one needs to determine 'additionality': what effects do the measure generate that would not have been generated without it. In principle, what one needs for that is longitudinal data over a sufficiently long period to trace effects, both for firms that participate in the programme and for firms that do not participate but are similar in other respects. That can and has been done, but only rarely.

In spite of the recent rhetoric against focusing on R&D and against the linear model of innovation, old practices still dominate the actual use of indicators (Hawkins 2007). There are three reasons for this. First, R&D as input and patents as output are more easily measured than many other, modern indicators (see Kleinknecht et al. 2002). Second and third, there are path-dependencies in both policy – with an ongoing focus on R&D – and in research, with an ongoing focus on familiar measurements of productivity and of impacts of R&D (Arundel 2007). In policy, R&D allows more grip for government than many other factors, such as

Table 1.1 Empirical Measures of Innovation

Inputs	Skill or educational levels of employees or entrepreneurs
	Science and engineering graduates
	R&D (formal and informal, internal and external to a firm)
	Innovation expenditures
	Use of ICT (hard- and software)
	Use of licences
	Purchase of machinery, instruments
	Information from trade fairs, conferences, press etc.
	High-tech industries
	Creative class and industries
	Risk capital, particularly seed capital
Process: linkages, marketing, organisation, strategy	Linkages with knowledge institutes
	Alliances, networks, chains of innovation, social capital
	Marketing, customer relations, design
	Supplier relations, co-makership
	Organisation, knowledge management
	Communicative skills
Outputs	Patents (in US, EU or elsewhere; applied for or obtained)
	Publications
	Graduates from education
	Licences sold
	Trademarks
	Prototypes
Impacts	Citations (of publications, patents)
	Renewal of goods or services due to development or application of new (to a firm) or recent technologies (adoption)
	Renewal not necessarily based on technological knowledge (e.g. management, marketing or organisational renewal), 'social' innovation
	Number or percentage of turnover of goods or services new to the firm / country / industry
	Improved processes, reduced costs
	Profit, market share growth
	New market entry
	High quality employment

collaboration. The ongoing one-sided pre-occupation with R&D is also reflected in the EU's Lisbon and Barcelona agendas.

A narrow focus on technological innovation and on innovation inputs, such as R&D spending and science and engineering graduates, could be especially misleading for innovation policy in small open economies like the Netherlands.

First, although the Netherlands may be lagging behind with respect to such things as the number of science and engineering graduates, business R&D expenditures, and high-tech manufacturing (see chapter 2), it may be leading the indicators on creative class, patents, and new designs, which may reveal more about the particular nature of the Netherlands economy than about the 'innovation gap' that it should bridge. Second, for a small open economy it might be more effective to improve the international networks and the absorptive capacity of the actors in the national economy than to aim at shifting the technology frontier (Weterings et al. 2006; Nahuis and Van de Ven 1999).

Within one week the following two headlines were published in the same Dutch newspaper (*Trouw*): "Innovation in the Netherlands lags behind neighbouring countries" (July 10, 2007) "Multinationals lead the Netherlands to a third place on the European innovation list" (July 16, 2007). These headlines may be puzzling to many: is the glass half full or half empty, or are we talking about different glasses? The latter seems to be the case: The Netherlands is lagging behind mainly in private R&D expenditures and in entrepreneurship, while it has been the third most productive patent applicant in Europe (after Germany and just behind France). This shows that innovation is a process with many phases and a concept with many faces that can thus be measured in multiple ways.

1.4 INNOVATION AND THE ECONOMY

The construction of new combinations is seen as a driving force of economic welfare. However, much that counts as innovation has dubious added value.

Two leading shaving equipment manufacturers – Gillette and Wilkinson Sword – spend billions of dollars on R&D to develop ever more advanced shavers. Has the evolution from one-bladed razors to eight-bladed razors really delivered a societal benefit that covers all these R&D investments? According to free-market ideologists, the answer is yes, because these two firms still seem to be profitable. According to Gillette, the key is to "provide benefits people think are worth paying for". However, one might wonder whether all the energy and creativity of all the thousands of talented highly qualified researchers could be allocated to more productive activities.

Innovation can also lead to the ruin of culture and nature. As Schumpeter already recognized, there is also 'creative destruction' involved in innovation: new knowledge, technologies, markets, firms, and employment make old ones obsolescent, and adjustments in demand and supply can be very painful.

In the earlier example of digital cameras, there was creative destruction of the production and use of chemical films and their processing, and hence of corresponding manufacture and retailing. There were wide repercussions for the media, publishers, advertising, design and artists. It led to cameras in mobile telephones that turn passers by into potential news photographers, who may have to shift their profession.

Radical innovations are generic in the sense that they generate waves of related and subsequent innovations, which enhance their application; and incremental innovations that widen their application, make improvements, and differentiate them into varieties. While radical innovations are the most dramatic, incremental innovations are by far more frequent, and in their accumulation have the greatest impact. Indeed, it is through incremental innovations that radical innovations have most of their economic impact. Here, the distinction between minor innovation and the diffusion of a major innovation becomes blurred. One might envisage this as a pyramid, with radical innovations at the top, which fan out into a broad basis of applications and incremental innovations. Incremental innovations, being in the majority, mostly require limited creation of new knowledge, and hence limited R&D, building on existing knowledge, and entail minor variations upon an existing theme, with adjustments in styling, branding, distribution, added services, etc. Thus, it is a misconception that all innovation requires R&D and a linkage to knowledge institutes. However, such linkages are required in the diffusion of new knowledge and technology, and for the inspiration of new knowledge. Even radical innovation is not necessarily based on R&D and new scientific knowledge, particularly if it is an innovation in organisation, marketing, or distribution.

An example is self-service, which has revolutionized retailing, not only in the type of service offered, but also in creating the opportunity for an enormous increase of scale, in large self-service stores. From retailing it spilled over to wholesaling, cafés, restaurants, and even hotels.

As a result, when complaints are made concerning lack of R&D, lack of application of knowledge generated by universities, and lack of connections between university and industry, the question is how much one would expect is needed and useful. What is the benchmark?

While economic value is not a precondition for all innovation, innovation is needed for economic reasons, also for developed Western nations, if they want to maintain their prosperity in the face of changing conditions in the global economy. As will be shown in chapter 2, total factor growth in the EU has slowed down, as a result of, among other things, slow down of the increase of the volume of the working population and of the volume of capital. In fact, due to an increase in age, the share of working people in the population is declining. A major determinant of total factor productivity is innovation, and for ongoing growth that will have to increase, in order to compensate for lack of growth of the volumes of labour and capital. This line of argument is further developed in chapter 2. There is a shift of growth to emerging nations (e.g., China, India), which offer lower wages and are bridging their gap in development. This is not necessarily a threat. There might even be a net benefit from this aspect of globalisation for the Netherlands, due to lower prices of goods produced in lower-wage countries, opportunities for exports to emerging markets, and opportunities for the Netherlands to function as a gateway into Europe for imports from emerging economies (Suyker and De Groot 2006). However, development is accompanied by structural change, where for a number of industries activities are shifting to emerging economies. Opportunities for innovation should be grasped to fill the gaps.

With globalisation, industrial value chains are increasingly fragmented or 'splintered', with different links of chains being distributed across the globe (Gereffi 1999; Dicken 2003). The front end of the chain, in R&D and design, may still remain in a developed country, while production takes place elsewhere, after which marketing, distribution, and user support may again be located at 'home', to a greater or smaller extent. Thus, it is out of date to think in terms of domestic versus foreign industries, and one should think in terms of segments of industries that need to be connected to other segments, across the globe. Nowadays, this fragmentation of value chains applies not only to manufacturing but also to services, in what has been called the "second unbundling" (Went 2007). Even some provision of distribution hubs , a traditional specialty of the Netherlands, has now been outsourced to China (Wright 2007).

There may no longer be a future for a full-fledged motor car industry in the Netherlands, but there may still be a future for a cluster of parts, say the 'front end' of a car, or its carriage and suspension system, provided it is well connected to, say, the German car industry. In the formerly extensive Dutch textile industry, most production has been outsourced to low wage countries, but fashion and design (Amsterdam) and logistics (Rotterdam) are still in the Netherlands.

This has an important policy implication: when instituting a programme for developing economic activity, there is a tendency to frame it in national terms, if only because in the EU a single nation is limited in what it can arrange in subsidised programmes across different EU nations. This bias towards national activities carries a risk of tying the initiative to the country while the opposite should occur: a loosening of local ties to connect local activities to parts of fragmented value chains that are located abroad. This is one of the reasons why we will argue for more openness to the world in national innovation policy.

We can no longer assume that all high value-added activities, such as R&D, will remain in developed countries. The arguments are familiar: there are hosts of engineers and other well-educated workers entering the global economy in less developed countries. What, then, will not easily dissipate to emerging economies? What may help us to survive is the condition, increasingly recognised in recent economic thought, that innovation is a network phenomenon, arising from interaction between a variety of firms, knowledge institutes, and public authorities, together fashionably and pompously labeled as 'the triple helix', embedded in local conditions of infrastructure and institutions, which cannot so easily shift to other locations, and may not easily be imitated. This is one reason for current policy attention to 'regional innovation systems' and 'local clusters'.

According to Meeus and Faber (2006: 67), in the Netherlands more than 90% of innovations are achieved with contributions from external parties. In much research over many years (see e.g., Wever and Stam 1999, Pittaway et al. 2004), it has been shown that firms maintain relationships first of all with customers, suppliers, and even competitors, more than with public and private knowledge institutes. Before this is taken as evidence for some kind of failure in the knowledge system, recall that much of the large volume of incremental innovation hardly requires new scientific knowledge.

Conceptually, a reason for attention to collaboration is the increasing recognition that innovation requires learning by interaction (Lundvall 1988, Nooteboom 2000). It is not quite clear, however, why, precisely, that is the case. It is clear that it has something to do with variety as a source of Schumpeterian 'novel combinations', but it is not clear how that works, cognitively and socially. Analysing that is one of the objectives of this volume. Conceptual analysis of collaboration for innovation is given, in particular, in chapter 3, which looks at the more cognitive aspects, and chapter 7, which looks at the governance of collaboration. Analysis of inter-organisational networks and regional clusters or innovation systems is given in chapters 10 and 11. The focus on knowledge and learning, developed in chapter 3, connects this volume to the 'Lisbon strategy' to make the EU the most dynamic and competitive knowledge-based economy in the world by 2010, and to the aim of the Innovation Platform to make the Netherlands one of the five leading knowledge economies of the world.

1.5 WHAT MARKETS?

Before turning to innovation policy, the question for many is why we need policy in the first place. Why wouldn't markets work for innovation? Arguments for innovation policy usually arise from market failures. But what are those? And are there also other failures? Here we take the view from institutional economics that there is co-evolution of markets and institutions: markets require institutions – as has been shown in the difficult transition of former communist economies to market economies – and market processes shift and create institutions.

Market failures

In traditional economic versions of innovation policy, the main orientation is one of laissez faire. The market should be allowed to do its work of achieving optimal allocative efficiency, with incentives from competition, and the focus of policy is on the reduction of barriers to firm entry, growth, and exit, in competition policy. The changes brought about by innovation can require adjustments by policymakers. Their prime task is to restore an optimal market structure. In such a perspective, results of technological change are supposed to be known, as antitrust authorities intend to impose a specific (optimal) state of affairs. The only rationale for state intervention is market failures, such as those connected with 'public goods', characterised by non-rivalry (one's use does not exclude another's use) and non-excludability of use, and external (dis)economies, such as pollution. Public goods are goods for which an increase in one person's consumption does not reduce its availability for others ('more for you means no less for me'). Examples of public goods are national defence, monuments, street lighting, lighthouses, or radio and television broadcasts. In addition to these most often 'national public goods', there are also public goods whose benefits extend beyond national boundaries, such as peace, the environment, biodiversity, health, scientific and technical knowledge, financial stability, and shared technical standards (see Acocella 2005).

In particular, a problem for innovation lies in possible non-appropriability of rewards for innovation, due to spillover of the results to imitators, which yields an argument to postpone competition by protection of property rights. In traditional economics innovation is generally seen as the introduction of new more efficient production technologies. Firms are assumed to immediately obtain gains associated with the new technology once they have decided to implement it. As a consequence, the only motive to innovate lies in the perspective of benefiting from a monopoly rent that is at least transitory.

Within the traditional economics framework, "the best way to understand market failure is first to understand market success, the ability of a collection of idealised competitive markets to achieve an equilibrium allocation of resources which is Pareto optimal" (Ledyard 1989: 185). This particular definition of market success and hence of market failure is nothing but a reading of the first fundamental theorem of welfare, according to which: "if there are enough markets, if all consumers and producers behave competitively, and if an equilibrium exists, then the allocation of resources in that equilibrium will be Pareto optimal" (ibid.). Therefore, "market failure, the inefficient allocation of resources with markets, can occur if there are too few markets, non-competitive behaviour, or non-existence problems" (ibid.). The main objective of public policy in this perspective is to realize optimal (static) efficiency: optimal allocation of given resources.

Often failure in financial markets is assumed when it is hard for certain innovators to acquire capital for investments. However, lack of private financing might well be the effect of rational choices: private financers perceive the market prospects of certain innovations to be too limited due to obstacles to commercialisation. Another frequently used argument in favour of policy (i.e., public investments in R&D) is underinvestment in private R&D due to limited appropriability of the returns to R&D. A conventional argument in market failure reasoning is that knowledge is a semi-public good. Because of the problem of non-appropriability, government should take action to overcome underinvestment in new knowledge through the provision of R&D subsidies or the establishment of property rights. An example of government intervention is the support of new technology-based firms, which is assumed to stimulate R&D investments and to improve public returns on a longer term when new technology is developed and commercialised in many new applications and sectors of the economy (see chapter 5).

These conventional market failure arguments provide a basis for public intervention in innovation policy. It is based on a linear model of innovation that focuses on R&D infrastructure and technology transfer, as if these will automatically lead to innovation. Sometimes targeted industrial policy is introduced, based on the assumption of market failure, in that public intervention stimulates economic growth not otherwise likely to occur. A standard argument in this line of reasoning is that knowledge is characterised by increasing returns to scale. For that reason, investments in public R&D, technology transfer, and education are expected to foster economic growth (Hall 1994). These increasing returns might even give rise to selective government intervention, for example through subsidies for industries characterised by this 'imperfect competition'. With this intervention government can ensure that a larger part of the supposed 'excess returns' is earned by domestic firms. The basic idea underlying such a strategic trade or industrial policy (Krugman 1987) is that in the presence of increasing returns and imperfect competition, firms in some industries may be able to earn excessive returns, and that a country can ensure that the firms earning these excess returns are domestic rather than foreign. This kind of policy could easily lead to 'locational tournaments' in which national governments compete with each other – in a zero sum game – to attract foreign direct investment.

Markets for discovery

Most people, also those outside mainstream economics, would agree that it is important that "markets are open, that they facilitate and create incentives to challenge established positions and that they eliminate activities which are no longer viable in the prevailing environment" (Metcalfe and Ramlogan 2005: 230). Stimulating innovative entrants (outsiders) to compete with (obsolete) incumbents (Boone and Van Damme 2004, Aghion et al. 2006), or 'backing the challengers' is a key policy instrument (Jacobs and Theeuwes 2004).[1] While markets may fail by undersupply of innovation, policymakers may also fail. They are susceptible to mistakes (due to information problems) and may be prone to capture by private interests (Olson 1965). Some authors have argued that government policy fails, giving rise to 'deadweight' and 'substitution' effects (Jacobs and Theeuwes 2004a, 194; Santarelli and Vivarelli 2002). Deadweight effects occur when the beneficiary of an innovation subsidy is a firm which would have started or innovated in any case. In other words, there is a lack of 'additionality' of public funding. See for innovation Irwin and Klenow (1994), Wallsten (2000), Cornet and Van de Ven (2004)[2]; and for new firm formation: Santarelli and Vivarelli (2002). Substitution effects lead to more distortions, since the subsidy is not only a social waste but also implies the substitution of potentially more efficient new firms or innovation projects by less efficient ones. In the presence of a – publicly funded – incentive, the firm adjusts its own capacity not on the basis of either passive or active learning in a market, but as a consequence of the artificial support brought about by the subsidy received. See for innovation Niosi (1995); and for new firm formation Maggioni et al. (1999), Santarelli and Vivarelli (2002).

Outside mainstream economics, however, there is a notion of a market as a 'discovery process' rather than as an efficient mechanism for allocation of scarce resources to given goals. This notion goes back to Austrian economics, in particular Hayek (1945).[3] Knowledge is differentiated and distributed in society. In chapter 3 we will use the notion of 'cognitive distance' between people. New economic knowledge is best discovered by actors in a competitive process. In this view, centralisation is inferior to a market system because the central actor attempts to make decisions on the national economy despite inferior information. Because the market system relies on the efficient exploitation of bits and pieces of information held by different actors dispersed over society, a decentralised market system can make most decisions better than a centralised one. As a result, in a decentralised market society, people can make decisions about the exploitation of innovation on the basis of idiosyncratic information gathered through their life course (Shane 2000). Why would policymakers be better informed about making the right choice than professionals in the marketplace and people with access to local knowledge?

What sets this Austrian approach apart from the traditional one, is that it regards uncertainty and information asymmetry as fundamental to the process of innovation. Entrepreneurs make conjectures about new combinations that are uncertain – that is, one cannot know (or even calculate the probability) *ex ante* whether these conjectures will be correct (Knight 1921). Several types of uncertainty can be distinguished: for example technical, market, and competition (Shane 2004: 205). The entrepreneur does not know in advance if the good or service he is producing will work, and, if so, if it can be produced at a cost less than the price at which it will be sold (technical uncertainty). The entrepreneur also does not know if demand will exist for the product, and, if so, if customers will adopt in large enough volumes, quickly enough, and at a high enough price, to make the effort profitable (market uncertainty). Finally, the entrepreneur does not know if she will be able to appropriate the profits from the exploitation of the opportunity or if they will dissipate to competitors. This uncertainty will only be resolved with entrance into the market (Rosenberg and Birdzell 1986: 257-258), hence the description of the market as a discovery process.

Consistent with this aspect of the Austrian view, but going beyond it, is the more recent approach inspired by evolutionary economics and institutional economics. Evolution is driven by processes of variety generation, selection, and transmission of what survives selection. Here, innovation is conceived of as a process of research, learning, and selection, which results in the appearance of new productive options that bring about a modification of the environment itself. The notion of selection points to the – often forgotten – function of competition in economic growth that it eliminates obsolete forms of economic activity, burying the economically dead (the 'destruction' part of Schumpeter's creative destruction). This function is not to be taken for granted: consider the difficulty often experienced by policymakers in getting rid of programs that are obsolete or that have simply failed. Thus defined, innovation is a sequential process, which takes (and can change) form, content, and direction at each successive step of its implementation. Firms do not know *ex ante* whether it is profitable to innovate.

"Indeed the answer to this question for any single firm depends on the choices made by other firms, and reality does not contain any provisions for firms to test their policies before adopting them. Thus there is little reason to expect equilibrium policy configurations to arise. Only the course of events over time will determine and reveal what strategies are the better ones" (Nelson and Winter 1982: 286).

This approach takes the radical uncertainty of innovation seriously, which pulls the rug out from under rational choice analysis and analysis in terms of equilibria. The efficient operation of markets is limited by the uncertainty of conditions and outcomes of innovation, limited insights in demand and supply, transaction costs, and the endogenous change of preferences.

Transaction costs are costs of contact (search, evaluation, distribution), contract (negotiation, agreements, contracts) and control (monitoring, haggling, litigation). Internet and related technologies (credit cards, bar codes, electronic tracking and tracing, database management, intelligent software agents) have drastically reduced such costs and have enabled novel and transformed relationships between producers, suppliers and customers. That is an important part of the technological background of globalization, in particular the 'splintering' and fragmentation of value chains across the globe.

While there may be forces that tend towards equilibrium, in the diffusion of innovations, equilibria are never approached due to ongoing change. Under the radical uncertainty of innovation, and consequent lack of foresight, one needs the view of an evolutionary process where a variety of trial and error is submitted to a selection environment of markets and institutions. One of the reasons equilibria are not reached is that the selection environment co-evolves with innovations: innovations affect the institutions upon which selection is based.

In such evolution one can get locked into path-dependencies. Institutional set-ups limit and enable evolutionary processes, and this varies across countries and regions. In other words: history and context matter. A policy that is effective in one setting of time and place may not be so in another. Universally valid instruments of policy are an illusion.

The evolutionary perspective recognises the possibility of system failures[4] that may obstruct the processes of variety generation, selection, and transmission, which need to be tackled by policy. This deviates from the more radical laissez faire view of Hayek. In this context, coordination among the firms is not only "highly beneficial to the economy" (Baumol 2001: 727), but a necessary condition for innovative investment to be carried out. Practices usually perceived as anti-competitive can be tolerated by antitrust authorities, at least so long as they do not lead to actual market failures. In order to gain (dynamic) efficiency (innovative choice), coordination among firms may be required (market imperfections), but should not lead to abusive market power that would block innovative initiatives. From this perspective, innovative choice consists not so much in the choice between given alternatives (whether based on complete or incomplete information) as in a search for coordination. What matters is no longer the 'rationality' of the choice between known alternatives. It is the 'viability' of the process through which a different alternative is brought about: a viability that depends on how coordination problems are dealt with step by step. In this light, competition is not only aimed at equalising supply and demand in a given market and technological environment, but "has also to adapt both structure and technology to the fresh opportunities created by expanding markets" (Richardson 1975: 353).

The evolutionary perspective, and its connections to theory of knowledge and learning, will be further discussed in chapter 3.

Variety and effects of scale

The importance of variety for innovation has implications for arguments of scale. There is a persistent inclination to think that large scale is efficient. In innovation policy this leads to arguments for 'focus' and 'mass' in research and development by concentrating efforts in a given discipline or field of research or in a single large institute or university, research programme, or project. And indeed, there are several valid arguments of economy of scale. For example, some areas of research require expensive instruments or installations, or specialised support and collaboration, whose cost can be justified only when they are used on a sufficiently large scale (in terms of number of researchers, students, or patients). Thus, there are good arguments for collaboration, in joint use of facilities, between the three technical universities in the Netherlands, for example. However, there are also diseconomies of scale, as the Dutch have learned from their experience of increased concentration and scale of education and health care provision, which came at a large cost. To some extent the advantages of scale may be utilised without incurring disadvantages, by having decentralised, highly autonomous operational units that share resources for support, such as housing, administrative support, ICT services, recreational and catering facilities, and libraries.

As we will argue in more detail in the chapter on entrepreneurship (chapter 5), many small, independent units allow for more variety of experimentation, at a lower cost of failure per unit, which is bound to be frequent in innovation, since the units are small. This yields more innovative output at lower cost of failure than concentration in a large effort in a single large unit. Large scale can lead to coordination failures, loss of motivation and, in particular, loss of variety.

If there are n people, then there are n(n-1)/2 possible direct links between them. If everybody talks to everybody, communication explodes, with the square of n, together with its costs, confusion, interpretive distortion of information, and delay, which crowd out effective work. This is a potential problem particularly, perhaps, in the Netherlands, with its deliberative tradition of the 'polder model'. The classic solution is to institute a hierarchy, with communication only between successive levels in the pyramid. Hierarchy entails delay and cumulative distortion and reduction of information, which isolates the decision making apex of the pyramid, resulting in inappropriate strategy, and frustrates the base which forms the front of the organization, and the main source of information needed for improvement and innovation, from contacts with customers, suppliers and technology. Decoupling of the system into separate, autonomous, specialized units that are nevertheless mutually dependent yields a problem of obstacles for coordination, novel combinations and diffusion between them.

Upon closer inspection, in the policy debate arguments of scale often turn out not to be proper arguments of scale, defined as more of the same, but arguments of scope, defined as configuring different, complementary activities. This can be closer to objectives of innovation than scale, in allowing for variety of insights and a potential for 'novel combinations'. It is not always clear when an argument is a bad argument of scale or a valid argument of scope. Note however, concerning arguments of scope, that in science the tapping of variety and the crafting of novel combinations occurs in the process of scientists moving around the world to conferences and seminars, and following it up with joint research, or participating in, say, an EU R&D programme. They do not necessarily have to be co-located for longer periods of time. However, it can be a good idea to institute a substantial research institute or university with a variety of specialisations within a discipline, or a variety of disciplines that yield perspectives for novel combinations, and attract the best talent in the world for each of the parts. An important argument may then be that this helps to attract the best talent, and contributes to what we will later (in chapter 3) call a 'knowledge ecology'. On the other hand, it is generally not a good idea to concentrate all efforts in a given field or specialisation in a single location. It may be an excellent idea to set up an elite institute for finance in Amsterdam, but not to concentrate all finance departments at a single university.

Discrepancies and agreement

Between mainstream economics and the Hayekian/evolutionary view there are not only deep-level discrepancies but also surface-level agreements. In the Hayekian/evolutionary view, the stuff out of which 'market failures' are made from the perspective of mainstream economics – such as asymmetric information, radical uncertainty, cumulative knowledge, path dependence, lack of equilibrium, and rigidities – are from an evolutionary perspective the stuff from which markets and innovation are made.

In view of differentiated and distributed knowledge it is nonsensical to assume that prices carry all the information needed for markets to work. Prices do not tell producers how technology works, nor what different qualities of products different consumers want. Prices do not tell users how products fit into their particular needs and user practices, and what the switching costs between different products are. Transaction costs, such as search costs, are not imperfections that one can tag onto an otherwise perfect market, but are endemic, part of how markets work. Markets require coordination to deal with differentiated and dispersed knowledge and transaction costs. Those are not market imperfections but part of markets, and the basis for discovery. Also, product differentiation is not an exception but the rule.

Gasoline in itself is as homogeneous as a product can get. Yet it is differentiated by means of logos and colour schemes of gas stations, projecting different images of being sportive, 'green', science based, etc., and with discount, bonus and gift schemes, loyalty cards, goods sold at gasoline stations, etc.

As argued by Anderson (2006), Internet and related technologies have enormously expanded opportunities for product differentiation. Much market thinking is still oriented towards the idea that only large volume sales are efficient, in view of the per unit cost of packaging, transportation, advertising, shelf space in shops, and search by consumers. However, since Internet started to dramatically reduce or even eliminate those costs, enormous opportunities have opened up to radically differentiate products, since very small volumes of sales now become economically feasible, opening up very small market niches. This 'fattens the tail' of the distribution of sales: very small volumes of demand that previously could not be served now can be.

What mainstream economists call rigidities to some extent are not only unavoidable but indispensable for markets to work. One example is that markets need communication and communication needs a certain stability of meanings, whereby new meanings tend to be ignored. Similarly, for technology to work one needs technical standards and those also need sufficient stability. In other words, to some extent efficiency requires orthodoxy. Interactions in markets require a certain reliability of behavioural expectations, on the basis of rules or norms of conduct. That entails rigidity. More generally, institutions, defined as enabling conditions for action, can work only if they cannot instantly be changed by that action. But enabling conditions are inevitably also limiting conditions. Hence institutions inevitably yield rigidities. Another example is that in many cases, investment for productivity and innovation require investments that are specific, i.e., that cannot (efficiently) be employed for alternative uses, and require a certain stability of activity to recoup the investment, or else they will not be made, thus sacrificing opportunities for productivity and innovation. In other words, they cause switching costs and when it is costly to switch one needs some stability, i.e., rigidity.

Furthermore, in the new perspective intense price competition can eliminate the slack in resources and time that are needed for exploration and coordination required for innovation. There should not be excessive competition that hampers the viability of the process of change. This will be discussed in more detail in chapter 3. In sum, what in mainstream economics are seen as virtues of market mechanisms are seen here as possible obstacles, and what are seen there as obstacles for markets, are here seen as features of their functioning.

"the imperfections identified in the market failure approach, therefore, can be viewed in a different perspective, as integral and necessary aspects of the production and the dissemination of knowledge in a market economy" (Metcalfe 1998: 114).

Now we are facing a puzzle. On the one hand this perspective says that what mainstream economics sees as imperfections of markets are not imperfections but the very stuff markets are made of. The crux of markets is not optimal allocation given perfect information, but the utilisation of profit opportunities from imperfect information and differences in knowledge and competence. It also says that markets can be too perfect, in the sense that extreme price competition can eliminate conditions for innovation. On the other hand, it agrees with the old view that markets are needed and that some market failures can obstruct innovation. The paradox can be resolved.

First, in the new view the notion of public goods and services still applies. If knowledge is dispersed, asymmetrical, and cumulative, yielding limited absorptive capacity, this means, in effect, that it does not spill over so easily. This means that it becomes more excludable, and its returns become more appropriable, so that the market failure concerning knowledge as a public good has less force, and less policy may be needed, in that respect. In other words, limitations of market operation through rigidity in the form of imperfect spillover limits market failure that discourages investment in knowledge. However, there can still be spillovers to a greater or lesser extent, and appropriability of innovation can still be a problem.

Second, in the new view one can still maintain that markets function better than central planning (see Hayek 1945). The reason may be different – based on the idea of dispersed knowledge, rather than optimal allocation – but the conclusion is still the same. In other words, the coordination required should be left as much as possible to decentralised actors that have the local knowledge.

Third, one may still agree that competition is needed. Here again, the reason may be different, based on the idea of selection in an evolutionary process, to select from the many trials and errors that emerge from a diversity of views and knowledge, rather than efficient allocation of scarce resources, but again the conclusion is still the same. One can still agree that entry barriers that obstruct entry of innovators and 'challengers' are to be prevented or lowered. Not so much to increase efficiency of allocation, but to give room for innovation. In evolutionary terms: to open up the selection environment.

Fourth, one may still agree that while certain rigidities are inevitable and indeed enable markets, they can become excessive. We should look for optimal or temporary, and hence 'flexible' rigidities (Dore 1986) that provide the basis for investment for productivity and innovation while not eliminating variety generation and selection, maintaining openness to new ideas, products, processes and institutions. This issue of how to combine rigidity and flexibility, which is a manifestation of the ancient problem of stability and change, will re-appear later in this book, in chapter 3, as the problem of combining exploration and exploitation. It is arguably the most fundamental issue in innovation policy and research.

1.6 INNOVATION POLICY

Science and technology policies emerged in the mid-twentieth century, triggered by the challenges of World War II and the subsequent Cold War (e.g., the space race) (Lundvall and Borras 2005). These policies were focused upon universities, research institutions, technological institutes, and R&D laboratories (see Bush 1945). In subsequent technology policy, emerging in the 1960s, attention was widened from knowledge institutes to sectors of the economy using and producing technologies, and linkages between them and knowledge institutes. A classic issue that is still with us, is to what extent universities should produce knowledge for industry or remain autonomous. In some EU countries technology policy has led to the promotion of industrial complexes, connecting public users and segments of industries, and 'national champions' in selected industries. Later, such industrial policies were seen to evoke too much public involvement in private interests and to detract too much from market mechanisms. Since the 1990s, policy was further widened to include additional actors in the 'innovation system' such as, in particular, entrepreneurs, but also users, and the public with its perceptions of technology. Also, policy opened up to non-technological aspects of innovation – such as design, marketing, and organisation – and innovation in and by services.

According to Lundvall and Borras (2005: 612):

"The major reason for innovation policy becoming more broadly used as a concept was the slow-down in economic growth around 1970 and the persistence of sluggish growth as compared to the first post-war decades. The reasons for the slow-down in the growth in 'total factor productivity' were, and still are, not well understood but there was a feeling that it had to do with a lack of capability to exploit technological opportunities".

In the Netherlands in the 1950s policy was oriented towards reduction of unemployment and regional backwardness, for the sake of equity (Raspe and Van Oort 2007). In the 1960s policy was more oriented towards growth potential, but in the 1970s, during an economic slump, policy again became more defensive, protecting firms and regions. In the 1980s, based on a report from the Netherlands' Scientific Council for Government Policy (WRR 1980), policy became more offensive again, orienting itself towards 'picking winners' in new growth areas, focusing on technological change, and innovation, in 'arrowhead' sectors. In the 1990s the strategy shifted, with more attention given to knowledge, networks and clusters, in the transition from an industrial to a knowledge economy, with a role for the government as a broker in relationships. More recently, attention has shifted from picking winners, with the recognition that it is at odds with the unpredictability of future success, as indicated by an evolutionary perspective, to 'backing winners' that have demonstrated their viability in market success (AWT 2003).

Radical, path-breaking innovation is rare, and most innovative activity lies in incremental innovation, diffusion, and imitation. A key question that is asked in policy is where we should choose to lead, creating radical innovation, and where to follow. Should we 'pick winners' or 'back winners', or are both choices problematic? If radical, path-breaking, paradigm-switching innovations are indeed rare, relative to incremental innovation, diffusion, and imitation, one should expect a small country like the Netherlands to be a leader, in the sense of creating those radical innovations, only in few cases, compared to a mass of imitations and improvements. Why would one want to be a leader rather than a follower in the first place? In early stages of innovation, which carry an emphasis on product innovation, profits are higher, allowing for higher real wages, while in later stages competition increases, there is a shift to process innovation, and prices fall (Freeman and Perez 1988). As a follower one benefits only from the latter, while as a leader one benefits from both. This issue is taken up further in chapter 3. There may be other advantages to being a leader, such as being able to set a technical standard that fits established assets and competencies, while as a follower one may have to accept a less congenial standard.

Given that one cannot be a leader everywhere, where, then, should one choose to be a leader and where a follower? Is such a choice possible, or is leadership an emergent phenomenon that cannot be planned? We note that also followers need to build and maintain absorptive capacity, needed to implement what leaders have created. In other words, even as a follower one would still need to have a basis of knowledge and skill, in research and education, which is sufficiently broad not to miss out on opportunities to be a follower, and sufficiently deep to be a fast follower. Nevertheless, the question remains where to be a leader and where a follower.

Note that the requirement to develop absorptive capacity, in order to be a follower, yields an argument for the policy of protecting an economy from world trade until it has built up absorptive capacity, because otherwise it will loose out on both leadership and followership. Here, absorptive capacity includes not only cognitive capabilities, based on education and research, but also physical and institutional infrastructure, health care, basic industries supporting transport and communication, and a variety of business services needed, among other things, to reduce transaction costs. After World War II this build-up occurred in Europe with the aid of the Marshall Plan. China and

India are growing now, and are developing from followers into leaders, after many years of protective policies to build up a basis of absorptive capacity.

Here, a perceived urge arises for policymakers to make a choice, on the basis of expected unique strengths in the future ('picking winners'), or to enhance strengths that have been proven in the market ('backing winners'). As noted above, the former approach was taken, in the Netherlands, in the 1980s, with a choice of 'arrowhead' sectors, and the latter policy is taken now, with the choice of 'key areas'. Picking winners is now recognised as going against the unpredictability of future success, emphasised by an evolutionary perspective. But for backing winners the question is why they should be backed if they are winners. If they have been proven in the market, surely this means that they are generating profits for their own expansion. We will pick up the issue of targeted industrial policy again in chapter 12.

Whither innovation policy?

The Austrian and evolutionary views have yielded a perspective of 'innovation systems'. Originally this was a purely descriptive, analytical category: innovation arises from interactions, of both competition and collaboration, between a multiplicity of individual and corporate actors, in science, industry, and government, and interest groups, conditioned by historically and regionally or nationally specific institutions (shared habits, norms, routines, established practices, rules, and laws). Innovation requires not only science, technology, and entrepreneurship, but also finance, education, training, and a variety of institutions concerning property rights, standards, competition, and disclosure. Next to these supply factors, advanced users are also of the utmost importance. Organisation, within and between organisations, conditioned by institutions, plays a central role. Innovation is conditioned by institutions but also causes institutional change. For recent summaries of the system view, see Borras (2003) and Chaminade and Edquist (2006). The system view has been adopted by many innovation scholars, the OECD, and in the Netherlands by the Dutch Advisory Council for Science and Technology (AWT) and, more recently, by the Innovation Platform instituted by the Netherlands government in 2004. One policy implication of the system view is that policy should be based on an 'integral perspective', taking into account diverse actors, institutions, and linkages between them (Edquist 1997; AWT 2003, 2007).

Systemic coherence of factors arises on the level of the entire innovation system, but also in parts of it. Take finance. Venture capital is needed to supply capital for entrepreneurial ventures that banks are too risk averse to supply. Even venture capital mostly fails to provide 'seed capital' at the early start of new ventures. For this one has to fall back on friends, family or 'business angels', who themselves are mostly successful entrepreneurs who have cashed in on their own successful ventures. They are not in the position to coach firms along the full trajectory of development, and require venture capital to take over or pitch in after the early start. Venture capitalists, in turn, want to exit after the venture has proven itself, and require stock markets to capitalize successful ventures, or private equity funds or firms that acquire them. Venture capitalists need to acquire their funds from institutional investors (pension funds, insurance companies), banks and private investors. Thus there is whole supply chain of finance. All the contracting involved requires appropriate legal forms and services. Hellman (2000) reports that in the US the SBIC programme, instituted to help private markets to supply capital to new entrepreneurial firms, played an important role in getting the venture capital business under way. Thus there is a whole web of firms and institutions that makes the finance subsystem work. It is an important part of innovation policy to make it work.

Innovation is a distributed phenomenon, even from the single firm's viewpoint. As a matter of fact, most innovations are the result of new forms of coordination among several firms and institutions rather than of the independent actions of single dominant innovating firms. In this light, particularly important is how the innovating firms acquire, accumulate, and develop knowledge other than scientific and technical knowledge which is material to innovation, (namely) knowledge about the specific characteristics of customers and markets, which in turn has wider connections to knowledge about economic, social, and regulatory changes (Metcalfe 2000). Such coordination among firms and institutions in the innovation process is a core element of the innovation system literature, with its core assumption that a number of organisations (such as research institutes, educational facilities, and financial organisations) provide complementary inputs essential to the innovation process (Edquist 1997).

The distributed, network nature of innovation is particularly pronounced in services, which often perform a linkage function, and in the increasing intertwining of manufacturing and services, in the combination of technology, design, marketing, distribution, organisation, the involvement of users and suppliers, and learning by interaction. In the past we were inclined to see innovation as technology, and a given technology as being related to a given industry, so that innovation policy is prone to be given the form of industrial policy. Now, we see innovation as a phenomenon of chains and networks that run, in principle, crisscross through all industries. This has important implications for policy, in particular the debate on 'key areas', which will be taken up in chapter 12.

Metcalfe (2003) argues that (innovation) system failure should be taken as a starting point for policy intervention, instead of conventional policy, which is preoccupied with market failure and optimal policy. Again, this is not to say that the relevance of market failures for underpinning innovation policy is denied. On the contrary, poor access to information, for instance, should be tackled by policy intervention, but it requires additional policy actions to be effective. The objective of innovation policy is to encourage and facilitate the generation, application, and diffusion of new ideas. The government might directly intervene, through the supply of R&D, education, and capital that match the need of local firms, and which increase the absorptive capacity[5] and innovative capability of firms. Public policy can also stimulate the effective transfer of knowledge through various mechanisms, such as spin-off dynamics, labour mobility, and collaborative networks (see chapter 11).

For innovation, there are system failures of many kinds. One is the condition of inconsistencies between different parts of a system innovation, or clashes of interest and stalemates among multiple stakeholders involved. Examples are given in chapter 6. System failures also include institutions that often were functional in previous stages of development but now hinder radical innovation. Examples are distribution channels, forms of organisation, standards, physical infrastructure, educational structures and programmes, political views, social practices, modes of thought, and ways of looking at the world, which developed to suit earlier changes of technology, products, tastes, and styles. Strong forces of social legitimation enforce conformity to established views and practices. In evolutionary terms, the selection environment, with its market structures and institutions, in due course moulds itself to suit successful breakthrough innovations that have diffused, and then can form a powerful obstruction for the next radical innovation. In other words, there is co-evolution between markets and institutions. Systemic coherence between different elements of the selection environment obstruct piecemeal deviations.

A problem for policy is that often market and system failures are case-specific, and are difficult to cover with a generic, one-size–fits-all policy. For example, when an innovation is competence-destroying, or when it does not fit the installed base of assets of powerful incumbents, the conservative force of established interests is clearly greater than when such conditions are not present. Some innovations entail more externalities, and thus require more government intervention, than others. Some innovations require more change of standards, educational systems, distribution channels, or forms of organisations than others. When price elasticity is low, established firms are less inclined to go for cost-reducing innovations, since they will yield few extra sales, whereas new entrants that can gain market share are more motivated to enter with the newest technology (Langlois and Robertson 2005), so that there is more need to make room for entrants.

Thus, there are good arguments against laissez faire, and in favour of specific policy interventions. However, in time, among some innovation scholars the innovation system perspective has developed from a purely descriptive category to a normative one, with ambitions not only to eliminate system failures but also to design and govern the system. Recognition of systemic coherence developed into ambitions for systematic design. While we accept that system failures occur, and government should address them, we are very wary of top-down system designs.

The perspective of innovation systems is still very general and little-developed in terms of the causalities involved in the processes of interaction between the various actors, as is recognized in the literature (Borras 2003, Chaminade and Edquist 2006). It is a frame of a painting that mostly still has to be painted. It might even be said that the innovation systems approach is a rather vague framework within which to speculate on some possible relationships between hypothetical actors at a vaguely specified level of abstraction, and that it consequently has only very limited utility for guiding innovation policy; theory has even been led by policy, not the preferred other way around (see for example Lovering 1999; Hers and Nahuis 2004).

1.7 OUTLINE OF THE BOOK

We proceed with a macroeconomic analysis of innovation as a source of productivity, and we confirm the idea that we need to delve further into the micro-level (chapter 2). For that, we start with an analysis of knowledge and learning (chapters 3 and 4), and proceed with an analysis of entrepreneurship (chapter 5), obstacles to innovation (chapter 6), institutional conditions for trust to support learning by interaction (chapter 7), forms of organisation (chapter 8), conditions for innovation within organisations (chapter 9), network structure as organisation between organisations (chapter 10), and properties of regional systems of innovation (chapter 11).

Each chapter ends with implications for policy. These policy implications are primarily for public policy, but secondarily also for policies of firms and other organisations in the private and public sector. The latter implications, oriented towards industry, are still relevant for public policy with a view to the diffusion of best practices, which applies to organisational innovation no less than to technological ones. The final chapter integrates the insights from the preceding chapters and discusses policy implications.

The occasion for the production of this volume was a study on innovation policy produced for the Dutch government, and thus has a Dutch orientation and flavour, both in the selection of illustrative cases and in the reference to existing innovation policy. However, the analysis of micro-foundations, and types of policy are, we expect, more generic and also of interest to policymaking elsewhere.

In a little more detail, the further content of thos volume is as follows:

In chapter 2, Gerard de Vries presents a macroeconomic diagnosis of the Netherlands, in comparison with other EU countries and the US. As is customary in macroeconomics, the core of this is an analysis of total factor productivity and an attempt to understand its development. The conclusion is threefold. First, the declining growth of total factor productivity indicates the need for enhanced innovation. Second, there are some indications that the Netherlands could perform better in a number of relevant dimensions of innovation. Third, apart from this broad diagnosis, macroeconomic analysis does not help us much further in the design of innovation policies, and for that we must proceed further into micro-foundations, in the analysis of processes of interaction between relevant agents, in the innovation system.

In chapter 3, Bart Nooteboom discusses, criticises, and extends the evolutionary perspective of innovation. While that perspective is very useful, and yields some important policy implications, it is also limited in that it does not provide a theory of discovery or invention. To compensate for this, he brings in elements from theories of cognition that underpin learning as a process of interaction. The need and difficulty of combining exploration and exploitation is identified as a key and pervasive issue. A model is applied of how exploration and exploitation may build upon each other, in a 'cycle of discovery'. Insight into processes of innovation by interaction are further developed by means of the notion of 'optimal cognitive distance'. The chapter ends with a call for an innovation policy that is open in four dimensions: open with respect to collaboration, with open communication; open for surprises and changes of direction during innovation projects; open to new players ('challengers'); and open to global linkages.

In chapter 4, Gerrit Kronjee and Bart Nooteboom discuss the creation and application of knowledge, in particular the role of universities, science policy, and the relations between university and industry. They discuss the necessary openness of innovation to variety and surprise in R&D. They argue that while efforts to apply scientific knowledge, and to indicate priority themes for research are valid, autonomy of fundamental university research should be preserved. There is a natural division of labour between universities and institutes of higher vocational education, with the latter focusing more on applied research, and they should obtain the means and the authority to do so. Institutes of higher vocational education are a natural ally for knowledge for smaller low-tech firms. For the interaction between university and large firms and high-tech small firms ideas are proposed to further activities of exploration between them, in a 'third space' of activities, to support industry in activities of exploration and for universities to test their ideas and to gain inspiration for fundamental research.

In chapter 5, Erik Stam discusses the nature of entrepreneurship and its relation to innovation. He addresses the necessary openness of innovation to outsiders and challengers. An overview is given of theory and empirical research on the effects of entrepreneurship on innovation and economic growth in OECD countries. An in-depth study is made of entrepreneurship and innovation in the Netherlands from an international and historical perspective. This study shows that the annual number of new firms has increased spectacularly (almost tripled) during the period 1987-2006. However, the study shows more weaknesses than strengths with regards to entrepreneurship. First, a large part of the new firms seem to be self-employed individuals who continue with the same activities (mainly in the construction and services sectors) they they had executed as employee before. This is not likely to improve the (product) innovativeness of the economy. Second, on average small- and medium-sized enterprises (SMEs) have become less and not more innovative in the last decade (1999-2007). The percentage of innovative SMEs is much lower than the EU average. Third, the Netherlands are lagging behind internationally with respect to ambitious entrepreneurship. The low number of ambitious entrepreneurship seems to be especially worrying, as such entrepreneurship is a strong driver of national economic growth. The chapter continues with the role of entrepreneurship in innovation policy. Several specific types of entrepreneurship – technology start-ups, spin-offs and corporate venturing, and high growth start-ups – are considered.

In chapter 6, Leo van der Geest and Lars Heuts discuss four Dutch cases in which innovation ran up against obstacles, and the policy implications. Thus, like chapter 5, this chapter addresses the issue of openness of systems and institutions to novelty. The cases are: energy from windmills and other alternative sources, an electronic patient dossier, a 'whisper' coach, and energy neutral houses. These cases give insightful examples of obstacles to innovation in general, and of system failures in particular, where there are several stakeholders that each have excellent reasons not to make the move they would need to make in order to set the system going. Later, in chapter 12, this results in the ideas of a 'deblocking brigade' and an 'ombudsman for entrepreneurs'. The deblocking brigade should help to unblock difficult stalemates among stakeholders in ways that only the government can do. An ombudsman for entrepreneurs should help entrepreneurs through the density of rules and regulations, and monitor complaints against obstacles from inconsistent or excessively complex institutional arrangements, and against obstacles from vested interests.

In chapter 7, Bart Nooteboom considers the governance of collaboration for innovation, within and between organisations, in particular the meaning, dimensions, conditions, and limits of trust. He addresses the openness of innovation to collaboration, within and between firms, and the corresponding need for open communication. He considers the role of government in facilitating trust and collaboration, and the implications of governance for the structure of networks and regional 'clusters'. One conclusion is that a mentality of excessive control has developed, in both public and private organisations, for the sake of 'accountability', that is detrimental to trust and excessively risk-averse, de-motivating, and constrains discretion of professional judgement as well as innovation. Another conclusion is that while in the emerging network economy there is a growing demand for go-betweens to facilitate collaboration, government should be careful in assuming that role.The development of a new branch of commercial services is to be preferred. Yet another conclusion is that policy should switch from a rhetoric of maximum flexibility of relationships, in labour and corporate governance, to a perspective of optimal flexibility that allows for minimum stability of relations to evoke investments in mutual understanding and trust.

In chapter 8, Bart Nooteboom and Robert Went discuss an important aspect of 'social innovation', in particular the relation between innovation and organisation. They further develop the issue of openness of innovation with respect to relationships within and between organisations. They look at both organisation *for* innovation and innovation *of* organisation. They consider conditions for work, forms of organisation to meet the central challenge of combining exploration and exploitation, platforms for serving customers, the role of users in innovation, and 'open source communities'. They also look at the scope that this, in particular the notion of platforms, yields for innovation in government services. Much of this is enabled by ICT and Internet. They see a large and largely untapped potential for improvement of quality and increase of productivity. This untapped potential might explain the lag in productivity behind the US, identified in chapter 2. In policy to promote social innovation one should beware that also the urgency and relative ease of exploitation tends to crowd out attention and resources for exploration.

In chapter 9, Neil Anderson and Rosina Gasteiger give a survey of the applied psychology literature on organisational conditions for creative teams. They address the openness of innovation with respect to collaboration and labour within organisations. For example, they look at the role of things such as work stress, autonomy, intrinsic motivation, trust, and type of leadership. This indicates that key dimensions of high quality of labour favour innovation, which yields scope for a coalition between employers and employees. A policy implication of this is that labour should be closely involved in innovation policy, in both firms' and public policy.

In chapter 10, Marius Meeus, Leon Oerlemans, and Patrick Kenis discuss interorganisational networks for innovation. They address the issue of openness of innovation to collaboration between organisations, and the trade-off between openness and closure. They summarise the literature on relevant dimensions of network structure and their effects on innovation. They then try to trace what instruments of innovation policy affect which of the relevant dimensions of network structure. One conclusion is that few instruments appear to impact those dimensions, which opens possibilities for further design of policy.

In chapter 11, Ron Boschma discusses the evolutionary view in economic geography, and in particular the role of 'related variety' in regional growth. Among other things, he addresses the necessary openness of agglomerations to outside linkages, with other agglomerations, and to entry and exit of actors. He concludes that regional innovation policy needs to account for the region-specific context because it provides opportunities but also sets limits to what can be achieved by public policy. In doing so, it should neither apply 'one-size-fits-all' policies nor adopt 'picking-the-winner' policies. Instead of copying best practice models or selecting winners, policy should take the history of each region as a basic starting point, and identify regional potentials and bottlenecks accordingly. To avoid regional lock-in, it is crucial that public policy is open to newcomers and new policy experiments.

In chapter 12, Bart Nooteboom and Erik Stam give an integrated survey of conclusions. Innovation policy should take into account the radical uncertainty of innovation, and the impossibility to predict and plan success. It should tap into the variety of ideas and initiatives in society. This is not, however, a traditional laissez faire policy.. Policy should condition, enable, and stimulate sources of variety, in ideas and ventures, remove obstacles, and coordinate where necessary. In addition to traditional arguments of market failure there are also system failures that require intervention. But in intervention we should take into account government failures. These principles and the four dimensions of openness are used to discuss policy implications. In particular, the following major issues are addressed:

- How can we manage the combination and tension between exploitation and exploration, and stability and change, in research and in the relations between university and industry, in the organisation of firms and government, and in inter-organisational collaboration and networks?
- Can we determine where, in innovation, a country should be a leader and where a follower. What are the merits and drawbacks of the current policy of 'backing winners'? Can such choices be made while maintaining openness to outsiders and challengers?
- How can we further improve openness for entrepreneurs, and deal with obstacles, in market and system failures? How can this be done while taking into account government failures?
- How can we provide openness for surprise in innovation projects and processes? In government-funded projects and programmes, can we switch from *ex ante* safeguards that lock away innovation to *ex post* accountability that allows for changes of direction?
- How can we organise openness for collaboration and communication, in alliances and networks? What is the role, and what are the limits of trust? How to combine trust and transparency? What are the effects of network structure?
- How can we combine the strengths of the local with the need for openness to the non-local and the global, in organisations, networks, regions, and countries?

NOTES

1 It is not clear, however, what defines 'the challengers': are these all new firm entrants, also including foreign direct investments, or only new firms that have reached a substantial size and are really able to challenge incumbents?

2 However, evaluations of the Dutch WBSO program (R&D subsidies) found that this triggered additional private investments of 2 to 94 percent (Brouwer et al. 2002; De Jong and Verhoeven 2007).

3 We adopt only part of Hayek's perspective: the heterogeneity, dispersion and local nature of knowledge. We do not adopt Hayek's view of laissez faire. What Hayek neglects is the social, interactive nature of knowledge, and the resulting importance of interaction, which encounters obstacles and system and institutional failures that often require government action.

4 The OECD (1998: 102) has defined 'systemic failures' as mismatches between the components of an innovation system. More specifically, it refers to a situation in which organisations, institutions, or interactions between elements of the innovation system are inappropriate or missing (Edquist 2001).

5 As noted earlier, investment in education/human capital might be necessary to improve the absorptive capacity of firms, in order to use new technologies in the production process (Bovenberg and Theeuwes 2004). A lack of investments in education can be seen as institutional failure, instead of market failure (Bovenberg and Theeuwes 2004).

REFERENCES

Acocella, N. (2005) *Economic Policy in the Age of Globalisation*, Cambridge: Cambridge University Press.

Aghion, P.H., R. Blundell, R. Griffith, P. Howitt, and S. Prantl (2006) 'The Effects of Entry on Incumbent Innovation and Productivity', NBER Working Paper 12027.

Amabile, T.M. (1983) *The Social Psychology of Creativity*, New York: Springer.

Anderson, C. (2006) *The Long Tail: Why the Future of Business is Selling Less of More*, New York: Hyperion.

Argyris, C. and D. Schön (1978) *Organizational Learning*, Reading, MA.: Addison-Wesley.

Arundel, A. (2007) 'Innovation Survey Indicators: What Impact on Innovation Policy?' in: OECD, *Science, Technology and Innovation Indicators in a Changing World: Responding to Policy Needs*, Paris: OECD, p. 49-64.

AWT (2003) *Backing Winners*, Den Haag: AWT.

AWT (2007) *Balanceren met beleid*, Den Haag: AWT.

Bateson, G. (1972) *Steps to an Ecology of Mind*, Chicago: University of Chicago Press.

Baumol, W. J. (2001) 'When is Inter-firm Coordination Beneficial? The Case of Innovation', *International Journal of Industrial Organization*, 19: 727-737.

Baumol, W. J. (2002) *The Free-Market Innovation Machine: Analyzing the Growth Miracle of Capitalism*, Princeton: Princeton University Press.

Boone, J. and E. Van Damme (2004) 'Marktstructuur en innovatie' in: Jacobs, B. and J.J.M. Theeuwes (eds.) (2004) *Innovatie in Nederland: De markt draalt en de overheid faalt*, Amsterdam: Koninklijke Vereniging voor de Staatshuishoudkunde, 71-92.

Borras, S. (2003) *The Innovation Policy of the European Union: From Government to Governance*, Cheltenham: Edward Elgar.

Bovenberg, L. and J. Theeuwes (2004) 'Arbeid en Innovatie' in: Jacobs, B. and J.J.M. Theeuwes (eds.) (2004) *Innovatie in Nederland: De markt draalt en de overheid faalt*, Amsterdam: Koninklijke Vereniging voor de Staatshuishoudkunde, 137-160.

Brouwer, E., P. den Hertog, T. Poot and J. Segers (2002) *WBSO nader beschouwd: Onderzoek naar de effectiviteit van de WBSO*, The Hague: Ministerie van Economische Zaken.

Bush, V. (1945) *Science: The Endless Frontier, A Report to President Roosevelt Outlining his Proposal for Post-war U.S. Science and Technology Policy*, Washington, DC: United States Government Printing Office.

Chaminade, C. and C. Edquist (2006) 'From Theory to Practice: The Use of the Systems of Innovation Approach in Innovation Policy' in: Hage, J. and M. Meeus (eds.) *Innovation, Science and Institutional Change*, Oxford: Oxford University Press.

Cornet, M. and J. van de Ven (2004) 'De markt voor nieuwe technologie' in: Jacobs, B. and J.J.M. Theeuwes (eds.) (2004) *Innovatie in Nederland: De markt draalt en de overheid faalt*, Amsterdam: Koninklijke Vereniging voor de Staatshuishoudkunde, 35-67.

De Jong, J. and W. Verhoeven (2007) *Evaluatie WBSO 2001-2005. Effecten, doelgroepbereik en uitvoering*, Zoetermeer: EIM.

Dicken, P. (2003) *Global Shift: Reshaping the Global Economic Map in the 21st Century*, London: Sage.

Dore, R.P. (1986) *Flexible Rigidities: Industrial Policy and Structural Adjustment in the Japanese Economy 1970-80*, London: Athlone Press.

Dosi, G. (1982) 'Technological Paradigms and Technological Trajectories: A Suggested Interpretation of the Determinants and Directions of Technical Change' *Research Policy*, 11: 147-162.

Dosi, G. (1984) *Technical Change and Industrial Transformation*, London: MacMillan.

Edgerton, D. (2007) *The Shock of the Old: Technology and Global History Since 1900*, Oxford: Oxford University Press.

Edquist, C. (ed.) (1997) *Systems of Innovation: Technologies, Institutions and Organizations*, London: Routledge.

Edquist, C. (2001) 'Innovation Policy – A Systemic Approach' in: Archibugi, D. and Lundvall, B.A. (eds.) *The Globalizing Learning Economy*, Oxford: Oxford University Press, 219-238.

Feinstein, J. S. (2006) *The Nature of Creative Development*, Stanford: Stanford University Press.

Freeman, C. and F. Louca (2002) *As Time Goes By: From the Industrial Revolutions to the Information Revolution*, Oxford: Oxford University Press.

Freeman, C. and C. Perez (1988) 'Structural Crises of Adjustment: Business Cycles and Investment Behaviour' in Dosi, G., C. Freeman, R. Nelson, G. Silverberg, and L. Soete, L. (eds.) *Technical Change and Economic Theory*, London: Pinter, 38-66.

Gereffi, G. (1999) 'International Trade and Industrial Upgrading in the Apparel Commodity Chain', *Journal of International Economics*, 48: 37-70.

Gault, F. (2007) 'Science, Technology and Innovation Indicators: The Context of Change' in: OECD, *Science, Technology and Innovation Indicators in a Changing World: Responding to Policy Needs*, Paris: OECD, chp. 1, 9-23.

Hall, P. (1994) *Innovation, Economics and Evolution: Theoretical Perspectives on Changing Technology in Economic Systems*, Hemel Hempstead: Harvester Wheatsheaf.

Hayek, F. (1945) 'The Use of Knowledge in Society', *American Economic Review*, 35(4): 519-530.

Hawkins, R.W. (2007) 'University Research in an "Innovation Society"' in: OECD, *Science, Technology and Innovation Indicators in a Changing World: Responding to Policy Needs*, Paris: OECD, p. 171-191.

Hellman, T. (2000) *Venture Capitalists: The Coaches of Silicon Valley*, Stanford: Stanford University Press.

Hers, J. and N. Nahuis (2004) *The Tower of Babel? The Innovation System Approach Versus Mainstream Economics*, The Hague: Ministry of Finance.

Innovation Platform (2006) 'Kennisinvesteringsagenda 2006-2016: Notitie van de werkgroep kennisinvesteringsagenda van het innovatieplatform' Den Haag: Innovatieplatform.

Irwin, D. and P. Klenow (1994) *High Tech R&D Subsidies: Estimating the Effects of Sematech*, Cambridge, MA: NBER.

Jacobs, B. and J.J.M. Theeuwes (eds.) (2004) *Innovatie in Nederland: De markt draalt en de overheid faalt*, Amsterdam: Koninklijke Vereniging voor de Staatshuishoudkunde.

Kleinknecht, A., K. Van Montfort, and E. Brouwer (2002) ,The Non-trivial Choice between Innovation Indicators', *Economics of Innovation and New Technologies*, 11(2): 109-121.

Kline, S.J. and N. Rosenberg (1986) ,An Overview of Innovation' In: R. Landau and N. Rosenberg (eds.) *The Positive Sum Strategy: Harnessing Technology for Economic Growth*, Washington DC: National Academic Press.

Knight, F. H. (1921) *Risk, Uncertainty and Profit*, Boston, MA: Hart, Schaffner & Marx.

Krugman, P. (1987) 'Is Free Trade Passé?', *Journal of Economic Perspectives*, 1:131-144.

Ledyard, J.O. (1989) 'Market Failure' in: Eatwell, J., M. Millgate, and P. Newman (eds.) *The New Palgrave: Allocation, Information and Markets*, London: Macmillan, 185-190.

Levinson, M. (2006) *The Box: How the Shipping Container Made the World Smaller and the World Economy Bigger*, Princeton: Princeton University Press.

Lovering, J. (1999) 'Theory Led by Policy: The Inadequacies of the "New Regionalism"', *International Journal of Urban and Regional Research*, 23 (2): 379-395.

Lundvall, B.A. (1988) 'Innovation as an Interactive Process: From User-Producer Interaction to the National System of Innovation' in: Dosi, G., C. Freeman, R. Nelson, G. Silverberg, and L. Soete (eds.) *Technical Change and Economic Theory*, London: Pinter, 349-369.

Lundvall, B.A. and S. Borrás (2005) 'Science, Technology, and Innovation Policy' in: Fagerberg, J., D.C. Mowery, and R.R. Nelson (eds.) *The Oxford Handbook of Innovation*, Oxford: Oxford University Press, 599-631.

Maggioni, V., M. Sorrentino, and M. Williams (1999) 'Mixed Consequences of Government Aid for New Venture Creation: Evidence from Italy', *Journal of Management and Governance*, 3(3): 287-305.

March, J. (1991) 'Exploration and exploitation in organizational learning', *Organization Science*, 2(1): 101-123.

Meeus, M.T.H. and J. Faber (2006) 'Interorganizational Relations and Innovation: A Review and a Theoretical Extension' in: Hage, J. and M. Meeus (eds.) *Innovation, Science and Institutional Change*, Oxford: Oxford University Press.

Metcalfe, J.S. (1998) *Evolutionary Economics and Creative Destruction*, London: Routledge.

Metcalfe, J.S. (2003) 'Equilibrium and Evolutionary Foundations of Competition and Technology Policy: New Perspectives on the Division of Labour and the Innovation Process' in: P. Pelikan and G. Wegner (eds.), *The Evolutionary Analysis of Economic Policy*, Cheltenham: Edward Elgar, 162-190.

Metcalfe, S. and R. Ramlogan (2005) 'Competition and the Regulation of Economic Development', *Quarterly Review of Economics and Finance*, 2(3): 215-235.

Mokyr, J. (1990) *The Lever of Riches: Technological Creativity and Economic Progress*, Oxford: Oxford University Press.

Nahuis, R. and J. Van de Ven (1999) 'Endogenous Growth in a Small Open Economy: Any Policy Lessons?', *The Economist*, 147: 353-360.

Nelson, R.R. and S.G. Winter (1982) *An Evolutionary Theory of Economic Change*, Cambridge, MA and London: The Belknap Press.

Niosi, J. (1995) *Flexible Innovation*, Montreal and Kingston: McGill-Queens University Press.

Nooteboom, B. (2000) *Learning and Innovation in Organizations and Economies*, Oxford, Oxford University Press.

OECD (1998) *The OECD Jobs Strategy: Technology, Productivity and Job Creation, Best Policy Practice*, Paris: OECD.

OECD (2005) *Oslo Manual: Guidelines for Collecting and Interpreting Innovation Data*, 3rd edition, Paris: OECD/Eurostat.
OECD (2007) *Science, Technology and Innovation Indicators in a Changing World: Responding to Policy Needs*, Paris: OECD.
Olson, M. (1965) *The Logic of Collective Action*, Cambridge: Cambridge University Press.
Pittaway, L., M. Robertson, K. Munir, D. Denyer, and A.D. Neely (2004) Networking and Innovation: A Systematic Review of the Evidence, *International Journal of Management Reviews* 5-6(3-4): 137-168.
Raspe, O and F. van Oort (2007) *Ruimtelijk-economisch beleid in de kenniseconomie* (*Spatial economic policy in the knowledge society*), The Hague: Ruimtelijk Planbureau.
Richardson, G.B. (1975) 'Adam Smith on Competition and Increasing Returns,' in: Skinner, A., and T. Wilson (eds.) *Essays on Adam Smith*, Oxford: Oxford University Press.
Rosenberg, N. and L. Birdzell (1986) *How the West Grew Rich*, New York: Basic Books.
Santarelli, E. and M. Vivarelli (2002) 'Is Subsidizing Entry an Optimal Policy?', *Industrial and Corporate Change*, 11: 39-52.
Schumpeter, J.A. (1934) *The Theory of Economic Development*, Cambridge, MA: Harvard University Press.
Schweizer, T.S. (2004) *An Individual Psychology of Novelty-Seeking, Creativity and Innovation*, Rotterdam: ERIM.
Shane, S.A. (2000) 'Prior Knowledge and the Discovery of Entrepreneurial Opportunities', *Organization Science*, 11(4): 448-472.
Shane, S. (2004) *A General Theory of Entrepreneurship: The Individual-Opportunity Nexus*, Cheltenham: Edward Elgar.
Sternberg, R.J. (1996) *Cognitive Psychology*, Fort Worth: Harcourt Brace College Publishers.
Suyker, W. and H. de Groot (2006) *China and the Dutch Economy; Stylised Facts and Prospects*, CPB Document 127, The Hague: CPB.
Wallsten, S. (2000) 'The Effects of Government-Industry R&D Programs on Private R&D: The Case of the Small Business Innovation Research Program', *Rand Journal of Economics*, 31(1): 82-100.
Went, R. (2007) Globalisering is onbekend en omstreden (Globalization is unknown and contested), *Economisch Statistische Berichten*, July 10, 10-14.
Weterings, A., O. Raspe, and F. Van Oort (2006) *De atlas van kennis en innovatie*, The Hague: Ruimtelijk Planbureau.
Wever, E. and E. Stam (1999) 'Clusters of High Technology SMEs: The Dutch Case', *Regional Studies*, 33(4): 391-400.
Wright, R. (2007) 'Western Retailers Shift their Supply Chain Tasks to China', *Financial Times*, March 27.
WRR (1980) *Plaats en toekomst van de Nederlandse industrie*, The Hague: Wetenschappelijke Raad voor het Regeringsbeleid.

2 INNOVATION AND MACROECONOMICS

Gerard de Vries

2.1 ECONOMIC GROWTH, PRODUCTIVITY AND INNOVATION

2.1.1 INTRODUCTION: THE IMPORTANCE OF ECONOMIC GROWTH

The functioning of welfare states profits immensely from a fair measure of economic growth. It enables them to cope with structural changes of their economies and labour markets, to address new challenges such as those stemming from the ageing of the population and climatic changes, and to enact reforms to modernise its social arrangements. Furthermore, economic growth provides the means for realising the demands of the population for more economic progress and more welfare. Therefore, it is interesting to assess the extent to which innovation can contribute to the emergence of economic growth.

2.1.2 THE CONTRIBUTION OF INNOVATION TO ECONOMIC GROWTH

Innovation is considered an important factor of economic growth, and more often than not, as the single most important factor. William Baumol thinks that it is obvious that "a reliable stream of innovation is the most important requirement for the remarkable long-run economic growth that has been experienced by the industrialised economies in the past two centuries" (Baumol 2004: 186). In the same vein, Paul Romer is convinced "beyond any doubt" that "discovery, invention, and innovation are of overwhelming importance in economic growth" (Crafts 2004: 205).

However, the role of innovation is seldom given its full due because its meaning is limited usually to that of economic growth. Its importance in general with regard to welfare is even greater. Of the seven major factors that affect happiness only two concern economic values: income and work (Layard 2005: 63). The other five are family relationships; community and friends; health; personal freedom and personal values, which can all profit immensely from innovation as well. The scientific progress in the medicine and health care sector may be of vital importance in many cases. The same can be said of the social sciences, which help us to understand and improve human relationships.

The assertion that innovation is the "mainspring of economic growth" (*The Economist* 2007: 29) has been the subject of a great deal of research. From an historical point of view, economic growth is a rather recent phenomenon. Over the past two centuries, per capita income has risen 14-fold world-wide. Prior to 1820, economic growth was quite limited and parallel to the increases in population, which left per capita income roughly unaltered (Maddison 2004: 29). Over the past two centuries, not only economic growth, but also increases in capital and education were unsurpassed. The question of which factors caused this rise in per capita income seems to be a salient one. Growth accounting was introduced to investigate the contributions of the various factors. If one supposes that the traditional factors, labour and capital, show constant returns, one can thus calculate what the rate of economic growth would have been without the contributions of increases in the labour force and in capital. By subtracting the weighted growth rates of the capital stock and the labour force (weighted by the share of wages for labour and one minus this share for capital) from the growth rate of GDP, the result is a residual that can not be attributed to traditional factors. This residual is the rate with which the so-called total factor productivity or multiple factor productivity increases, because it is caused by an increase of their productivity and not of their volumes.

This method of eliminating the roles of the growth of capital stock and labour can be refined by also taking into account the increase in the quality of the traditional factors. Thus, the impact of improved or longer education and of improvements in machinery and equipment is eliminated as well. The resulting total factor of productivity (TFP) can be distinguished from the 'crude' or 'unrefined' TFP, which does not specify the improvements in quality. Because TFP is a residual, which is derived after the exhaustive application of available statistical knowledge, Abramovitz has called it "a measure of our ignorance" (Verspagen 2005: 490). Because TFP is the result of all of the factors that influence economic performance other than labour and capital, each of these factors might be given preference. Many authors have attributed the growth of TFP, largely or wholly, to technological change (Helpman 2004: 34-5), but there is no evidence for this claim (Crafts 2004: 216).

In order to give an impression of the main explanatory sources behind the growth of TFP, one can use Denison's pioneering work and Maddison's studies as guides. See table 2.1.

The authors have three sources in common: b, c and e. The catch-up effect results from adopting technologies and business methods used in more productive and

Table 2.1 Sources of TFP growth

	Denison	Maddison
a	Advances in knowledge	
b	Catch-up effect	Catch-up effect
c	Improved allocation of resources	Structural effect
d		Foreign trade effect
e	Economies of scale	Scale effect
f	Unexplained	

Source: Baily and Kirkegaard 2004: 37; Crafts 2004: 210.

better performing countries and industries. The structural effect occurs with a shift of employment from low productive sectors to higher productive sectors. The benefit of scale effects occurs when extensive production enables the application of technologies that are only feasible for mass production. Compared with the growth of productivity that is realised within sectors, the growth that results from shifts between sectors is currently rather small (Erken *et al.* 2007, Table A.2).

One may notice that cultural, political, fiscal and cyclical factors are not listed among the sources presented by Denison and Maddison, although it is clear that TFP must register any impact these factors may have on the growth of GDP.

This section is devoted to the assessment of the importance of innovation for economic growth. It is therefore crucial to learn to what extent the categories presented in table 2.1, are related to innovation. This question enters a difficult area, which concerns the conceptual, analytical and size dimensions of innovation, which makes a short digression on these issues permissible. The subject of innovation has been analysed by economic theorists, econometric researchers and statisticians. The founder of the theory of innovation is Joseph Schumpeter who distinguished five types of innovation; the introduction of new products and new methods of production; the opening of new markets; the development of new sources of supply for inputs; and the creation of a new market structure in an industry (OECD/EC 2005: 29)

Econometric research has concentrated on the quantification and interpretation of TFP, the studies of which resulted in the categories mentioned in table 2.1, which are rather different from those of Schumpeter. Statisticians have developed comprehensive frameworks to comprehend the sometimes elusive elements of innovation. The Oslo Manual that was developed by the OECD and EC, has reached its third edition, with the previous one focussing mainly on technology and manufacturing. The third edition provides three extensions: the role of linkages in innovation processes; the inclusion of services and the acknowledgement of organisation and marketing as a source of innovation (OECD/EC 2005b: 11). The Oslo Manual defines innovation as "the implementation of a new or significantly improved product (good or service), or process, a new marketing method, or a new organisational method in business practices, workplace organisation or external relations" (OECD/EC 2005b: 46).
Besides the Oslo Manual, three other manuals have also been developed especially for Research & Development (Frascati Manual), patents (Patents Manual) and human resources (Canberra Manual) (OECD/EC 2005b: 23). The EC developed a survey based on the Oslo Manual in order to promote the comparability of innovative activities in its member states (Smith 2005: 163).

The various concepts and definitions of innovation and their measurements need to be brought in accordance because it does not make sense to attribute an impact based on a broadly defined concept of innovation to a narrowly selected innovative activity. Another problem is that much of the statistical information is related to inputs because these can be better monitored than outputs. However, as the ultimate goal of innovation is improved total production, inputs are poor substitutes for outputs because of the often deficient relationship between innovative inputs and their intended outputs (Economist Intelligence Unit 2007: 7).

Although the definition of the Oslo Manual mentions 'newness' as a consistent requirement for an innovation, it is also clear that this novelty does not have to be of a radical nature, nor does it mean new in the sense of nonexistent prior to the introduction of the innovation. For this reason it seems clear that both the catch-up effect and the scale effect mentioned in table 2.1 may be classified as resulting from innovation.

2.1.3 THE CONTRIBUTION OF LABOUR TO ECONOMIC GROWTH

A country's population contributes to its economic welfare by the extent to which it is employed or self-employed. Several factors determine this extent:
- the share of the population participating in the active labour force;
- the number of hours of work for a full-time job per week and per year;
- the share of part-time work;
- the number of years people are employed over the course of a lifetime.

Economic welfare is measured in terms of GDP per capita, it therefore follows that labour contributes to increases in economic welfare only when the labour force grows faster than the population as a whole. Although this type of increase has a lasting effect on economic welfare, the increase itself can only be realised once for all. The opportunities for fostering economic growth by higher participation, as opposed to innovation, are limited over the long term.
The labour participation rate in the Netherlands is among the highest in Europe, which means that the probability of any further increases seems rather limited. Meanwhile, however, the participation levels of women and the elderly are in terms of full-time equivalents (much) lower than those in the best performing countries, which means that some improvements can be made in these areas (WRR 2007: 45-8).

Considering the second and third factors mentioned above, an increase seems to be highly dependent on the preferences of the people involved. It seems that the percentage of people in the EU (15) who prefer to work fewer hours is much higher than the percentage of those who prefer to work longer hours (CPB/SCP 2006: 67). The conclusion is that labour volume, notwithstanding the fact that there is still some room for improvement, is not a factor that is likely to increase economic growth substantially.

This can be illustrated by noting that all four factors, mentioned earlier in this section, on balance, actually have had a negative impact on economic growth in the Netherlands (and a minor positive impact in the EU-15 area) (EC 2006: 23).

2.1.4 SOME FIGURES

As has already been noted, economic growth as a source of economic prosperity requires an increase in the output per work hour. The various factors which constitute the growth of labour productivity are summarised in table 2.2., which reveals how, by eliminating the three different contributions, the growth of refined TFP can then be derived from the rate of economic growth.

Table 2.2 Decomposition of labour productivity growth

	Economic growth
Minus:	Contribution of increases in the number of hours worked (1)
	Growth of labour productivity
Minus:	Contribution of the growth of capital stock (2)
	Growth of unrefined TFP
Minus:	Contributions of better education, and an increase in the quality of capital (3)
	Growth of refined TFP

The main factors that explain the growth of TFP:
- Catch-up effect
- Structural effect
- Scale effect

Of course, it would be very interesting to learn which forces are behind the three main factors mentioned at the end of table 2.2. This will require a switch from macro to micro-economic analysis, because these forces lie within the entrepreneurial domain. The growth of labour productivity is presented in table 2.3.

Table 2.3 Labour productivity per hour (total economy, annual percentage changes)

	EU-15	USA
1913-1950	1.6	2.5
1950-1970	4.7	2.8
1970-1990	3.1	1.4
1990-1995	2.2	1.1
1995-2003	1.5	1.9

Source: Smid 2005

Based on the data presented in various studies concerning TFP, it can tentatively be indicated that in the first half of the previous century (some major) OECD countries showed (refined) TFP growth averaging 0.5 percent annually (Crafts 2004: 210), while in the third quarter, the percentages ranged from 0.5 to 3.5 (Crafts 2-4: 210; Baily and Kirkegaard 2004: 37), with the fourth quarter showing growth rates averaging 1 percent (Crafts 2004: 210; Kets and Lejour 2003: 9; Gomez-Salvador *et al.* 2006: 18).

Comparison of these figures with the ones in table 2.3 that correspond in period, suggests that 30 to 60 percent of labour productivity growth can be explained by TFP growth, the remainder being due to increased levels of education and increases in capital stock and its quality (cf. also Crafts 2004: 211; Helpman 2004: 33-4). This margin of 30-60 percent is not uninteresting because it seems that the share of TFP is higher when the growth rate of labour productivity is higher. This is rather plausible, when one assumes that the increases in levels and duration of education, and the increases in capital stock and its quality, may have had a fairly stable impact during the long period concerned.

2.1.5 CONCLUSIONS

Economic welfare in the near future will be mainly dependent upon the growth of labour productivity, and not on labour utilisation. Moreover, labour productivity is dependent on both education and the increases in the quantity and quality of labour, on the one hand, and the growth of TFP, on the other. The higher the growth rate goal, the more crucial the growth of TFP becomes. Furthermore, it is very likely that, in view of the many factors discussed and summarised in table 2.2, innovation is the most important single factor explaining economic growth. With the growth of TFP attributed, to a large extent, to innovation, policy measures to foster economic growth should ultimately be directed at the strengthening of the innovative capacities of the economy.

2.2 PRODUCTIVITY IN THE EU, US AND THE NETHERLANDS

The conclusion that innovation is most likely the most important single factor determining economic growth drawn in the previous section, seems to be more relevant for the Netherlands than for the rest of the OECD area as a whole. Two arguments support this observation. First, the contribution of labour to economic growth has changed more drastically in the Netherlands than elsewhere and, second, the prospect for growth based on innovation is poorer. In this section, the Dutch case will be discussed vis-à-vis the EC and the US.

2.2.1 INTERNATIONAL COMPARISON

Table 2.4 presents growth data for the US, some major EC countries and the Netherlands, with Finland as one of the outstanding performers GDP growth is often considered the most important measure of economic performance. Average growth rates seem to be the best in the US, Finland and the Netherlands. Economic growth divided by population growth yields the per capita GDP growth rate, the most common key figure for increases in economic wealth. It appears that the per capita figures tend to converge. If we subtract the contribution of a higher participation rate from GDP per capita, GDP per employed person results,

Table 2.4 Growth data (annual percentage changes)

	GDP growth			GDP per capita			GDP per employed person			GDP hourly
years	70-80	80-90	90-00	70-80	80-90	90-00	70-80	80-90	90-00	90-00
US	3.2	3.2	3.2	2.1	2.2	2.2	0.8	1.4	1.9	1.6
Germany	2.7	2.2	1.6	2.6	2.0	1.3	2.6	1.7	1.5	1.8
France	3.3	2.4	1.8	2.7	1.8	1.4	2.7	2.1	1.3	1.7
UK	1.9	2.7	2.3	1.8	2.5	1.9	1.7	2.0	2.0	2
Finland	3.5	3.1	2.2	3.1	2.7	1.8	2.5	2.4	2.9	3
Netherlands	2.9	2.2	2.9	2.1	1.6	2.2	2.6	1.3	0.8	1.7

Source: OECD 2003a, Tables A1.1-3; Figure 1.3

the figure for the Netherlands drops drastically. Finally, when subtracting the contribution resulting from an increase in hours worked per person, the hourly GDP emerges.[1]

In table 2.4, the final column clearly reveals a narrow margin, with all of the countries' showing figures within the range of 1.6 to 2 percent, with the exception of Finland. Going from the third column to the final one, it isn't clear why the higher growth rate of the Netherlands, when compared to Germany, France and the UK, results in the lowest hourly GDP rate in the Netherlands among the four.

Figure 2.1 GDP per capita and hour worked in the EU-15 relative to the US, 1960-2006 (US=100)

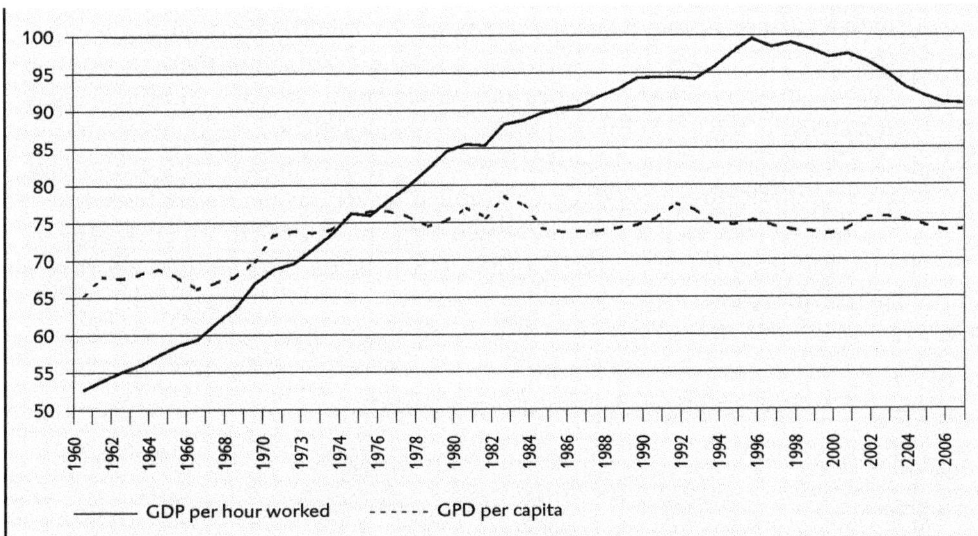

Source: The Conference Board and Groningen Growth and Development Centre, Total Economy Database, January 2007, http://www.ggdc.net

Presenting growth rates as averages for the past decades hides the breach that has emerged since 1995. Until 1995, GDP per hour worked increased more in the EU-15 than in the US. Since 1995, however, the EU-15 increase has been less than that in the US, as shown in graph 2.1 and discussed in detail below. Meanwhile, it becomes clear from table 2.4 that the Dutch growth record over the 1990s was due to a population increase and increased labour participation, and not to productivity, which ranks among the lowest among the countries investigated.

Countries can be compared with respect to both increases in productivity and level of productivity. When measuring the latter, it appears that a number of European countries have productivity levels (well) above that of the US (see table 2.5).

Table 2.5 Productivity per hour (US=100) in 2002

Austria	88-101
Belgium	107-111
Denmark	89-101
France	107-113
Germany	92-105
Ireland	104-109
Netherlands	102-106
Norway	121-125

Note: The authors refer to three sets of data H. This table presents two figures for each country to cover their range.
Source: Bourlès and Cette 2005

Bouclès and Cette (2005) argue that, in order to make the figures more comparable, two corrections must be made. First, Europeans work (when considering both regular full-time and part-time work weeks) shorter hours than workers in the US, which lowers the US level because of the law of diminishing returns. Second, most European countries have employment rates that are (much) lower than those in the US. The most productive people will be hired first, which means that higher participation levels leads to lower levels of productivity. Taking into account these differences gives the so-called 'structural' hourly productivity figures. See table 2.6., which only presents the countries that, after the correction, still show higher productivity levels than the US. Data for the Netherlands are also presented.

Table 2.6 Structural hourly productivity (USA=100 in 2002)

Ireland	99-105
Netherlands	89- 93
Norway	100-114

Source: see table 2.5

It can be concluded that the superior productivity level of the Dutch economy compared with the American is due largely to the more favourable composition of its smaller labour force. Moreover, as a result of the recent slowdown in productivity growth, the level of productivity in the Dutch market economy has declined to below that of the American economy, even when structural values are disregarded (Inklaar *et al.* 2006, table 4).

This slowdown in Europe dates from 1995, as is shown in figure 2.1, and has been discussed in numerous articles. Recent figures from the Groningen Growth & Development Centre show that, notwithstanding the actual developments (recording the lowest rate of productivity growth in more than a decade in the US in 2006) there has been a remarkable reversal of positions between the EU-15 and the US (see table 2.7).

Table 2.7 GPD per hour worked (average annual percentages changes)

	1980-1995	1995-2006
EU-15	2.3	1.4
United States	1.3	2.2

Source: GGDC/TCB Total Economy Database, January 2007, www.ggdc.net

It is clear that while the EU led in the first period when compared to the US, it has now dropped below the US in the second, more recent period. This reversal is even more detrimental to the Netherlands, which already had the lowest growth rates in the period 1980-1995 (see table 2.8).

Table 2.8 GPD per hour worked (average annual percentage changes)

	1980-1995	1995-2006
Germany	2.4	1.7
France	2.5	1.8
UK	2.6	2.0
Finland	3.0	2.5
Netherlands	1.7	1.5
Average of 15 EU countries	2.3	1.8

Source: see table 2.7

These figures show that the Dutch economy had the lowest productivity rates of any of the major economies as well as a lower figure than the average for the 15 EU member states.

There have been investigations to clarify the causes of this reversal in the productivity figure rankings for the EU and the US, and, before turning our attention to the Dutch case, some of the reasons that emerged will be discussed in some detail in the next section.

2.2.2 EXPLAINING THE PRODUCTIVITY GAP BETWEEN THE EU AND US

In this section the productivity gap between the EU and the US will be discussed. Because the institutional arrangements and the size of the public sector in these two areas differ so greatly, it may seem necessary for the EU to not try to duplicate the US strategy. However, a comparison might indicate that there is room for improvements in the EU.

'European Paradox'

This paradox concerns the situation that, while the EU has a better record in the area of scientific output than the US, it does not benefit from this fact because it is unable to convert this scientific excellence into profitable innovation. Dosi *et al.* (2005), however, have argued that there is no such paradox. In terms of top publications; expenditures for R&D; share of tertiary education; and the budgets available for these students, the US always outperforms the EU.

The Role of the ICT

Much importance has been attached to the role of information and communication technology (ICT) to explain the production gap. The EU allegedly had three disadvantages in this respect when compared to the US: it produces less ICT, it uses less ICT and it is slower in reaping the benefits from ICT. These concerns have received ample attention from Van Ark and other researchers at the Groningen Growth and Development Centre. In the Netherlands, an entire issue of the economic biweekly ESB was devoted to this subject (Van Ark and Inklaar 2005, note 1 and ESB 2006). However, the impact of ICT is not as evident as it may seem. Van Ark and Inklaar show that ICT production in the US yielded more productivity than in the EU, but this is also the case for market services (Van Ark and Inklaar 2005, fig. 4). Furthermore, the sectors which caused most of the productivity gap are on average not characterised by a large share of ICT capital in total capital services (*ibid.*; figs. 5 and 6). In addition, the slowdown of productivity growth in the EU's non-ICT sectors was greater than in other sectors (Van Ark 2005a: 14; Gomez-Salvador *et al.* 2006, fig. 5). However, on the other hand, successful ICT applications seem indispensable for non-technological innovations (Van Ark 2007: 61).

Service sector

The service sector currently accounts for more than 70 percent of GDP in the Euro area (EA). This sector consists of three components: trade and transportation (ca. 20%), financial and business services (ca. 30%) and governmental services (ca. 20%). Increases in labour productivity in this sector have been very limited compared to that of the manufacturing sector as well as when compared to the US. Productivity growth actually decreased in the EU while increasing in the US, and this might explain (part of) the productivity gap. The most striking difference occurs in two sub-sectors: wholesale and retail trade (ECB 2007). Larger retail outlets like Wal-Mart, Home Depot and Best Buy make the exploitation of economies of scale, and a fuller use of ICT possible (Gordon 2004; ECB 2007). Here, it should be noted that the Netherlands have pursued a deliberate policy of limiting the location of superstores, thus depriving itself from advantages of increasing scale. Another conspicuous difference is that the US has twice as many highly skilled workers in market services than in the EU (Inklaar *et al.* 2007, table 5).

And lastly, governmental services constitute an obstacle as increases in production in this sector are difficult to determine. It is not easy to measure the output of, say, a ministry. Government sector output in many OECD countries is commonly measured by its inputs. This, of course, obliterates any increase in productivity. Hence, countries with larger public sectors tend to have lower productivity levels than countries with smaller public sectors. This method of measuring the output of the government sector has been regarded as unsatisfactory , especially when the likelihood of real productivity increases is very strong. For example, with the introduction of ICT in administrative services, there is much evidence that there are clear productivity gains. For this reason the UK, notwithstanding the obvious problems it has encountered, has switched from measuring inputs to measuring outputs (Atkinson 2005). In the Netherlands, productivity increases in the public sector are supposed to equal the wage drift in this sector, which is ¼ percent for 2007 and 2008 (CPB 2007: 110).

R&D

As R&D spending in percentages of GDP is substantially higher in the US than in Europe, it has been suggested that the productivity gap could be lessened by stimulating R&D in Europe (Daveri 2004: 31). However, the relationship between R&D activities and successful innovations is far from clear and have not been persuasively demonstrated (Jaumotte and Pain 2005b: 25, table 13). To innovate, the productive *use* of new technology is crucial, and R&D is not necessarily the right vehicle for this (Van Ark 2005b: 85). Most of the productivity growth derived from R&D is obtained from R&D that originates abroad (Guellec and Van Pottelsberghe 2001: 113). The importance of foreign R&D is not incompatible with another finding that on average almost half of successful product innovations are imitations (Jaumotte and Pain 2005b: 25).

Thus, it seems that the productivity gap that the Dutch economy appears to suffer from more than the EU member states on average (see table 2.8), cannot be explained, so it seems, by pointing to macroeconomic causes. And even if a distinct explanation was offered, the question remains what the best remedy would be.

2.2.3 DUTCH INNOVATIVE POTENTIAL IN AN INTERNATIONAL PERSPECTIVE

Ranking a country's innovative potential is a hazardous enterprise. Ultimately, it is the contribution to economic growth that matters, but the channels through which this contribution is realised has neither been definitively identified nor quantified. For this reason, comparative statistics continue to rely heavily on innovation inputs and innovation enablers. Patents are often considered innovation outputs, although it is their successful and profitable application that actually matters. Moreover, trademarks, design registrations and copyrights are also used as means to protect intellectual property. Nevertheless, the Economist Intelligence Unit has developed a ranking based primarily on patents, innovation inputs and innovation enablers. Their index shows the Netherlands dropping from ninth to 13th place in the forecast for the period 2007-2111 (2007: 13).

In this respect, the European Innovation Scoreboard (EIS) is also important. It compares the innovation performances of the EU countries and also includes countries such as the US, Japan, Switzerland, Norway and Iceland. Here the Netherlands ranks 11th, and does not belong to the category of innovation leaders. Instead it is categorised as an 'innovation follower.' The scoreboard consists of 25 indicators, divided into five subcategories (see table 2.9).

Table 2.9 Dutch ranking in the subcategories of the EIS

Innovation drivers	10
Knowledge creation	10
Innovation and Entrepreneurship	18
Applications	18
Intellectual property	8

Source: EIS 2006, www.proinno-europe.eu/inno-metrics.html

Information about the three countries with the highest score on each of the 25 indicators is also presented. In this way the European member states can qualify for 75 recognitions as best performer. For these 75 places for European innovation leaders, the Netherlands only qualifies only once (EIS, table 2).

Table 2.10 Dutch ranking in sciences and research

Share of HRST occupations	4
Of which are researchers	20
Scientific articles	5
R&D expenditures	15
Attractiveness of foreign R&D location	9

Source: OECD 2006: 93; OECD 2005a: 159; OECD 2006: 23; OECD 2006: 135

Other indicators concerning human resources in science and technology (HRST) can be sampled from OECD sources (see table 2.10).

Although the Netherlands ranks fairly high as an R&D attractive country, it comes in 20th place as a current foreign R&D location (OECD 2006: 134). The high costs of R&D personnel, the limited co-operation between firms and knowledge institutes and the lower than average availability of private R&D capital may serve as the primary reasons for this discrepancy (OECD 2005b: 110). The inability to attract sufficient R&D investments from abroad, accounts for 40 percent of the lower level of private R&D in the Netherlands when compared to the rest of the OECD (Erken and Ruiter 2005: 78).

The Netherlands, together with Denmark, Ireland and Luxemburg, has the highest GDP per capita (Sapir 2003: 101), so one would expect better scores in the area of innovation indicators. Moreover, when the Lisbon agenda is also taken into consideration, the lack of progress is even more remarkable. The Netherlands scores best in the sciences, which has led the OECD to describe 'a Dutch paradox' meaning "an excellent record in knowledge *creation*, but a mediocre record in innovation activity, which is defined as the successful *development* and *application* of new knowledge in new product and/or processes" (OECD 2005b: 104).

2.3 INNOVATION IN THE MARKET ECONOMY AND INNOVATION POLICY IN THE NETHERLANDS

2.3.1 INNOVATION IN THE DUTCH MARKET ECONOMY

The Dutch bureau of statistics (CBS; Statistics Netherlands) collects plenty of data concerning innovation, which is published annually in *Kennis en economie* (Knowledge and Economy). Only 24 percent of all Dutch firms employing at least 10 workers, can boast having had an innovation in the period 2002-2004. Looking at the various sectors it appears that this figure is roughly similar for the service sector. As this figure of 24 percent is exceeded in the manufacturing sector (40%), it is clear that this implies lower figures in the remaining sectors. Product innovation occurs more frequently than process innovation. While firms involved in both the innovation of their products and their processes, form a minority (CBS 2006: 129-31). In most cases, product innovation leads to a broader assortment of products and services offered, or new markets, or an improvement in the quality of the services and/or products. Process innovation leads to more flexibility or a greater capacity, or generates lower costs or a higher level of effectiveness (CBS 2006: 153). If one were to concentrate solely on the novelty of a product, the share of sales of products 'new to the firm' in terms of turnover is about 24 percent in the manufacturing sector, 14 percent in the services sector, with both figures being below the EU average, 29 percent and 19 percent respectively (OECD 2005a: 166).

If one were to equate non-technological innovation with social innovation, then one can conclude that social innovation occurs at a slightly higher rate than technological innovation. Social innovation may involve internal organisation, new marketing concepts, or improved customer and supplier relationships (SER 2006: 27). It is intensely related to 'tacit knowledge', which includes skills, customs, intuition, and cultures within the involved firms (Weehuizen and Soete 2001: 43). Research indicates that service sector firms that are actively engaged in social innovation, experience the highest productivity increases (CPB 2002: 148).

It has been observed that better education is a decisive factor for fostering innovation. Because of this skill levels have received renewed attention. Top performers, or those with the highest levels of skills, seem to be particularly important when it comes to increased productivity. The Netherlands is in the top ten countries in the world, when countries are ranked according to the levels of the skills of their labour force. However, when the ranking is focussed on the top five percentage of the labour force with the highest skill levels, the Netherlands falls out of this elite and ranks only as thirteenth (Minne *et al.* 2007: 25, 47).

In the Netherlands, R&D expenditures as a percentage of GDP is below both the EU-15 and OECD average (CBS 2006: 79). As has already been mentioned, this is in part explained by the Netherlands' inability to attract a sufficient inflow of foreign R&D. However, a larger factor is the very structure of the Dutch economy, which is less knowledge-intensive than the average OECD economy. As Erken and Ruiter rightly stipulate, it is very doubtful that this is an explanation, which might serve as a justification. On the contrary, so it seems. In general, it is the less-advanced economies that need to undergo an intensification of R&D activities, rather than merely being resigned to their predicament (Erken and Ruiter 2005: 78-79).

2.3.2 INNOVATION POLICY IN THE NETHERLANDS

In assessing the Dutch system of innovation, the OECD highlights a number of positive features: the high quality of its scientific research, the high rate of patenting, the rich endowments with HRST and the positive framework conditions resulting from generally sound economic policies. However, on the negative side, the OECD emphasises the low level of private R&D, the less than optimal interaction between industry and academia, insufficient innovative entrepreneurial activity and the limited ability to commercialise scientific output (OECD 2005a: 153-4).

Dutch innovation policy, in some respects, addresses the problems listed above. The Ministry of Economic Affairs spends some 1 billion Euros on innovation, or about 0.2 percent GDP (EZ 2006: 7). About 60 percent of this is spent on a large number of items related to either supportive measures aimed at facilitating innovation or to specific industries to strengthen their innovation efforts. About 40 percent is spent on WBSO, a fiscal incentive for innovation, i.e., a tax deduction for R&D staff wages in excess of the standard wage costs (EZ 2006: 33-37). Furthermore, there are numerous smaller measures that have been developed, which has led, according to the OECD, to a plethora of instruments (OECD 2005b: 114). A comparable budget is also spent on research and science by the ministry of Education, Culture and Science (OC&W 2006: 161-64). This budget is allocated to a large number of institutes and organisations for research or for the channelling of this money to research projects based on various criteria, including the degree of excellence of the research itself.

Although the abundance of instruments may reflect the manifold demands the various parties involved in innovation have, a lack of transparency may be the result. The Dutch Ministry for Economic Affairs has, for this reason, taken steps to reduce the number of instruments and to improve the ease of access to the various instruments for each target group (EZ 2006: 129). Tax incentives to promote R&D, like the WBSO, may be an effective measure as many studies have shown (OECD 2003b: 24). The same can be said for the innovation voucher which has been introduced to enable the SME in particular to use external resources for technological innovation (Cornet *et al.* 2007: 48).

The desirability of stimulating promising innovative enterprises ('backing winners') has been questioned frequently. Generally speaking, government funding can replace private payments. If the government pays for (or contributes to) investments which would have been made in the absence of public funding by the private sector, its impact might be very limited. The final result might be even a negative one because of the distortion being caused by taxes. OECD case studies indicate that the percentage of (partial) substitutions range from 25 percent to 80 percent (OECD 2006a). This percentage only increases when successful firms can profit from this kind of funding. Another OECD study strongly suggests that R&D in the private sector is not affected by tax relief (OECD 2003b, fig. 1).

A sometimes implicit evaluation of Dutch innovation policy can be derived from prevailing opinions. Dutch economists, in their role as policy advisors, are greatly concerned about the lack of productivity growth in the Netherlands. They have thus recommended more competition (REA 2005; CPB 2005; KVS 2004: 207) although there is a clear indication that competition should not be promoted beyond its optimum (Boone and Van Damme 2004). They also believe that there should be less regulation because some regulations frustrate innovation and entrepreneurship (REA, KVS) and that more attention should be paid to education, science and research (CPB, REA, Van Wijnbergen 2004). That part of the present policy mix that can be aptly labelled as 'backing the winners', has been duly criticised, and should, in the view of its critics, be transformed into a policy of 'backing the challengers' (Jacobs and Theeuwes 2004; REA). An original, albeit singular explanation for the slowdown of the Dutch productivity growth has been offered by Naastepad and Kleinknecht who regard the Dutch wage moderation as its main cause (2004).

2.4 CONCLUSIONS

2.4.1 THE URGENCY OF REINFORCING THE GROWTH POTENTIAL OF THE DUTCH ECONOMY

There are three major reasons that denote that there is an urgency for the strengthening of Dutch growth potential:
- absence of factor accumulation as a source for economic growth
- globalisation
- specific characteristics of the Dutch innovation system.

Factor accumulation

The ageing of the population in the coming decades is, to varying degrees, relevant for all of the EU countries. The consequences for the Netherlands are especially severe as the growth of the Dutch economy has until now to a large extent been based on a steadily increasing participation rate. The labour force has increased from about 40 percent to 48 percent of the population over the two last decades. In the coming decades, this percentage, whatever the envisioned scenario, will continue to decrease (Huizinga and Smid 2004, figure 4.6). Capital accumulation will also exert a limited influence because the rate of its growth will be less than that of the GDP (*ibid.*, table 3.5).

Globalisation

The Dutch economy has always profited from economic openness and the CPB forecasts that it will continue to benefit from international liberalisation. As the OECD method of ranking countries according to their 'globalisation-readiness indicator' suggests, this depends not only on the capacity to adapt to changing circumstances but also on one's capacity to innovate (Bergeijk 2007: 5-7).
In the second half of the twentieth century, the Netherlands made large gains as it caught up with the advanced technological economy of the US. This situation has changed drastically because most of the EU members states – including the Netherlands – have been unable to keep pace with US increases in productivity But it is not only the US that the European countries have to compete with. Asian nations like South Korea, Taiwan, Singapore, Malaysia and China currently have the highest growth rates per capita GDP. Some of these countries actually have more university degrees per capita in the natural sciences and engineering departments than EU countries. They also spend more on R&D and acquire more patents (Fagerberg and Godinho 2005: 527-320). Thus, it is understandable that the Esko Aho Report 'Creating an Innovative Europe,' warns that "Europe and its citizens should realise that their way of life is under threat" (OMC 2007: 11).

Dutch situation

Dutch productivity growth is substantially lower than both US rates, and that of the EU-15 country average, as can be observed in tables 2.7 and 2.8 above. "No comparable OECD member country evolved along a growth path that was less productivity-driven," than the Netherlands writes the OECD (2005a: 149). Moreover, a large number of other details show that the Netherlands ranks well below its expected position considering its GDP per capita. Figure 2.2 shows that the Netherlands cannot even keep abreast with most of the EU averages, let alone the countries exercising best practices.

Figure 2.2 Innovation

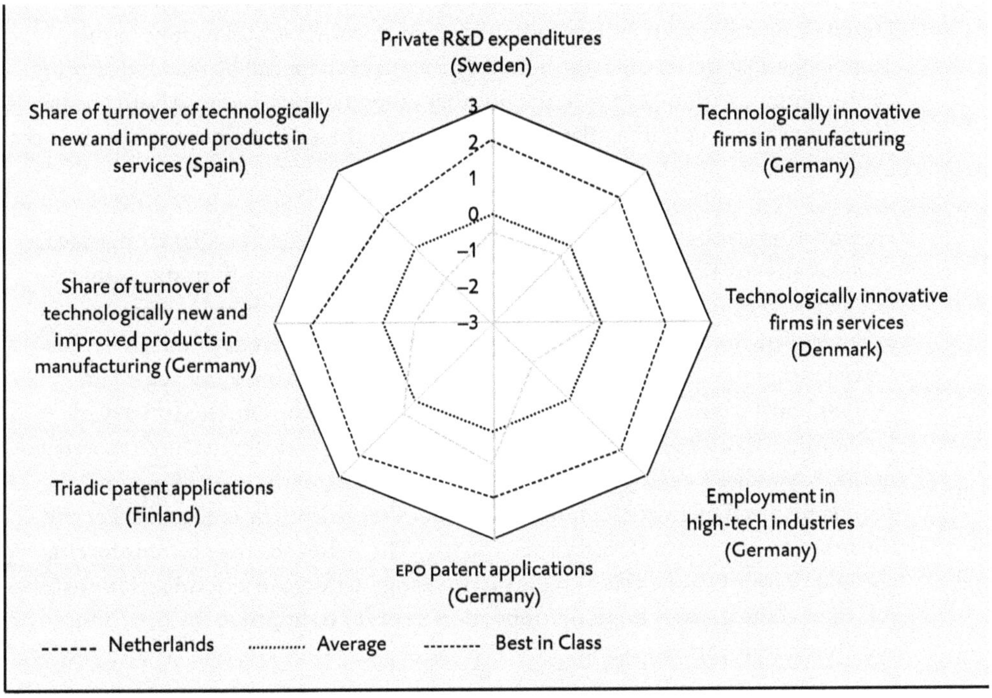

Source: CBS (2007): 23

Figure 2.2 presents some selected features of innovation systems in the EU. The dotted line shows the best performers, the regular inner circle reveals the EU averages, and the light irregular octagonal line represents the scores for the Netherlands. Clockwise, and starting at the top, the rays show: business sector R&D expenditures (1); technologically innovative firms in manufacturing (2); the same in services (3); employment in high tech sectors (4); patents submitted to EPO (5); submitted triadic patents (6); share of technologically (re-) new(ed) products in total sales in manufacturing (7); and the same in services (8). The Netherlands appears to perform below the EU average in six out of the eight categories.

2.4.2 MACROECONOMIC ANALYSIS AND INNOVATION POLICY

Macroeconomic analysis has been invaluable in determining the causes of economic growth and in locating the factors that contribute to it. The importance of innovation as the main driving force for economic progress is generally acknowledged. However, the emergence of innovation, to a large degree, remains enigmatic: "we know much less about why and how innovation occurs than what it leads to" (Fagerberg 2005: 20). In these circumstances, macroeconomists tend to rely heavily on favourable macroeconomic conditions, as well as on education and science as has already been mentioned in section 2.4.1. Although harmless, these no regret policies scarcely address the actual shortcomings of the Dutch innovation system. As the Economist Intelligence Unit's report and the "knowledge paradox show, it is not primarily a lack of knowledge which hampers the Dutch innovation system. It is rather a lack of a profitable application of this knowledge that needs the attention. A similar comment can be made regarding the recommendation to reduce the number of regulations, because both the product market and sectoral regulations are less strict than the EU average" (ECB 2007: 81-2).

However, as Jacobs and Theeuwes (2004: 197) have noted, it is "far from easy to develop effective instruments" for innovation policy. These limitations may be due to the very nature of the macroeconomic discipline, which is "usually unable to determine the direction the causality flows from one event to the other," as Lewis, rather harshly, believes (2004: 7). One can, however, agree with Jan-Willem Oosterwijk, the former Secretary-General of the Dutch Ministry of Economic Affairs. He noted that although the need to reinforce Dutch macroeconomic basics has largely been dealt with, there is currently a need for a reform agenda inspired by microeconomic challenges regarding the labour market and the innovation system (Oosterwijk 2007: 5) The better strategy for reinforcing the Dutch innovation system seems to investigate micro-economic relationships relevant to the development of innovation in order to improve the functioning of the Dutch innovation system.

NOTE

1 On the basis of GDP/population = GDP/employed person x employed persons/population, and GDP/employed person = GDP/hours worked x hours worked/employed persons.

REFERENCES

Ark, B. van (2005a) 'Does the European Union need to revive productivity growth?', *Research Memorandum* GD-75, Groningen: Groningen Growth and Development Centre.

Ark, B. van (2005b) 'In search of the silver bullet for productivity growth', *International Productivity Monitor* 10, Spring.

Ark, B. van and R. Inklaar (2005) 'Catching up or getting stuck? Europe's trouble to exploit ICT's productivity potential', *Research Memorandum* GD-79, Groningen: RuG.

Ark, B. van (2007) 'Enhancing productivity requires more than ITC alone' in T. Wolters (ed.) *Measuring the New Economy*, Amsterdam: Elsevier.

Atkinson Review: Final Report 2005, Basingstoke: Palgrave.

Baily, M.N. and J.F. Kirkegaard (2004) *Transforming the European Economy*, Washington: Institute for International Economics.

Baumol, W.J. (2004) 'Four Sources of Innovation and Stimulation of Growth in the Dutch Economy', in Gelauff et al. (eds.).

Bergeijk, P. van, 'Globaliseren zonder angst', *ESB dossier*, 2007 July.

Boone, J. and E. van Damme (2004) 'Marktstructuur en innovatie', in Jacobs and Theeuwes (eds.).

Bourlès, R. and G. Cette (2005) 'A Comparison of Structural Levels of Productivity in the Major Industrialised Countries', *OECD Economic Studies* 41, 2.

CBS (2006) *Kennis en economie*, Voorburg: CBS.

CBS (2007) *Het Nederlandse ondernemingsklimaat in cijfers 2007*, Voorburg: CBS.

Cornet, M., M. van der Steeg and B. Vroomen (2007) 'De effectiviteit van de innovatievoucher 2004 en 2005', *CPB Document* 140, The Hague: CPB.

Crafts, N., 'Quantifying the Contribution of Technological Chance to Economic Growth in Different Eras: A Review of the Evidence' in Heikkinen and Van Zanden (eds.)

CPB (2002) *CEP 2002*, The Hague: CPB.

CPB (2005) *MEV 2005*, The Hague: CPB.

CPB (2007) *CEP 2007*, The Hague: CPB.

CPB/SCP (2006) *Europese tijden*, The Hague: CPB/SCP.

Daveri, F. (2004) *'Why is there a Productivity Problem in the EU?'* 205, July, CEPS Working Document.

Dosi, G., P. Llenera and M. S. Labini (2005) *Evaluating and Comparing the Innovation Performance of the United States and the European Union*, Expert report prepared for the *TrendChart* Policy Workshop.

EC (European Commission) (2006) *European Economy, Special Report No 4*, Brussels: EC.

ECB (2007), 'Competition in and Economic Performance of The Euro Area Service Sector', in *ECB Monthly Bulletin* 2007, May: 73-87.

Economist, The 2007, August 4.

Economist Intelligence Unit, The (2007) *Innovation: Transforming the way business creates*, London.

Erken, H. and M. Ruiter (2005) *Determinanten van de private R&D-uitgaven in internationaal perspectief*, Den Haag: Ministerie van Econimische Zaken and Dialogic.

Erken, H., P. Donselaar and S. Raes (2007) 'An International Comparison of Productivity Performance: the Case of the Netherlands', forthcoming, 2007.
ESB (2006) 'ICT en economische groei', *ESB dossier* September 2006.
EIS, *European Innovation Scoreboard 2006* (www.proinno-europe.eu/inno-metrics.htlm).
EZ, Ministerie van Economische Zaken (2006) *Begroting 2007*, The Hague: Sdu.
Fagerberg, J. (2005) Innovation: Guide to the literature, in J. Fagerberg, D.C. Mowery en R.R. Nelson (eds.). *The Oxford handbook of innovation,* Oxford: Oxford University Press: 1-26.
Fagerberg, J. and M. M. Godinho (2005) 'Innovation and Catching-Up', in Fagerberg et al.(eds.).
Fagerberg, J., D.C. Mowery and R.R. Nelson (eds.) (2005) *The Oxford Handbook of Innovation,* Oxford: Oxford University Press.
Gelauff, G., L. Klomp, S. Raes and T. Roelandt (2004) *Fostering Productivity*, Amsterdam: Elsevier.
Gomez-Salvador, A. M., M. Stocker and J. Turunen (2006) 'Labour Productivity Developments in the Euro Area', *Occasional Paper Series* 53, Oct., Frankfurt am Mainz: ECB.
Gordon, R.J.(2004) 'Why was Europe left at the station when America's productivity locomotive departed?', *NBER Working Paper Series* 10661.
Guellec, D. and B. van Pottelsberghe de la Potterie (2001) 'R&D and productivity growth: panel data analysis of 16 OECD countries', *OECD Economic Studies* 33, Parijs: OECD.
Heikkinen, S. and J. Luiten van Zanden (eds.) (2004) *Explorations in Economic Growth*, Amsterdam: Aksant.
Helpman, E. (2004) *The Mystery of Economic Growth*, Cambridge, Mass.: Harvard University Press.
Huizinga, F. and B. Smid (2004) *Vier vergezichten op Nederland*, The Hague: CPB.
Inklaar, R., M.P. Timmer and B. van Ark (2006) *Mind the Gap! International Comparisons of Productivity in Services and Goods Production*, Groningen: Universiteit van Groningen.
Inklaar, R., M.P.Timmer and B. van Ark (2007) *Market Services Productivity Across Europe and the US*, Groningen: GGDC, University of Groningen.
Jacobs, B. and J.J.M. Theeuwes (eds.) (2004) *Innovatie in Nederland*, Amsterdam: KVS.
Jacobs, B. and J. Theeuwes (2004) 'Innovatie in Nederland, De markt draalt en de overheid faalt', in Jacobs en Theeuwes (eds.).
Jaumotte, F. and N. Pain (2005a) 'From Innovation Development to Implementation: Evidence from the Community Innovation Survey', *Economics Department Working Papers* 458, Paris: OECD.
Jaumotte, F. and N. Pain (2005b) 'Innovation in the Business Sector', *Economic Department Working Papers* 459, Paris: OECD.
Kets, W. and A. Lejour (2003) 'Sectoral TFP developments in the OECD', *CPB Memorandum* 58, 18 February.
KVS, Koninklijke Vereniging voor de Staathuishoudkunde (Royal Netherlands Economic Association) (2004), see Jacobs and Theeuwes.
Layard, R. (2005) *Happiness. Lessons from a New Science*, London: Penguin.

Lewis, W. W. (2004) *The Power of Productivity*, Chicago: University of Chicago Press.
Madison, A., 'When and Why Did the West Get Richer than the Rest?' in Heikkinen and Van Zanden (eds.).
Minne, B. et.al. (2007) *Excellence for Productivity*, The Hague: CPB.
Naastepad, C.W.M. and A. Kleinknecht (2004) 'The Dutch productivity slowdown: the culprit at last?' in *Structural change and Economic Dynamics* 15.
OC&W, Ministerie van OC&W (2006) *Begroting 2007*, The Hague: Sdu.
OECD (2003a) *The Sources of Economic Growth in OECD Countries*, Paris: OECD.
OECD (2003b) *Tax Incentives for Research and Development: Trends and Issues*, Paris: OECD.
OECD (2005a) *Innovation Policy and Performance, A Cross-Country Comparison*, Paris: OECD.
OECD (2005b) *OECD Economic Surveys, Netherlands* 2006,2, Paris: OECD.
OECD (2006) *OECD Science, Technology and Industry Outlook*, Paris: OECD.
OECD/EC (2005) *Oslo Manual, Third Edition*, Paris: OECD.
OECD (2006a) *Government R&D Funding and Company Behaviour*, Paris: OECD.
OMC, OMC Policy Mix Review Report (2007) *Country Report The Netherlands.*
Oosterwijk, J.W. (2007) 'Maatwerk en modernisering in Europa', ESB 12 January 2007.
REA, Raad voor Economische Adviseurs (Council of Economic Advisors) (2005) 'Innovatie en economische groei', *Brief* TK 30 385, 1.

Sapir, A. (2003) 'An Agenda for a Growing Europe', *Sapir Report* 2003, July.
SER (2006) *Welvaartsgroei voor iedereen, Thema Sociale Innovatie*, The Hague: SER.
Smid, B. (2005) 'De arbeidsproductiviteit op lange termijn in historisch en internationaal perspectief', CPB *Memorandum* 107, The Hague: CPB.
Smith, K. (2005) 'Measuring Innovation', in Fagerberg et al. (eds.).
Verspagen, B. 'Innovation and Economic Growth', in Fagerberg et al. (ed.)
Weehuizen, R.M. and L. Soete 'Nederland kennisland, let op uw zaak', in Gradus et al. (eds.).
Wijnbergen, S. van (2004) 'Groei, innovatie en overheidsbeleid', in Jacobs and Theeuwes.
WRR (2007) *Investeren in werkzekerheid*, Amsterdam: Amsterdam University Press.

3 LEARNING, DISCOVERY AND COLLABORATION

Bart Nooteboom

3.1 INTRODUCTION

In chapters 1 and 2 one conclusion was that for a foundation of innovation policy we need to move beyond macro analyses of factor productivity and into the innovation system, with micro-level analyses of actions and interactions of the actors involved, in the context of institutions and institutional change. R&D does not directly yield innovation, but a pool of ideas and emergent technologies, products, and practices, which require testing, design, commercialization, marketing, and the development of production. This requires entrepreneurship and organisation, internally within firms and externally, within networks of firms. R&D may also serve to explore new developments and opportunities, build up absorptive capacity, knowledge on sources of knowledge and contacts for future collaboration. The innovation process is not linear, but entails feedback from testing, as well as trial and application to the generation of new knowledge. In this chapter, we attempt to contribute to further insight on the micro level of actions and interactions, from the perspective of learning by interaction, which is also central in the literature on innovation systems (Lundvall 1988).

As in the literature on innovation systems, an important source of inspiration here is an evolutionary perspective, which recognises the fundamental importance of uncertainty and unpredictability, in the emergent nature of innovation, and the crucial role of diversity as a source of novelty. This in itself has important policy implications, as it points to the limitations of rational planning and design of innovation. However, we also recognise that learning is inherently social, and hence we should derive insight also from theory of cognition, communication, language, and motivation, and these do not necessarily conform to details of Darwinian evolution as found in biology. In other words, while cultural evolution may derive important insights from neo-Darwinian theory, it is not necessarily isomorphic to it.

This chapter outlines and elaborates a 'cyclical model' of innovation, where the development of new knowledge and new practices (exploration) and the improvement of existing practices (exploitation) build upon each other. A key theme is how such a combination can be achieved, in spite of fundamental differences between exploitation and exploration, and between the conditions of markets and organisation that they require. Among other things, the analysis offers a deepening of our understanding of why collaboration furthers innovation and the positive role of differences in insight ('cognitive distance').

The analysis will have implications for policy concerning universities and relations between universities and industry (chapter 4), entrepreneurship (chapter

5), organisation for innovation (chapter 8), inter-organisational networks (chapter 10), and regional innovation systems (chapter 11).

3.2 EVOLUTION AND LEARNING

According to the evolutionary perspective, new forms of life emerge without prior, goal-directed design, in a process that is cumulative, and path- and context-dependent, because development builds upon what has developed before. That was a revolutionary insight in the field of biology, and presents a formidable challenge to thinking in management and public administration, including innovation policy. A tenacious, deep-rooted intuition of good management is that managers know where we should go (goals), the means for goal achievement, and causal relations between means and goals, from which they can issue directives (planning) for execution. But the question is how in an uncertain and volatile world they could know all that. According to an evolutionary perspective, new forms of life, and analogous to that innovations, can emerge through selection in an 'ecological niche'. This comes about as a result of survival pressures on a variety of more or less blind trials whose success is fundamentally unpredictable, followed by a broad transmission and adoption of what turns out to be successful. The maintenance of sources of diversity and mechanisms of selection, and the recognition that success cannot be predicted are both crucial for innovation policy. Management and administration become the facilitation of the creation of variety, selection, and transmission.

In the economy, creation of variety lies in entrepreneurship, invention, and basic research. Selection lies in experimental testing within firms, competition in the ecological niche of an organisation, and regulation. Transmission lies in education, training, and imitation. This connects with the notion of the market as a discovery process, as suggested by Hayek, where perceptions, knowledge, and ideas are widespread, diverse, and rich in local specificities, across a wide variety of actors that interact to generate Schumpeterian 'novel combinations'.

According to evolutionary theory, the development of new forms of life (species) is stimulated by a change of ecological niche ('niche selection'), as a result of external events, such as a disaster or climate change, or because of migration to a different but existing niche.[1] Analogous to that I will argue that innovation is stimulated by the application of products, knowledge, and technology in new environments (countries, markets, industries, technologies, or organisations) that are sought or imposed. Modern evolutionary theory also recognises 'niche creation', where species have an impact on their selection environment, in co-evolution with it. Analogous to that I will argue that innovations affect markets, and that while markets require institutions, they also produce institutions, which can provide obstacles to further innovation. Therefore, for innovation, one may need to escape to an outside niche. Thus niche creation can create a need for the selection of another niche. These considerations form an important part of the basis for an argument for creating or maintaining openings, in innovation policy.

While the importance of selection processes is consonant with the importance of markets and competition in a traditional economic perspective, the path- and context dependence of evolution stands in contrast with economic claims of universalistic structures and mechanisms, and similar universalistic pretensions in innovation policy. In particular, local and regional agglomerations and clusters of activity have emerged from an interplay of local specificities, deliberate policies, quirks of nature, chance events, institutional and political conditions, and outside developments, in historical processes. As a result it is very hazardous to infer general 'recipes' for activities that can be transplanted elsewhere. This point is further taken up in chapter 11.

In political thought, the Austrian/evolutionary perspective has emerged, although not under that banner, in the notion of 'experimentalism' or 'directly-deliberative polyarchy' (Gerstenberg and Sabel 2002). Here also, the idea is that local diversity of knowledge and insight, and room for experimentation, form a basis for learning and development, here of institutions and government. This notion of experimentalism forms perhaps the most fundamental idea for the policy view taken in this volume. It yields a preference for government that resists the urge to conduct centralised, top-down planning, and sees its task as granting the autonomy, and creating conditions and boundary conditions for:

"... lower level actors to experiment with solutions of their own devising within broadly defined areas of public policy. In return they furnish central or higher-level units with their rich information regarding their goals as well as the progress they are making ..., and agree to respect in their actions framework rights of democratic procedure and substance ... With periodic pooling of results ... (that) reveals the defects of parochial solutions, and allows the elaboration of standards for comparing local achievements, exposing poor performers to criticism from within and without, and making of good ones (temporary) models for emulation" (Gerstenberg and Sabel 2002).

However, the evolutionary perspective also has its limitations (Nooteboom 2001). It is useful for a conceptual twist away from intuitions of innovation as central design and planning, and for the basic idea of variety, selection, and transmission of success as the motor of change. However, the further content of those notions in the economy is different from that in biology. In society, innovation emerges from cognition and learning. These are social-cultural processes with their own characteristics (Nooteboom 2001). These limitations of an evolutionary perspective are closely related to the limitations of a neo-Darwinian take on cultural evolution, with its notion of ideas as 'memes' in analogy to genes. It is enlightening to see concepts in a non-essentialist way as emergent, in a process where new ideas build upon old ones and in the process shift their meanings. It is also enlightening to recognise that ideas are subjected to survival or death in selection, by adoption, citation, rejection, or neglect by scientific and policy communities. It is also revealing to compare disciplines to species, to understand why difficulties of interdisciplinary research may have a similar function to cross-species infertility: cross-species mixing might eliminate the evolutionary process. It is also useful to note that if invention is to a large extent trial and error, then the detection of error is indeed needed in selection 'to clear up the mess' of the massive failure of blind trials.

In a strict rendering of universal Darwinism, it should be plausible to explain phenomena in terms of 'replicators' (genes in biology, memes in culture) carried by "interactors" (Campbell 1974; McKelvey 1982; Hull 1988) or "vehicles" (Dawkins 1983) that are subject to selection. To function as an interactor, an entity must have a reasonably cohesive and stable set of components. This is the *ecological* side of evolution (Baum and Singh 1994). Interactors carry replicators that in the *ontogenetic* development of interactors generate characteristics that affect their survival and the replication of their replicators. This generation of characteristics (in biology: gene expression) takes place in interaction, within the interactor, and between replicators and other features of the interactor, as well as with the environment of the interactor. Replicators may lie dormant until triggered by conditions. Note that it is not the replicators themselves that determine survival but the characteristics that they produce. Replicators from surviving interactors are replicated and re-combined, mostly within populations of interactors that partake of a common pool of replicators. This is the *genealogical* side of evolution, in the *phylogenetic* development of a species.

These conditions hardly apply in culture (as recognised, for example, by an eloquent promulgator of evolutionary thought such as Daniel Dennet (1995)). In cultural evolution, it is not clear what or who the interactors are. If ideas or concepts are the replicators, what are the vehicles and interactors? Are they the words or symbols in which ideas are expressed, coherent stories and theories, the media (journals, books, films, computer programmes, websites) that carry them, mental structures, or the people who generated the ideas or who adopt, reject, or ignore the ideas? 'Replication' is a dubious term for communication. Cognitive and linguistic theory show that in communication meanings of words get transformed as they are assimilated and embedded in brains, and in the process are transformed rather than replicated. In evolutionary terms, the implication is that transmission is also a source of variety creation. Meanings of concepts shift as they are combined with other concepts, and meanings are context-dependent. Similarly, identities of people shift as they connect with other people, and identities are context-dependent. The rate at which meanings of concepts 'mutate' is so high that the effectiveness of the selection process is in doubt. As generally recognised, in culture transmission is Lamarckian: new meanings acquired during life are included in what is transmitted. Ideas mix between different carriers, and hence the notion of their 'lineage' is problematic. While in biological evolution lineage only branches out, in culture lines of lineage may come together again. While in biology extinct species cannot be revived, in culture, old, discarded concepts can be revived.

During the oil crises of the 1970s, when oil quadrupled in price, the concept of sailing ships was revived, but in a new form, with computer-directed aluminium sails. The idea was not, however, put into practice, since oil prices fell again.

While genes are malleable in gene expression, in the sense that the characteristics they generate depend on interaction between genes and the setting in which they operate, meanings of words appear to vary much more radically as a function their position in a sentence, a story and the context of action. The co-evolution of ideas and their selection environment is much closer than co-evolution in biology is: since people use communication, novelty and niche creation are much faster than in biology, and with communicative strategies and political maneuvering people generating ideas and innovations can more easily create their own selection environment. This enhances the importance of niche creation as a possible obstacle to ongoing innovation.

Within disciplines or fields of research, new schools of thought are often rejected or ignored by the established orthodoxy. Yet, regularly outsiders find a way out by setting up a society of their own that subsequently sets up a proprietary journal, with its own editors and editorial board, developing its own brand of orthodoxy. In the field of innovation an example is the Joseph A. Schumpeter Society, setting up its *Journal of Evolutionary Economics*, and in institutional economics an example is the European Association for Evolutionary Political Economy, setting up its *Journal of Institutional Economics.*

In particular, creativity and the sources of invention remain a black box, and are neglected in the theory of innovation. Partly, this is because, since Schumpeter, innovation is defined as the successful application and market introduction of novelty (cf. chapter 1). Invention is seen, or at least treated, as inexplicable. In biology, variation is blind, resulting from random mutations of genes and chromosome cross-over, in sexual reproduction. While invention is subject to radical uncertainty (in the Knightian sense), and is subject to much trial and error, there is creativity and a certain amount of intelligent inference of causes from phenomena, on the basis of experience and smart experimentation. In evolutionary terms, variety precedes selection but also arises from it. In other words, economic evolution, and by implication innovation policy, is not entirely devoid of intelligent design.

The tacit assumption, adopted by intellectual contagion from biological evolution, that invention and idea creation are blind or random and hence do not admit of systematic explanation, has led to a focus on application, in innovation and diffusion, and neglect of sources and processes of invention. That is highly undesirable and unnecessary. It is undesirable because without idea creation there is no innovation. What is worse, while invention is unpredictable, innovation, seen as application of something that has proven itself in the market is subject to rational choice, and innovation policy can fall back on either of two dominant modes of thought. One is the market as an efficient mechanism for allocating resources to the achievement of identifiable goals by known means. The second mode, when markets fail, is to fall back on rational selection, design and planning. The problem then is that market mechanisms or planning that are appropriate for the promotion of application may have an adverse effect on idea creation.

This development has been apparent also in knowledge management. Old or 'type one' knowledge management has focused on the efficient utilisation and sharing of existing knowledge, by the identification of potential users, the specification of roles and tasks, technical standards, meanings and rules, in which ambiguity is eliminated for the sake of efficiency. People realized that creativity, in the generation of new ideas, was also important, but since it was ill understood, that was postponed to an indefinite future, in the reasonable drive to first grasp the low hanging fruits of existing knowledge. As for creativity, one would turn to that later. The problem with this is that measures taken to utilize existing knowledge may obstruct the generation of new insights and opportunities, which would require an opening up to ambiguity and new meanings that would upset existing efficiencies, roles, tasks and positions. In short: lack of openness to surprise. As a result, knowledge management developed into an impasse that is well recognized, and is leading to attempts to develop knowledge management 'type two', to open up to new knowledge creation.

The neglect of idea creation is unnecessary because the possibility to understand it is not a priori hopeless. The idea that it is hopeless rests upon a misunderstanding. The assumption is that a new idea is by definition unpredictable, because if it were predictable it would no longer be a new idea. However, one can have some insight into processes of idea creation without the pretension to an ability to predict its outcomes. That is relevant for policy that is not geared to outcomes that are fixed in advance but to processes that yield surprise.

The question for a foundation of innovation policy now is what such insight in idea creation might be. For this, we need an excursion into theory of cognition. It is hardly surprising that for an analysis of innovation, in the 'knowledge economy' and in a 'learning society', some insight is needed in knowledge and learning.

3.3 COGNITION

Learning can mean the adoption of existing knowledge and competence (learning type 1), and it can mean the development of new knowledge (learning type two), also called 'experiential learning' in the literature on organizaisational learning. This distinction runs parallel to the distinction, in the innovation literature, between the adoption and the production of innovations. As we shall see, however, the adoption and production of innovations interact. In experiential learning, the development of new knowledge is grounded in experience, i.e., the application and adaptation of existing knowledge. It is important to understand how this interaction between application and creation of knowledge works. For policy, it has implications for how to stimulate not only the application of knowledge from universities, which is the current focus of innovation policy, but also, the other way around, how experience forms the basis for new scientific insights, which is neglected in policy.

Here we employ a constructivist, interactionist perspective on cognition (Berger and Luckmann 1967). The basic idea is that people perceive the world, and interpret and evaluate it, on the basis of mental categories that are constructed in interaction with other people. That implies that people see and understand the world differently to the extent that they have developed their cognition along different life paths, in different environments, with different experiences, in different practices, and that consequently people never have identical knowledge or views. There is, in other words, a greater or lesser 'cognitive distance'.

An individual's web of mental categories constitutes his 'absorptive capacity': the ability to form perceptions, interpretations, and understandings. That ability at the same time constitutes a limitation: we can only absorb what fits. However, in absorption, or assimilation, there is always to some extent reconstruction. Even memory is not 'retrieval' but reconstruction, conditioned by the contexts and triggers of memory. The Dutch (and the German) term for memory '*herinnering* (*Erinnerung*)', i.e., 're-internalisation', is literally an accurate term. Knowledge 'transfer' carries a misleading connotation since 'transfer' in fact entails reconstruction and transformation. Goods get transferred, ideas get transformed. Recall that this yielded an argument against a literal interpretation of the term 'replication' from evolutionary theory. In other words, learning type 1 can lead to learning type 2, and indeed to some extent always does.

These ideas go back to the work of developmental psychologists Piaget (1970, 1972, 1974) and Vygotsky (1962), who both proposed, with some important differences, that intelligence is 'internalised practice'. One can only 'assimilate' perceptions into existing cognitive frameworks, but assimilation entails processes that lead to change of cognitive structure ('accommodation'). Their insights, developed in the first half of the previous century, were based on observation and experimentation in the conduct of children. Interestingly, more recent developments in cognitive science, in the stream of "embodied cognition" (Lakoff and Johnson 1999; Damasio 1995, 2003), the idea is supported by neural research, e.g., in the work of Edelman (1987, 1992). That shows how neural-cognitive structures arise from action. There also, an evolutionary perspective is used, in 'neural Darwinism'.[1] This shows how tentative ideas, grounded in neuronal groupings, in competition with each other, can reinforce themselves and can dominate depending on the success of actions that they support, and how new ideas can arise from connections between old ones, in linkages between neuronal groupings.

The idea of intelligence as internalised practice is important for the analysis of the knowledge paradox discussed in chapter 1. It suggests that the issue is not only one of application of knowledge from universities, but goes both ways, meaning that practice can be a source of inspiring ideas. That is very different from programming scientific research on the basis of the perceived needs of industry. But it does suggest the need for interaction between theory and practice, on the basis of a certain mobility of researchers and developers, between university and industry, or for collaboration in joint projects. This yields our proposal for the facilitation of such encounters and collaboration between university and industry. The specification of reasons and ways to do this, as well as feedback from industry and universities on this idea, are discussed later, in chapter 4.

We propose that cognitive processes in the interaction between theory and practice also provide the theoretical basis for the 'cyclical innovation model' of interaction between invention, trial and testing, innovation, and diffusion that has replaced the old 'linear model' of the transfer of knowledge from research (Kline and Rosenberg 1986).

Below, we analyse how exploration and exploitation may build on each other, in a 'cycle of discovery'. There is a connection with the notion of "mode 1 and mode 2 science" (Gibbons et al. 1994). In mode 1, mono-disciplinary science and multidisciplinary application are separate domains, with their own methods and styles of thought, but they are connected, generally in the 'transfer' of knowledge from knowledge generation to application and in the programming of knowledge generation from practical priorities. In mode 2 science, generation and application of knowledge get integrated, in the joint, 'transdisciplinary' creation of knowledge between scientists and 'stakeholders', defined as parties that undergo or exert influence from the development and application of knowledge. Both sides have legitimate claims of knowledge, of different kinds, which need to be integrated. What our analysis here shares with that perspective is first of all the underlying constructivist, interactivist epistemology. Second, we share the idea that application contributes to knowledge creation. However, there is a difference. While we recognise that generation and application build on each other, we still recognise different stages, in a cycle of development, with different requirements and conditions concerning institutions, organisation, styles of thought and action, tolerance or appreciation of ambiguity, and uncertainty of goals, means, and the causal connections between. From application there is an essential step of disembedding knowledge from its context, so that it may be re-embedded elsewhere, in a novel application that yields new insights. In some accounts of mode 2 science (RMNO 2007) any such distinction seems to disappear, in a total merging of generation and application, in one single process, which would suggest that there is no longer any division of labour between universities and industry, and no distinction between invention, innovation, and diffusion. We reject that.

3.4 CYCLE OF DISCOVERY

With the idea that cognition is constructed from interactions of practices we arrive at an innovation process as a cycle or spiral of invention/idea generation, development, commercialisation, market penetration, diffusion, consolidation, and differentiation, which lead to the beginning of invention. This should explain how exploitation and exploration succeed each other and emerge from each other. Insight into this is only nascent, and requires further development.

One proposal (Nooteboom 2000) of a 'cycle of discovery' was originally inspired by the work of Piaget on the development of intelligence in children. Here it is applied at the level of personnel and teams within firms, and to the level of firms, products, and technologies within economies. How can such a shift of the level of analysis be justified? The claim here is that the cycle goes beyond empirical phenomena of child development. It represents a more general 'logic' of composition and break-up on the basis of experience, in an alternation of reducing variety of content, in the move towards consolidation, an opening up of variety of context, in generalisation, which leads on to a renewed opening of content, in novel combinations. The reduction of variety in consolidation is analogous to the evolutionary notion of niche creation; the escape to new contexts is analogous to niche selection. We propose that this yields general principles of a 'knowledge ecology'. In fact, we propose that it also applies beyond markets and business, in cycles of development and execution of government policy.

A basic idea of the cycle is that application of existing knowledge and competence in novel contexts (e.g., new applications of theory and technology, new markets for existing products, new jobs for people), called 'generalisation', leads to 'differentiation' of existing practice, for the sake of adaptation to the new selection environment.

The Dutch innovation of the 'Senseo' electric coffee machine, developed in an alliance between Philips Electronics (for the machine) and coffee producer Douwe Egberts (for the coffee pads used in the machine), has recently been differentiated with a bright red model, to appeal to emotions in the perceptions of new customers in the Brazilian market, which is not sold elsewhere, and with pads for making tea, in the same machine.

The new selection environment offers the room to deviate from the institutions that have formed themselves in the 'consolidation' of a previous innovation, the motivation for taking the trouble of adapting practice, to satisfy conditions of survival, and fresh insights from new conditions and demands, into new opportunities and into the limits of established practice. In the adaptation of a product or practice to new conditions, one first taps into earlier experience about how things might be done differently, going back to experiments, trials, and errors in an earlier round of innovation. One may also utilise ideas from existing partners in collaboration. If differentiation does not suffice in order to survive, or to profit from newly emerging opportunities, a further step is to allow oneself to be inspired by foreign practices encountered in the new environment, which appear successful or promising where one's own practice seems to fail. This leads to experiments of combinations of known elements from existing practice and new elements from unfamiliar, local practices, called 'reciprocation'. This yields hybrid practices. The history of technology offers many examples of the importance of hybrids in the transition to radical innovation (Mokyr 1990). The significance of hybrids, in the logic of discovery, is that they allow one to explore the potential of novel elements without yet surrendering the basic logic, structure, design principles, or architecture of established practice. The problem with hybrids is that they yield inefficiencies and inconsistencies in the system ('spaghetti'), with overlaps, redundancies, misfits, and 'work-arounds' to solve them. That then leads to more radical, architectural change, in Schumpeterian 'novel combinations'. The period of hybridisation gives insight into the elements one would most like to preserve, given their performance in the hybrid, and the directions in which one might think for novel principles of logic or architecture. Here, at this stage, small changes in design principles or basic logic can yield a saltation in the functioning of the whole. At the same time, the inefficiencies and contradictions of hybrids also form a stumbling block: they may be seen as evidence of failure and lack of perspective for the innovation. Progress then depends on the perseverance of the entrepreneur or inventor. Also, the inefficiencies of reciprocation and hybridisation are difficult to sustain under the pressures of competition. This frequently leads to failure – because problems do indeed prove to be insuperable or ongoing efforts and uncertainty cannot be sustained – but occasionally this leads to a breakthrough. The cyclical process of invention or discovery indicates how one can set out in exploration along a path of exploitation. Crucial for the process, in the stages of generalisation and reciprocation, is the opening to novel contexts ('niche selection'), with new challenges and opportunities, and openness in the form of curiosity and attention to unfamiliar practices and perspectives, and the willingness and opportunity to engage in experiments, and tolerance of the problems with hybrids. The cycle is illustrated in figure 3.1. In chapter 9 of this volume, Anderson and Gasteiger employ a cycle of convergent and divergent behaviour, where the first resembles the movement towards consolidation, in exploitation, and the second resembles the movement of exploration

Whereas previously multinationals used to internationalise for growth of sales, as a way out from saturated markets, now they increasingly use internationalisation as a process of discovery and a path to innovation. They appear to have caught on to the logic of the cycle. So far, the discussion of the cycle concerns the bottom half, in the transition from exploitation to exploration, which is new in the innovation literature. The top half is more in line with established innovation theory. Along the top half, in the emergence of a new idea or practice, in a novel combination, there is search for technical feasibility and commercial viability of a new technology or product and its optimal configuration, in the emergence of what in the innovation literature is called the 'dominant design'. This leads to what is labelled 'consolidation'. In that process, if a breakthrough to an invention succeeds, it faces the need to replace old practices, in Schumpeterian 'creative destruction'. Here, one runs into the problem that existing institutions, in the form of standards and regulations (technical, safety, commercial, fiscal, legal, administrative), market structures (distribution channels, installation, maintenance, repair), schooling and training, and established commercial positions, which form the existing selection environment, can block entry and change. In other words, in order to break through, innovation requires institutional change, and the analysis may contribute to the literature on institutional change. Our lack of understanding of innovation and of institutional change are connected. As a result, due to institutional barriers radical innovations can often break through only later, and initially can only succeed where they can be fitted into the prevailing order of existing institutions and market structures. They need to prove their worth and their potential more extensively before obstacles can be cleared. It is a well-known phenomenon that initially innovations do not find their application where their potential is highest but where the obstacles are lowest.

Figure 3.1 Cycle of discovery

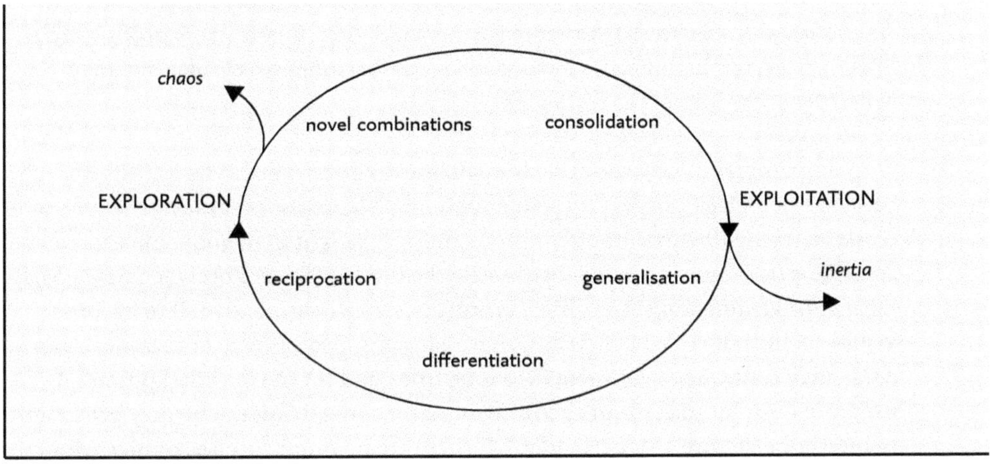

Source: Nooteboom (2000)

Thus, the steam engine was first used for pumping water out of coal mines, because there they did not have to replace any deeply rooted existing system. Only later [were steam engines] used in trains and steamships, in competition with coaches and sailing ships .

Hence, openness of markets for new product entry, with a critical attitude towards established interests and institutions is an issue for innovation policy. In this volume, the issue of institutional and commercial obstacles to entry is elaborated with four case studies, in chapter 6. One policy implication is that enabling entrepreneurs goes beyond helping them to find their way through the density of rules and regulations, and requires the gathering of insights as to how those may have to be changed to accommodate the shifts of innovation.

In the movement towards consolidation, goals, means, and causal relationships between them become clear. As also among potential users uncertainty decreases and familiarity with the novelty increases, demand increases, novel producers jump on the bandwagon of the emerging market, and price competition intensifies. Pressure on price yields pressures towards efficiency, on the basis of process innovation. In pressures towards efficiency, standard economic analysis applies. Market mechanisms are needed to ensure optimal allocation of scarce resources to known goals and means. In the drive towards efficiency, opportunities are taken to increase productive efficiency, by increase of scale, enabled by growing demand, which leads to concentration and the 'shake-out' of less efficient producers. Here, policy is oriented mainly toward removing barriers to entry, in competition policy.

As noted in chapter 1, the fall of profits, in the transition from product innovation to process innovation, in the development towards consolidation, yields an argument for trying to be a leader, in the early stage of innovation, because thereby one captures the high profits of early partial monopoly, before imitation sets in. As a follower, one enters at the stage of consolidation, where users profit from lower prices, but high profits have eroded. Furthermore, as a leader one may construct entry barriers to followers.

Ongoing progress along the cycle is by no means guaranteed. The cycle is not to be seen as a logically necessary sequence but as a heuristic that may generally work. In trying novel combinations one may get caught in ongoing uncertainty and chaos (see figure 3.1), unable to settle the inconsistencies between new goals, means, and connecting causalities. Prototypes may continually fail to become viable, technically or commercially. Rival designs, prototypes, or technical standards may continue to compete for a long time, and for the duration potential users are hesitant to commit themselves. After consolidation, one may get caught in inertia (see figure 3.1), particularly if there are no opportunities or incentives to escape to new contexts of application, or barriers to novel conditions being imposed from outside. In consolidation, institutions have shifted to accommodate the innovation, and once that has happened there are often strong pressures towards "isomorphism" (DiMaggio and Powell 1983), with strong pressures to conform, by 'coercion, mimesis (imitation) or normative pressures'. Existing institutions are also protected by vested economic interests, in their efforts to protect the installed base of both tangible and intangible investments, existing competencies and efficiencies (accumulated in learning by doing), and market positions. Therefore, innovation requires openness to novel contexts of application, e.g., global markets, or new users or suppliers, as arenas for exploration and sources of novel challenges. Stages of the cycle may be skipped, in a leap to novel combinations without much intervening differentiation or reciprocation. The process may not proceed beyond some stage. For example, differentiation, as a step in exploitation, may not proceed to reciprocation and novel combinations. In the example of the Senseo coffee machine development may not proceed beyond the addition of a red model for the Brazilian market and the use of filter pads for tea in addition to those for coffee.

Note that progress along the cycle is full of stress and potential conflict. In generalisation a stressful need arises to adapt existing practices, in order to survive in the novel context. In novel combinations innovators encounter stress in trying to have their innovation accepted, and established practices encounter the stress of creative destruction. In chapter 9, Anderson and Gasteiger report evidence that during innovation within organisations these stresses indeed arise, as both a motivator and a consequence of innovation.

Returning for a moment to the discussion of evolutionary theory, we note that the cycle offers a specification of how invention can remain subject to trial and error without being entirely blind, because it is rooted in a process of application and adaptation that is to some extent based on inference and imagination. It shows how exploration can emerge from exploitation, and how an apparent jump ('saltation') in form or function can arise from a more gradual step-by-step process leading up to it. Earlier in this chapter we noted that ecology (the functioning of 'niches') is an aspect of evolutionary theory. In our present analysis, aimed at taking discovery and learning into account, we propose that the cycle of discovery provides the basis for what one might call a 'knowledge ecology'. By this we mean a system where the cycle of discovery operates, more or less perfectly, depending on institutional conditions that inhibit or enhance the component processes of generalisation (opening up to new contexts); differentiation (deviation from established practice to survive in the new context); reciprocation (opening up to contributions from unfamiliar ideas or practices); experimentation with hybrids and new principles, interpretive schemes or architectures; convergence to a dominant design; and institutional change to accommodate the novelty. The top loop of the cycle, in figure 3.1, corresponds with niche construction and the lower loop arises from the selection of a new niche. We can then characterise innovation policy as guiding and enabling the knowledge ecology.

Crucial in the process is the opening to new contexts with new challenges and opportunities, opening to collaboration for the exploration of novel combinations, opening in the form of curiosity and attention to foreign practices, and the preparedness to engage in experiments with elements from those and with surprising hybrids. The Renaissance in Europe was accompanied by a lively interest and use of many things that could be found elsewhere (Mokyr 1990). This stands in contrast to China, for example, which from around 1400 lost its prior advantage by closing itself off to foreign influences. Perhaps this helps to explain why the first two industrial revolutions occurred in Europe.

Further application

Chapter 5 offers a systematic analysis of entrepreneurship, but here we briefly indicate the connection with the cycle of discovery. It is customary to distinguish between equilibrium breaking, Schumpeterian entrepreneurship that yields 'creative destruction', and 'Kirznerian' entrepreneurship, derived from Kirzner (1973), which finds new market niches for existing or adapted products, in a process of what economists call 'arbitrage,' and thereby tends towards equilibrium. We can recognise this in the cycle of discovery: the movement towards consolidation can perhaps be seen as equilibration, and the movement away from it as disequilibration. Instead of two kinds of entrepreneurship, we can identify a larger range of different types, all along the cycle of discovery (Nooteboom 2000). Thus, there are entrepreneurs who make a new idea technically feasible, commercially feasible, productively efficient, eliminate entry barriers, carry it into new markets or applications, differentiate it, bring in new elements, in hybrids, or bring together elements from different practices in new architectures and thereby produce new ideas.

Note that in the step of generalisation the actor who takes an existing product or practice into a new context is not necessarily an existing producer or practitioner. It may be an outside entrepreneur or user stepping in, adopting the product or practice with his own specific experience and perspective. This, however, may already happen prior to consolidation, so that exploration may set in when exploitation has not settled down yet. Entrepreneurs adopting the innovation will inevitably, and not necessarily deliberately, colour their use of it according to their perspective, and seeing that the product is on its way to widespread diffusion and consolidation, with an erosion of profit, may already differentiate it deliberately. What we are saying here is that disequilibration may take place even during equilibration, which seems to make nonsense of the very notion of equilibration. Why would entrepreneurs move towards equilibrium if they know that that will erode profits?

In this book, the main application of the cycle is to the emergence of new technologies, products and processes. However, the cycle also applies to experiential learning by organisations and people. Indeed, its origins lie in the latter, as mentioned earlier. There, new ideas are developed when existing ideas or practices are confronted with new challenges in novel contexts that are sought or imposed from outside (generalisation). In individual learning, this typically arises in discourse where one confronts one's ideas with those of others who think differently. In the processes of differentiation and reciprocation, existing ideas, interpretations, and meanings shift, in the development, by way of hybrids of old and new, the familiar and the unfamiliar, and via of new interpretive schemes and terminology. In the literature on organisational learning, this is called sense making (Weick 1995). Here, in exploration, discourse is based not primarily on logical analysis, but on 'storytelling', in which narration from specific experiences in specific contexts, shifts perspective and interpretation. Stories are means for reciprocation. Attempts at mutual understanding and reconciliation of differences yield the hybrids from which new shared perspectives may arise. It is in this interaction, between people in organisations, with shared goals, orientations, or practices, that individual and organisational learning connects. This interaction between people as a basis for learning is further analysed in the next paragraph.

Perhaps the cycle can also be applied to government, as a cycle of policy learning, in an elaboration of the notion of experimentalism or 'direct-deliberative polyarchy' indicated before (Gerstenberg and Sabel 2002). There is much talk of deliberative government as a source of policy learning, particularly for dealing with complex, 'untamed' problems (WRR 2006), where one does not yet have clear insight into goals, priorities, means, the causality between means and goals, and the resources needed. This includes uncertainty concerning possible outcomes, what might happen under what conditions, and uncertainty concerning the value or disutility of outcomes for different people. Here, one needs to tap into the diversity of experience and insight of the population. When this results in conclusions and policy, policy has to be implemented efficiently and transparently, which requires a clear distribution of tasks.

The link with the cycle of discovery may be as follows. To learn, existing policy thinking must be subjected to new contexts and perspectives (generalisation), which here entails that it is opened up to the variegated experience of citizens, in consultative deliberation. New elements arising from that opening are incorporated in experimental hybrids (reciprocation), and this experimentation yields further insight into what old and new elements to combine in what new architectures or perspectives, in exploration. After experimentation has developed into a 'dominant design', it is implemented and consolidated, in exploitation, with requisite changes in relevant institutions. This implementation requires and allows for tighter central coordination, enforcement, transparency, and accountability. A central task of government is to enable the overall cycle, in "metagovernance" (Soh 2007). After consolidation, execution may perhaps be delegated to markets or semi-public executive bodies. In sum, exploration occurs from bottom-up and exploitation from top-down. A potential problem with the latter is that the necessary connection between exploitation and exploration may be broken. Experience with implementation is needed to inform exploration. Therefore, people from executive bodies should also play a part in the process of citizen consultative deliberation, in a subsequent round of exploration.

3.5 MARKET FAILURES (WITH A VENGEANCE)

In the debate on innovation policy, the usual argument for public policy is that especially in innovation there are market failures. These failures concern the usual problems of externalities, public goods, merit goods, and entry barriers. For public goods the market does not work due to non-excludability and the non-rival nature of the goods. However, according to more recent insights the 'transfer' or 'spillover' of knowledge requires the absorptive capacity to do so, which arises from an accumulation of knowledge and experience. As a result, knowledge can be excludable to a greater or lesser extent. A more recent insight into obstacles to efficient markets arises from the notion of transaction costs; in the judgement of quality; costs of search, mutual understanding, and adaptation for producers and users; costs of coming to an agreement; obstacles to complete contracting and monitoring, due to asymmetric information, consequent costs of monitoring, haggling, re-negotiation, conflict, and litigation; and switching costs if a relationship breaks. The uncertainties of innovation increase transaction costs. Some of that will appear in later analyses of problems of collaboration, in chapters 7 and 8. The problem of entry barriers, and the need for public intervention to avoid them or break them down, came up in the previous section.

In spite of such market failures, however, in the 1990s policy rhetoric has increasingly shifted to the mobilisation of market forces, also in public goods and services, such as transport and health care, by means of deregulation, privatisation, the auction of franchises, and public-private arrangements with measures to deal with market failures by public setting of tariffs or targets, and other regulatory controls and openings for intervention. Currently market logic is moving into universities, with incentives for university scientists to commercialise their inventions by incentives of intellectual property, licensing, contract research, and consultancy.

However, Lester and Piore (2004) revealed less recognised and perhaps more fundamental failures of markets, which may not only fail to yield innovation but can actually break down sources of innovation. Lester and Piore propose that innovation consists of two distinct, different, and difficult to combine stages of 'analysis' and 'interpretation'. There is a strong link, not identified by Lester and Piore, between analysis and exploitation and between interpretation and exploration, whereby their analysis fits snugly into our analysis of exploration and exploitation. In analysis, ends, means, causal linkages between them, and requisite resources, are stable and clear, so that one can decompose a product or process into elements (hence the term 'analysis') and rationally optimise the allocation of scarce resources. Here, the allocative function of the market is central. This corresponds closely with the situation that we indicated, in figure 3.1, of 'consolidation', as a basis for exploitation. In sum, in consolidation markets work.

By contrast, in interpretation, ends, means, causal relations, and requisite resources are not clear, and one has to iterate between the interests, knowledge, competencies, and resources of a variety of stakeholders (potential customers, producers, suppliers, labour, regulators, citizens affected by externalities, etc.), with a corresponding variety of perceptions, views, norms, and values. Here, we are dealing with what have also been called "untamed problems" (WRR 2006). This requires what Lester and Piore call 'conversation', which requires openness, acceptance, and utilisation of ambiguity, confusion, and ways to deal with fears and suspicions and to build trust. This connects with the notion of sense-making in the literature on organisational learning (Weick 1995). It requires the opportunity, time, stamina, and patience to 'sort things out' and to build understanding and trust. Pressures of price competition tend to disable all that. Under harsh, pure price competition, there is fear of openness leading to loss of appropriability of profit from investments ('spillover'), pressure of price competition constrains the give and take needed to build trust, pressures on cost, and pressures of time, for short-term profits and for speed to market. Competition eliminates the 'slack' of resources and time needed to 'conduct the conversation', to accept and utilise ambiguity, and to build understanding and trust. Price competition yields an incentive to differentiate products, which is a form of innovation, but the point of it is that it reduces price competition. Price competition also yields an incentive to increase efficiency, which may be used to create the slack needed for exploration, but that works only when the efficiency advantage with respect to competitors is maintained, which entails limitation of competition.

Later, in chapter 7, we will argue that in exploration one needs to make relation-specific investments in mutual understanding and trust, which require a certain stability of the relationship. This leads to a plea for optimal, not maximum flexibility, in labour and organisation. These arguments appear to connect with the idea of "flexible rigidities" (Dore 1986): in early innovation a certain rigidity is needed, in the sense of a limit on price competition and a certain stability of relationships. However, rigidity should be optimal and temporary: providing sufficient stability but not so much as to obstruct the structural change and creative destruction that is the outcome of radical innovation.

Lester and Piore (2004) propose that innovation policy in the US suffers from a fundamental misinterpretation of what happened in the US in the 1990s. An increasing focus on market mechanisms, under globalisation, deregulation, and privatisation, has been beneficial in furthering efficiency and exploitation, and a dynamic of entry and exit, but has also broken down the opportunities in industry to conduct 'conversations for interpretation', or what we call exploration. This has led to a break-up of institutes that existed in forms between university and business, with participation of people from both, such as the famous Bell Labs (for their role, see also Chesbrough 2006), that provided hotbeds for exploration and spawned, among other things, the famous 'Silicon Valley'. As a result, industry now has to look more to universities to provide the 'public space' that allows for exploration. The fact that in the 1990s there was both an upsurge of innovative performance in the US, and a drive towards market mechanisms, has led to the mistaken interpretation that the first was caused by the second. In fact, that surge was built partly on the structures for exploration in industry that existed previously but was broken down in the 1990s, thus breaking down the wellsprings of future innovation. And now, the mistake is compounded by pressures to carry market forces into the universities, thus jeopardising the remaining sources of exploration. Europe, we may add, in its breathless admiration of US innovative performance, has been preparing to imitate this mistake along with factors of success.

For these and other reasons, we are wary of further commercialisation of university research. However, we do see the need to stimulate the utilisation of opportunities to apply university research, a process which has been referred to as 'valorisation' in Dutch policy debates. On the basis of our analysis of exploration and exploitation we add that the issue is not only that of moving university research (exploration) into commercialisation in industry (exploitation), but also, in the opposite direction, of generating inspiration for exploration from exploitation. We need to find ways for university and industry to interact in order for exploitation and exploration to build upon each other. This issue is taken up in chapter 4.

3.6 OPEN INNOVATION

The idea of 'open innovation' has gained widespread interest. It suggests that firms should not innovate in isolation, but in collaboration with others. Such collaboration would yield not only efficiency of scale or scope but, also of importance for innovation, a greater diversity and flexibility for Schumpeterian 'novel combinations'. Note that the key feature of the logic of the cycle of discovery was that novel areas of application of existing practices (ideas, products, processes) yield new pressures and insights for adjustments that start small, in differentiation, and are then extended, in reciprocation and novel combinations. This, we propose, yields the deeper significance of collaboration for innovation. In collaboration with suppliers, customers, firms in related markets, and sometimes even competitors, one fits existing ideas and practice into their different but related thinking and practice, whereby the dynamic of exploration by differentiation and reciprocation may be set in motion. This ties in with pleas to involve suppliers not only for efficient production, in 'co-makership', but also for development and innovation, in co-development. It also ties in with pleas to make use of lead users for development and innovation. See, in particular, Von Hippel's (2005) plea for the 'democratising innovation' by involvement of users.

This line of thought is further developed with the notion of "cognitive distance" (Nooteboom 1999). Cognition here is a broad notion, including both rational and moral dimensions, in both knowledge and value judgements. As noted earlier, the constructivist view of cognition implies that people have different cognition (perception, interpretation, evaluation) to the extent that they have developed their mental structures along different life paths, in different countries, markets, technologies, and forms of organisation. This yields greater or lesser cognitive distance.

Cognitive distance yields both a problem and an opportunity. The problem is that collaboration is more difficult under larger differences in cognition, because it is more difficult to understand each other (less requisite absorptive capacity) and because views on collaboration diverge. The opportunity lies in the fact that difference yields an opportunity to learn something new. Innovation is furthered by a balance between difference and agreement, at optimal cognitive distance: large enough to tell each other something new but not too large for understanding (Nooteboom 1999). Here we see the importance of diversity but also its limitations. In other words, innovation is favoured by variety that is still somehow related (see also 'related variety' as a driver of regional economic growth in chapter 11). In chapter 9, Anderson and Gasteiger present evidence that diversity and team heterogeneity have a positive effect on creativity within organisations, as well as a positive effect on cohesiveness and longevity of teams.

Figure 3.2 Optimal cognitive distance

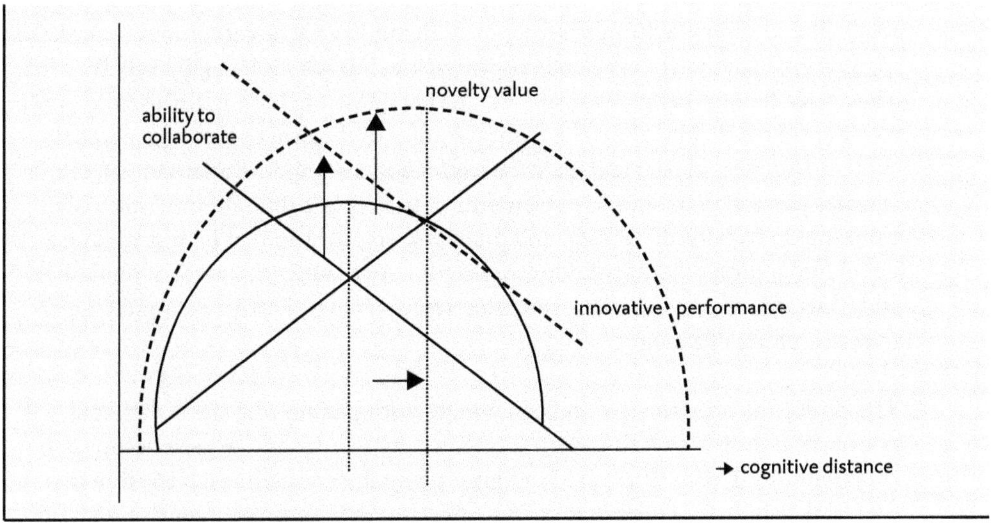

Source: Nooteboom 1999, Nooteboom et al. 2007

This is illustrated in figure 3.2. The downward sloping straight line[2] models the idea that it is more difficult to understand each other and to collaborate as cognitive distance increases. The rising straight line models the idea that novelty value increases with cognitive distance. If innovative performance equals the mathematical product of novelty value and the ability to utilise it, an inverted U-shaped curve (parabola) results, with a given level of optimal distance. When something is new but cannot be used, it is as useless, for innovation, as when something is easily used but not new.

An econometric corroboration of the model is given by Nooteboom et al. (2007), in an analysis of 994 alliances involving 116 firms during a period of 10 years. Patents were taken as a measure of innovative output, and the difference in technological profiles of firms, construed on the basis of patent data, were taken as a proxy for cognitive distance. Li et al. (2007) found a corroboration in an analysis of collaboration between suppliers and producers of 1,635 innovations of 550 Canadian firms during a period of 35 years. Innovative performance was measured as the number of radical innovations (new to the world) and cognitive distance was measured as difference of position in the classification of industries.

In exploration, the value of novelty is higher than in exploitation, resulting in a steeper slope of the line of novelty value. In exploitation the penalty of lack of understanding and ability to fine tune collaboration is higher, resulting in a steeper downward slope of the line for ability to collaborate. As a result, optimal distance is higher for exploration than for exploitation. For empirical evidence, see Nooteboom et al. (2007).

It has been widely recognised that collaboration is needed for innovation, but why is that, precisely? It is evident, especially from an evolutionary perspective, that collaboration yields more variety and flexibility for novel combinations than integration of activities in a single organisation, but why is it cognitively fruitful? The cycle of discovery gives an answer to that question. Interaction forces people to try and fit their ideas into the mental frames of the other person (generalisation), differences appear and yield a need for adjustment (differentiation), opportunities emerge to try and fit in elements of the other's thought into one's own thinking, in hybrids of thought and practice (reciprocation), which stimulate a novel integration of joint thinking and action. By the inventive use of metaphor and illustrations one can help each other to cross cognitive distance and trigger requisite shifts of thought.

The ability to communicate and collaborate with people who think differently is not fixed. It grows with the accumulation of knowledge and experience in such collaboration. This is illustrated in figure 3.2 by the dotted lines. If the line that represents ability to collaborate shifts upwards, then optimal cognitive distance shifts to the right, and corresponding performance shifts upwards. For empirical evidence, see Nooteboom et al. (2007). This demonstrates the economic advantage of the ability to collaborate with people who think differently: one can operate at a larger cognitive distance, achieving more innovation. A policy implication for the Netherlands is that for innovation it is crucial to restore former Dutch capabilities of appreciating and mobilising cultural diversity. The analysis also suggests the positive side of cultural diversity in Europe. As a matter of EU policy, we might aim for optimal variety: big enough for innovation but small enough for collaboration.

If the analysis arises on the level of individual cognition, how does it apply on the level of organisations? Organisations serve, among other things, for the sake of efficient collaboration, by limiting cognitive distance, with a certain cognitive 'focus', on the basis of organisational culture, but with the maintenance of sufficient distance for an internal source of innovation. In other words, firms have to make a trade-off between exploitation and exploration. Organisational focus by definition yields a form of myopia ('group think'), that needs to be compensated for with external contacts with actors (persons, organisations) with a different, complementary cognitive focus (yielding "external economy of cognitive scope", Nooteboom 1992). Thus we again obtain figure 3.2, but now on the level of organisations, where cognitive distance now applies to difference in the 'cognitive focus' of organisations.[3]

When individuals or firms maintain long-lasting relations, and those relations are also exclusive, i.e., are closed off from other, outside contacts, then after awhile cognitive distance will decrease and in the long run it will decrease to the point that the innovative potential of collaboration dwindles (see figure 3.2). For empirical evidence, see Wuyts et al. (2006). In their survey of studies of creativity and innovation within firms, Anderson and Gasteiger (chapter 9) also report inverted-U shaped effects of e.g., group cohesiveness and longevity. In view of this, for innovation it is necessary, to continually refresh the field of collaborators. Long-term relations do not necessarily have this effect. If actors also maintain other, outside relations that do not overlap among each other, so that they are continually refreshed with outside insights and impulses, even a long-lasting relation can maintain its innovative momentum. This ties in with what Burt (1992) called the bridging of 'structural holes' as a source of innovation, in the connection of otherwise unconnected groups. This is illustrated in figure 3.3.

Figure 3.3 Bridging structural holes

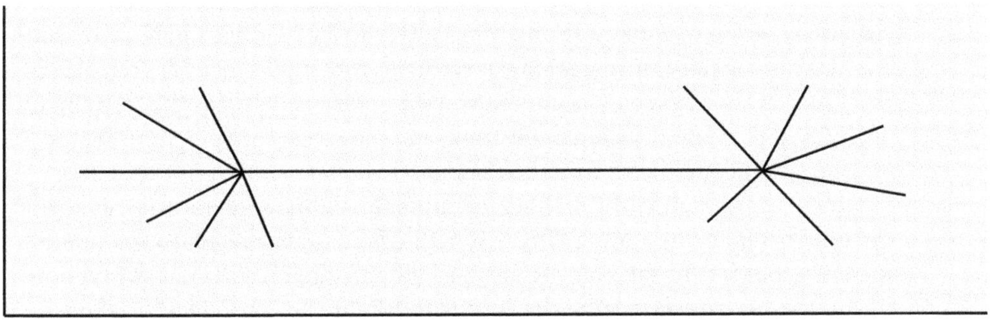

In sum, open innovation requires not only collaboration with others, but also the management of cognitive distance in collaboration, which requires openness for the entry of novel actors or connections across groups of collaborators.

The notion of optimal cognitive distance also has implications for utilising the cycle of discovery, in particular the step of generalisation. It is recommended to look for a new context of application, e.g., a new market for a product, at optimal cognitive distance (which includes cultural distance).

"A recent evaluation of the EU Framework programme for collaborative research (FP5 and FP6) illustrates the role of absorptive capacity (EU 2007). A survey among FP participants shows that in FP projects the distance to the core activities of the firm is larger than in an average R&D project. An analysis on the basis of the Community Innovation Survey (CIS) shows that participants in the FP programme do more R&D than non-participants. The survey shows that previous collaboration has a positive effect on outcomes, but the presence of new participants has a greater positive effect. All this can be interpreted from the perspective of the above analysis. In order to profit from the opportunity for innovation offered by distance to the core, firms need adequate absorptive capacity, which is formed by earlier R&D and experience in collaboration. However, previous collaboration can have a negative effect of reducing cognitive distance too much, and then it is better to have new participants to increase it again" (Polt et al. 2008).

3.7 CONCLUSIONS

An Austrian/evolutionary perspective in economics, together with a constructivist, interactionist perspective of knowledge and learning, with its appreciation of local specificity and diversity of cognition and meaning, as a source of learning, yields a perspective of 'experimentalism' that is also known in political thought. This is the perspective that forms the basis for innovation policy that is developed in this volume.

The evolutionary perspective suggests that especially the early stages of innovation, in exploration, are subject to radical uncertainty, are hence unpredictable, and an important task of innovation policy is to maintain sources of variety that yield unexpected outcomes. Maintaining such variety requires a policy to stimulate entry of outsider entrepreneurs. It confirms the importance of eliminating entry barriers to markets. The unpredictability of invention requires restraint of instincts to plan and program innovation, and the acceptance of failure of entrepreneurial ventures not as waste but as a necessary feature of innovation (see chapter 5). As a result of the problem of unpredictability, in innovation policy there is a paradox of accountability. For public funding one needs to account for spending, but setting targets ('deliverables') is paradoxical and clear *ex ante* criteria that are guaranteed to yield the desired results are not available. Yet, there is an inclination to fix verifiable targets, while in innovation one typically ends up somewhere else.

An illustration is the development of the famous Philishave of Philips Electronics. The development was initially aimed at a new dynamo for lighting on a bicycle, but what came out was a shaving apparatus.

There are several perverse effects. First, by specifying deliverables one locks up development in a possibly unfruitful direction. Second, one elicits strategic conduct to distort facts, falsely pretending that the original goal has been reached, in order to secure payment. It should be possible, in schemes to subsidise R&D, to allow for flexibility in outcomes, and pay for what has been achieved along the way and in the end, rather than for the achievement of outcomes determined a priori. This is complicated in the case of research consortia, as in the EU Framework Programme, where change along the way implies exit and entry of members of the consortium, with complications concerning contracts and shares in payment. However, the alternative is to continue with an unfruitful development with the wrong partners. A third perverse effect of the paradox of accountability is that to avoid the problem, one focuses subsidy schemes not on the difficult to predict exploration, but on the more predictable stage of exploitation. Thereby, innovation policy tends to confirm existing strengths rather than developing new ones. This last problem is compounded by an intellectual mistake, promoted by an uncritical adoption of the evolutionary perspective, that since invention is unpredictable, it is blind and random, so that a policy for invention is hopeless. In exploitation, laissez faire, to profit from market mechanisms, is mostly preferable. Here, we have tried to show that while invention is subject to much trial and error and remains unpredictable, it is not blind, but is based on experiential learning that employs a certain type of inference from failures and indications for the potential of new directions. For this we employed the idea of a cycle of discovery, where exploitation can lead to exploration.

An issue of fundamental and ill-recognised importance is a novel market imperfection in the conditions needed for exploration. In exploration ends, means, relations between them, and requisite resources are not yet clear. This requires open communication, abilities to mobilise ambiguity, to deal with confusion, and to spend resources and time in the iterations and trials needed to achieve clarity of ends and means. This, in turn, requires trust, but trust has its limits, especially under competition. This is taken up in chapter 7. Market forces may block such processes, disallowing the openness, slack resources, time and trust needed for exploration. However, competition is seldom extreme, pure price competition, due to transaction costs, uncertainties, and switching costs. As a result, combinations of exploration and exploitation, and of competition and collaboration, may be feasible.However, the problem is not only that competitive pressure on price may eliminate the resources and the slack needed for exploration in individual firms. Also, competition policy, instituted to prevent cartels, may forbid ongoing collaboration that excludes entry of new participants for the duration of the relationship. For exploration, relationships should have sufficient duration to evoke relation-specific investments involved in the development of the mutual understanding and trust needed to profit from cognitive distance, in the creation of Schumpeterian novel combinations.

Océ, a Dutch producer of photocopiers, was forbidden by the competition authorities to maintain durable relations for product development with its suppliers, on the ground that it limited competition.

The analysis of the cycle of exploration and exploitation, in combination with the notion of optimal cognitive distance, has many further implications for policy, in both its main lines and its details. For example, it has implications for our take on globalisation. Globalisation can form the basis for novel challenges and insights that contribute to innovation. For this we should seek and employ optimal cognitive distance, both within and beyond Europe. This requires an open-mindedness and appreciation for cultural diversity that is large enough for novelty but not too large to block collaboration. Local collaboration should be supplemented with linkages to communities elsewhere, preferably again at optimal distance. The capability to cross cognitive distance is a key competitive advantage, and it grows with experience, allowing one to deal with increasing cognitive distance, with an increased potential for innovation.

There are also implications for the mobility of labour, whereby application of existing knowledge and competence in novel contexts can open up paths to discovery. Thus, mobility is important not only for efficient allocation of competence, and for the diffusion of innovations, but also for discovery. The analysis also confirms the need for inter-disciplinarity as a source of innovation. Opportunities and obstacles for inter-disciplinarity are further analyzed in chapter 12.

The cyclical interaction between exploitation and exploration has implications for the interaction between university and industry. There should be paths leading from university exploration to exploitation in industry, but also, in the opposite direction, paths by which experience from exploitation in industry inspires exploration in universities. At the same time, we should recognise that exploration requires conditions that may be thwarted by market competition. This issue is taken up in chapter 4.

The problem of combining exploration and exploitation is arguably the most important problem for innovation management in and between organisations, in industry and government. What forms of organisation enable such combinations? That will be taken up in chapter 8. What forms of collaboration between firms, in alliances and networks, enable the combination of exploration and exploration? That will be taken up in chapters 9 and 10. What are the implications for 'regional innovation systems'? That will be taken up in chapter 11.

In sum, innovation requires openness, in four dimensions. First, openness to uncertainty in the innovation process, especially in exploration. Second, openness for cooperation with other organisations at an optimal cognitive distance. Third, openness for new innovative entrants. Fourth, openness to new areas of application, in new countries, industries, markets, and organisations. This stands in sharp contrast to much innovation policy, which in many cases creates obstacles to innovation in pre-conceived targets, established players, national programmes, key industries and technologies, isolated activities, and relationships at arms length that lack collaboration and openness of communication.

NOTES

1 Bateson (1972) also spoke of an 'ecology of mind'.

2 The straight lines are not necessarily straight, but simply, as a principle of parsimony, the simplest lines that decline or rise. When good theoretical or empirical reasons arise for non-linearity, it will be built in.

3 For an empirical study, cognitive distance between firms was operationalized in terms of differences in profiles of technological capabilities derived from the patent portfolios of the firms involved. See Nooteboom et al. (2007).

REFERENCES

AWT (2003) *Backing Winners*, The Hague: AWT.
AWT (2007) *Balanceren met beleid*, The Hague: AWT.
Bateson, G. (1972) *Steps to an Ecology of Mind*, Chicago: University of Chicago Press.
Baum, J.A.C. and J.V. Singh (1994) 'Organizational Hierarchies and Evolutionary Processes: Some Reflections on a Theory of Organizational Evolution', in: J.A.C. Baum and J.V. Singh (eds.) *Evolutionary Dynamics of Organization*, Oxford: Oxford University Press, 3-20.
Berger, P. and T. Luckmann (1967) *The Social Construction of Reality*, New York: Doubleday.
Borras, S. (2003) *The Innovation Policy of the European Union: From Government to Governance*, Cheltenham: Edward Elgar.
Burt, R.S. (1992) *Structural Holes: The Social Structure of Competition*, Cambridge, MA.: Harvard University Press.
Campbell, D.T. (1974) 'Evolutionary Epistemology', in D.T. Campbell, *Methodology and Epistemology for Social Science*, Oxford: Oxford University Press, 393-434
Chaminade, C. and C. Edquist (2006) 'From Theory to Practice: The Use of the Systems of Innovation Approach in Innovation Policy', in J. Hage, Jerald and Marius Meeus (eds.) *Innovation, Science and Institutional Change*, Oxford: Oxford University Press.

Chesbrough, H. (2006) *Open Innovation – The New Imperative for Creating and Profiting from Technology*, Boston: Harvard Business School Press.
Damasio, A.R. (1995) *Descartes' Error: Emotion, Reason and the Human Brain*, London: Picador.
Damasio, A.R. (2003) *Looking for Spinoza*, Orlando, FL: Harcourt.
Dawkins, R. (1983) 'Universal Darwinism', in D.S. Bendall (ed.) *Evolution from Molecules to Man*, Cambridge: Cambridge University Press, 403-425.
Dennet, D.C. (1995) *Darwin's Dangerous Idea: Evolution and the Meanings of Life*, New York: Simon & Schuster.
Dimaggio, P.J. and W.W. Powell (1983) 'The Iron Cage Revisited: Institutional Isomorphism and Collective Rationality in Organizational Fields', *American Sociological Review*, 48(2): 147-160.
Dore, R. (1986) *Flexible Rigidities*, Standord: Stanford University Press.
Edelman, G.M. (1987) *Neural Darwinism: The Theory of Neuronal Group Selection*, New York: Basic Books.
Edelman, G.M. (1992) *Bright Air, Brilliant Fire: On the Matter of Mind*, London: Penguin.
Edquist, C. (ed.) (1997) *Systems of Innovation: Technologies, Institutions and Organizations*, London: Routledge.
Gerstenberg, O. and C.F. Sabel (2000) 'Directly Deliberative Polyarchy: An Institutional Ideal for Europe?' in: C. Joerges and R. Dehousse, *Good Governance in Europe's Integrated Market*, Oxford: Oxford University Press, 289-341.
Gibbons, M., C. Limoges, H. Nowotny, S. Schwartzman, P. Scott and M. Trow (eds.) (1994) *The New Production of Knowledge: The Dynamics of Science and Research in Contemporary Societies*, London: Sage.
Hippel, E. von (2005) *Democratizing Innovation*, Cambridge, MA.: MIT Press.

Hull, D.L. (1988) *Science as Process: An Evolutionary Account of the Social and Conceptual Development of Science*, Chicago: University of Chicago Press.
Hurford, J.R. (2007) *The Origins of Meaning: Language in the Light of Evolution*, Oxford: Oxford University Press.
Kirzner, I.M. (1973) *Competition and Entrepreneurship*, Chicago: University of Chicago Press.
Kline, S.J. and N. Rosenberg (1986) 'An Overview of Innovation' in: R. Landau and N. Rosenberg (eds.) *The Positive Sum Strategy: Harnessing Technology for Economic Growth*, Washington, DC: National Academic Press.
Lakoff, G. and M. Johnson (1999) *Philosophy in the Flesh*, New York: Basic Books.
Lester, R.K. and M.J. Piore (2004) *Innovation – The Missing Dimension*, Cambridge, MA: Harvard University Press.
Li, Y., B. Nooteboom and W. van Haverbeke (2007) *The Effects of Industrial and Country Difference in Supplier Relationships on the Radicalness of Innovations*, University of Hasselt, Diepenbeek, België.
Lundvall, B.A. (1988) 'Innovation as an Interactive Process: From User-Producer Interaction to the National System of Innovation' in: G. Dosi, C. Freeman, R. Nelson, G. Silverberg and L. Soete (eds.) *Technical Change and Economic Theory*, London: Pinter, 349-369.
McKelvey, W. (1982) *Organizational Systematics: Taxonomy, Evolution, Classification*, Berkeley: University of California Press.

Mokyr, J. (1990) *The Lever of Riches: Technological Creativity and Economic Progress*, Oxford: Oxford University Press.
Nelson, R.R. and S. Winter (1982) *An Evolutionary Theory of Economic Change*, Cambridge: Cambridge University Press.
Nooteboom, B. (1992) 'Towards a Dynamic Theory of Transactions', *Journal of Evolutionary Economics*, 2: 281-99.
Nooteboom, B. (1999) *Inter-firm Alliances: Analysis and Design*, London: Routledge.
Nooteboom, B. (2000) *Learning and Innovation in Organizations and Economies*, Oxford: Oxford University Press.
Nooteboom, B. (2001) 'From Evolution to Learning and Language' in J. Foster and J.S. Metcalfe (eds.) *Frontiers in Evolutionary Economics*, Cheltenham: Edward Elgar, 41-69.
Nooteboom, B., W.P.M. van Haverbeke, G.M. Duijsters, V.A. Gilsing and A. Van der Oord (2007) 'Optimal Cognitive Distance and Absorptive Capacity', *Research Policy*.
Piaget, J. (1972) *Insights and Illusions in Philosophy*, London: Routledge & Kegan Paul.
Piaget, J. (1970) *Psychologie et epistémologie*, Paris: Denoël.
Piaget, J. (1974) *Introduction a lépistémologie génétique*, Paris: Presses Universitaires de France.
Polt, W., N. Vonortas and R. Fisher (2008) Innovation impact study, Brussels: OPOCE, forthcoming.
RMNO (2007) *Kenniscocreatie: samenspel tussen wetenschap en praktijk*, The Hague: RMNO.
Soh, C. (2007) 'Governance in Global and Regional Context: The Korean Case as a Study of Metagovernance', paper presented at the Korean-Dutch governance forum, Leiden, August 30.

Vygotsky (1962) 'Thought and Language' in: E. Hanfmann and G. Varkar (eds.), Cambridge, MA.: MIT Press.
Weick K.F. (1995) *Sensemaking in Organisations*, Thousand Oaks, CA: Sage.
WRR (2006) *Lerende overheid. Een pleidooi voor probleemgerichte politiek*, Amsterdam: Amsterdam University Press.
Wuyts, S., M. Colombo, S. Dutta and B. Nooteboom (2005) 'Empirical Tests of Optimal Cognitive Distance', *Journal of Economic Behavior and Organization*, 58(2): 277-302.

4 RESEARCH, HIGHER EDUCATION, AND INNOVATION

Gerrit Kronjee and Bart Nooteboom

4.1 INTRODUCTION

In this chapter we discuss scientific research and higher education. New knowledge in the fields of science and technology often has a positive effect on the wealth of nations. Consider, for instance, the discovery of radar that after military applications turned out to also facilitate transportation. However, new knowledge may also have non-economic social consequences. When knowledge promotes the attainment of goals like safety, public health, or a clean environment, that also is part of innovation. Many scientists are particularly motivated by such societal goals.

In this chapter we consider if and how innovation policy for scientific research and education for economic and other societal goals is possible. Such a policy could be designed to stimulate:

- the development of new knowledge by scientific research;
- the transfer of knowledge in education;
- the social application of that knowledge.

We first discuss the relation between knowledge and innovation. This is followed by an outline of conditions for the financing of research and of other important characteristics of the context of research in higher education and in the business community. Next we discuss more specific instruments for innovation policy. The final section formulates a policy recommendation on the generation and utilisation of knowledge in so-called 'third spaces'.

Much could be discussed under the rather general section-headings that follow. However, this chapter solely concentrates on (the implications for) innovation policy and thus leaves aside other aspects of science and education. Policies for science and education have many other dimensions that will not be discussed or weighed against goals for innovation policies. The following observations on innovation can therefore only be a limited (or partial) contribution to broader discussions on the future of higher education and scientific research.

4.2 THE RELATION BETWEEN KNOWLEDGE AND INNOVATION

4.2.1 EFFECTS OF KNOWLEDGE GENERATION

In search of innovation, the relation between scientific knowledge and the development of the economy gets much attention. This is both new and not new. Not new is the belief in progress, the expectation that knowledge leads to more economic prosperity. What is new is the degree of specificity, i.e., the search for specific scientific knowledge with economic utility, and the attempt to influence the acquisition and transfer of knowledge to obtain optimal effects.

As discussed in chapter 1, not all innovations are based on new knowledge. The economic significance of new technological knowledge is sometimes smaller than that of existing technology. The significance of new technological inventions is often exaggerated and is, even when important societal changes result, sometimes only experienced in the long run (see also Edgerton 2007). In addition, not all new knowledge will result in innovations. It is therefore wrong to expect that investments in scientific research will always have societal benefits before long. Nevertheless, sometimes new knowledge is a necessary condition for innovation, although it may take many years before such new insights lead to useful applications.

To be able to identify benefits, research on innovation in the longer term, focussed on specific discoveries, is necessary. To get a grip on the societal impact of more fundamental discoveries (such as e.g., fire, gunpowder, airplanes) we need a more historical-sociological approach. For less fundamental discoveries an economic-statistical approach may be adequate. In that way, it is possible for example to find out with a limited historical analysis how much time it takes before expenditures on research are followed by patents. According to an Italian study, for example, this takes in general four to five years, but much less – up to one year – in the ICT sector (Schettino 2006).

The traditional view of R&D and innovation is a linear one: R&D yields knowledge, which is applied to new products and processes. In reaction to this linear approach, a non-linear system approach according to the *Oslo Manual* (Gault 2006) has become fashionable, for example at Eurostat and in OECD research. In this innovation system approach, research and development activities are seen as no more than a segment in a range of innovative activities, and there are feedbacks between activities in the system. In fact, R&D refers to invention, not its use in innovation. As shown in chapter 1, many variables could be relevant as an indication of these activities; the limits of variables that could be incorporated in the measuring instrument have not yet been reached. For a survey of recent extensions of planned innovation indicators, see the 3rd edition of the *Oslo Manual* (OECD 2005).

However, the impact of public funding of R&D on innovation output is still difficult to prove by statistical/econometric means, for several reasons. One is that its impact often takes a long time to materialise, in terms of new products and processes that survive in markets. A second is that fundamental inventions have wide repercussions that are not all easy to trace (OECD 2007: chapter 7). A third is that while R&D yields new knowledge, its commercialisation requires many additional activities, in prototyping, testing, design, marketing, tooling, training, efficient production, and distribution, with requisite organisation within and between firms. A fourth is that it is often difficult to prove 'additionality', i.e., that the activities would not have taken place without the subsidy.

Nevertheless, some progress in the measurement of effects is being made, by adding variables for a variety of factors that affect outcomes, improved econometric technique (e.g., to take into account selection bias), and longitudinal research based, for example, on the accumulation of statistics in the EU Community Innovation Survey (CIS).

In a recent evaluation of the EU Framework Programme for collaborative research (Polt et al. 2008), Kleinknecht managed to show, in a comparison between successive sets of CIS observations (conducted every two years) that for smaller firms there was additionality, in that participation yielded an increase of R&D that exceeded the subsidy. However, in other research it was shown that overall in the program the commercialisation of outcomes was limited. This does not disqualify the programme, however, because its aim is pre-competitive R&D, and the generation of knowledge that is later used for commercialisation is a valid goal. It is, nevertheless, hard to prove that, when and how that happens.

However, the aim of this chapter is not to present an overview of the research results in this field. Whenever appropriate, we will refer to relevant research. Whatever insight is available will be used here to analyse directions for policy. As noted in chapter 1, in spite of the rhetoric against focusing on R&D and against the linear model of innovation, old practices still dominate the actual use of indicators (Hawkins 2007). One reason for this is that R&D as input and patents as output are more easily measured than many other, modern indicators. Second, in policy R&D allows more a more tangible, identifiable target for government than many other factors, such as collaboration, marketing, and organisation of firms. The ongoing one-sided pre-occupation of policy with R&D is reflected in the Lisbon and Barcelona agenda's of the EU, and in the Dutch policy of setting up 'technological top institutes'. In innovation research there is a bias to look only at firms that do R&D, ignoring other sources of innovation (Arundel 2007). Incorporation of non-technological commercial, organisational, social. and strategic factors requires a connection between the scientific literatures on innovation and management/organisation that is difficult to achieve.

A fundamental problem with the tenacity of the linear view is that the link back from application (exploitation) to the generation of ideas (exploration) is neglected. This has been discussed in chapter 3, with the proposal of a 'cycle of discovery'. The main implication of that cycle is that we should look not only at the application of science in practice, but also at its inspiration from practice. That will lead us to the proposal of a 'third space' between university and industry.

4.2.2 SCIENCE IN AN INNOVATION SYSTEM

One outcome of the system view is the so-called 'triple helix model'. After a brief presentation of this 'model', we will discuss its drawbacks.

"*The triple helix model.* It seems the innovation debate is characterized by a wish to negotiate about the direction of knowledge exploration and about the kind of knowledge that is transferred in education. That comes together in a 'triple helix' wherein science, government and business are made to fit and are inseparably linked together. A motive for this model is the observation that nearly all of scientific knowledge production could not be put in a simple dichotomy of fundamental and applied research. Most new knowledge is seen as being of a type 'Mode 2', where the boundaries between scientific disciplines dissolve. The role of traditional knowledge of 'Mode 1' is believed to be finished: 'The old paradigm of scientific discovery (Mode 1) – characterized by the hegemony of theoretical or, at any rate, experimental science; by an internally-driven taxonomy of disciplines; and by the autonomy of scientists and their host institutions, the universities – was being superseded by a new paradigm of knowledge production (Mode 2), which was socially distributed, application-oriented, trans-disciplinary, and subject to multiple accountabilities'" (Nowotny, Scott and Gibbons 2003).

Three questions can be raised about this assumed paradigm shift:

- Is it correct that knowledge nowadays is produced in a totally different manner than previously, i.e., that knowledge comes more directly out of exploitation and is less of an outcome of fundamental research?
- Is it right to say that the Mode 1 approach is no longer valid, given the fact that this approach is often the cornerstone of important innovations?
- To what extent is such a factual observation – to the extent that it is true – also a norm for future research and for investing in research; or: is it necessary to throw the disciplinary approach overboard?

It will be argued that a variegated negative answer to these questions is appropriate.

In the innovation system approach, knowledge is seen as part of an innovation *system* and subordinated to it. The next, perhaps initially unintended step, is then to move from an analytical, descriptive proposition that this is how in fact innovation works to a normative ambition for policy to stimulate innovations by *organising* cooperation between science, government, and business. In chapter 1 it was argued that planning may have an adverse effect on the creation of ideas. This becomes evident when we look at higher education and research in an academic environment.

The risk of Mode 2 is that it leads to a neglect of activities that are related to knowledge but that at first sight do not fit optimally in the innovation system. Such knowledge activities can nevertheless have a relation with other important societal functions apart from innovation. It is not advisable to organise education for the new generation primarily for the purpose of transfer of knowledge for innovation. Education has other functions too, like civil education. These other functions could also be important for the cycle of discovery. Apart from the cultural, social, and political value of education, the economic value of providing well-educated people may be larger than the direct economic value of research. It was emphasised in one of our interviews with industry that this academic function of providing well-educated people is far more important, even for industry, than the function of research.

The accumulation of knowledge through research partially takes place in an academic environment that for good reasons has characteristics other than those of a business environment. These include differences in recruitment and socialisation, career patterns, values that are seen as important, and activities that knowledge workers are paid for. These aspects are important for other functions of research in higher education, such as stimulating social responsibilities and educating citizens. Within the existing institutional differentiation researchers are not (entirely) oriented to (economically) innovative exploitation of their new knowledge. Related to the different functions of knowledge development and transfer, academic scientific research and the business world have their own different institutional contexts. The same is, by the way, the case with the arts – also important for innovation, but not discussed in this chapter.

Innovation policy should not in an attempt to incorporate science, business, and government into an innovation system, neglect other functions that require a different context. To do so would place processes of exploration at risk processes of exploration may be at risk, as argued in chapter 3. Serving the other functions could even benefit exploration for innovation. For example, educating the new generation produces inspiration for academic researchers in developing new scientific theories. That is one reason why at universities we want to maintain links between research and education.

According to Salter et al. (2000; Hawkins 2007), there are seven categories of benefits from research, as follows:
- Training skilled graduates,
- Increasing the stock of useful knowledge.
- Creating 'social knowledge' (i.e., knowledge about the socio-economic environment),
- Creating new scientific instruments and methodologies,
- Increasing capacity for scientific and technical problem solving,
- Creating new firms and licensing patents,
- Forming networks and stimulating interaction.

The thinking behind Mode 2 is valid in its attempt to take into account the interaction between exploitation and exploration, in a learning ecology, that was discussed in chapter 3. However, such *interaction*, in a knowledge ecology, should not be interpreted as *integration* in a single process and organisation. Exploitation and exploration require different conditions that should not be neglected. When approached in a mode of exploitation, explorations is killed, and so is the learning ecology. The challenge is to specify directions for policy that take this into account. We will argue later that in innovation exploitation is a task of industry, exploration is a task of the university, and that their interaction, which is indeed needed for both application and inspiration of knowledge, should take place in a 'third space' between them.

4.2.3 SOCIAL RELEVANCE AS A MEASURE FOR ACQUISITION AND TRANSFER OF KNOWLEDGE

> *As we know, there are known knowns.*
> *There are things we know we know.*
> *We also know there are known unknowns.*
> *That is to say we know there are some things we do not know.*
> *But there are also unknown unknowns.*
> *The ones we don't know we don't know.*
>
> DONALD RUMSFELD, 12 FEBRUARI 2002

There are many possible options in the development of knowledge, and a key question is whether knowledge development has the potential to result in innovations. In chapter 3 it was proposed that in principle it is not possible to know this in advance, but there may still be good reasons to try.

Among the outcomes that are difficult to predict, there are also negative outcomes. Some knowledge could lead to difficult questions. For instance in the medical field there is the question of how far we should go in using expensive medical technology that only marginally improves a patient's quality of life. The accumulation of knowledge produces societal progress but can also raise ethical problems, and the growth of certain knowledge can also lead to regrets.

As argued above, economic effects of research can be indirect, and there are other motives than economic ones for doing research. Research also has intrinsic value, just as the arts. It serves progress in science and contributes to the education of new generations. However, the fact that research has more than just economic functions should not be taken as an argument to stop paying attention to the social significance of research. A relevant concept in debates and analyses about innovation is the 'valorisation' of knowledge. In a knowledge society there is the continuous need to collect and transfer knowledge, and the valorisation concept points to this.

The social significance of research is not self-evident. As noted above, with fundamental research we have to wait and see whether it will bring social benefits. Nevertheless, there are social problems that have to be solved and that cannot wait for science to spontaneously offer a helping hand. Therefore, programming of at least a part of fundamental research is necessary. Such programming calls for a difficult differentiation between 'useful' and 'non-useful' knowledge for society. The usefulness of knowledge is derived from the way it is used; it is not possible to know or decide whether knowledge itself is useful or not. The problem is then how it is possible to make the right choices, knowing that we don't know what we don't know. Inevitably, most expectations will fail to be realised, and surprise is endemic. Yet, sometimes it is necessary to accept the gamble and make choices about aims and themes of research to help solve social problems, such as traffic safety, the effect of climate change, or the supply of energy. Even if only a small amount of research on such issues contributes to solutions, that research may still be important enough to decide to invest in it. In other words, for problems with a public goods nature, which industry does not take up out of its own interest, governments have to take the initiative even if success is unpredictable. Lack of success then is not failure but a consequence of inevitable uncertainty.

A special case: Research in the social and behavioural sciences

In principle it is not necessary that knowledge leading to an economic or broader social innovation has a basis in technology, or in the natural sciences. As also argued by the Dutch Council for Science and Technology (AWT 2007), research in the humanities or in the social and behavioural sciences can also result in innovations. Such research could result in changes in social relations, with the effect that problems are solved and economic progress is obtained. Historically one can think of research on the working conditions of the poor, which resulted in changes in laws, followed by a general increase in prosperity. The social sciences in particular investigate the social conditions in which the economy can make use of technological progress. These social conditions are not an invention or discovery, in the sense of discoveries in the natural sciences. After all, the repertoire for social interaction is limited; it will not be easy to find something new. The same kind of argument holds for the behavioural sciences, and possibly also for the humanities. Chapter 8 gives a discussion of non-technological innovation in the form of organisational innovation.

Often, social research is inherently scientifically relevant as well (Overlegcommissie Verkenningen 1996). There is no need to see a discrepancy here with fundamental research. The history of the social sciences teaches us that social research has often been an important force in the development of these sciences. See for instance the wide scope of social-psychological research during World War II, which resulted in a series of publications under the common title 'The American Soldier'. Or think of the Leyden Talent Project, a Dutch research program in the 1960s that tried to promote the upward mobility of children from lower social-economic classes.

The importance of social relevance of research, in both technology and social science, could be emphasised in research that is financed by the so-called 'second money flow' (see the next section). Here, research should be sufficiently problem oriented and geared toward the solution of social problems, such as e.g., ensuring mobility or protection from flooding, without being so specific in advance, in trying to determine the content, approach, and feasibility of research projects. Finding a balance between specificity and feasibility on the one hand and the promotion of problem-oriented research with the possibility that it will lead to surprising results (serendipity), is essential. This connects with the recommendation, in chapter 3, for openness in the sense of allowing for changes of direction in innovation projects, and for letting ideas for research emerge from the bottom up, from the diversity of experience and views in the research community.

4.3 SOURCES OF FINANCE

Table 4.1 below presents an overview of money flows funding research in the Netherlands. It turns out that government is a major funder of scientific research in the higher education system, i.e., at the university level. Research institutes are second in the reception of funding. However, the business community spends more on research than government, mainly to fund research activities within companies. This business research is not part of the public programming of research. Figure 4.1 provides insight into the money flows of funding for scientific research in the Netherlands].

Table 4.1 Funding for R&D in 2005 in the Netherlands (in billions of euros)

	Destination			Total
Source	Higher education	Research institutes	Business	
Government	2.1	0.9	0.2	3.2
Business	0.2	0.2	4.0	4.4
Private, non- commercial	0.1	-	-	0.1
Abroad	0.1	0.1	0.9	1.1
Total received	2.5	1.2	5.1	8.8

Source: Ministerie van Onderwijs, Cultuur en Wetenschap, Kerncijfers 2002-2006

Government funds are important for the financing of scientific research in the Dutch higher education system. Academic research is financed and programmed via three channels: fundamental research with the first and second money flow, and applied research with the third money flow, which includes financing by non-governmental institutions . The funding in the second money flow, administered by the Netherlands Organisation for Scientific Research (NWO), is partly allocated in thematic programmes established by NWO, partly to proposals based on ideas generated by researchers, and partly to successful researchers ('person-directed funding'). There has been a tendency to increase the share of thematic and person-directed funding.

Figure 4.1 shows that government departments, e.g., health, agriculture, transport, and economic affairs, also fund research at universities and technical institutes. There is a certain tendency to concentrate research on a certain field (agriculture, environment, transport) at a certain university (e.g., the University of

Figure 4.1 Money flows from government to science

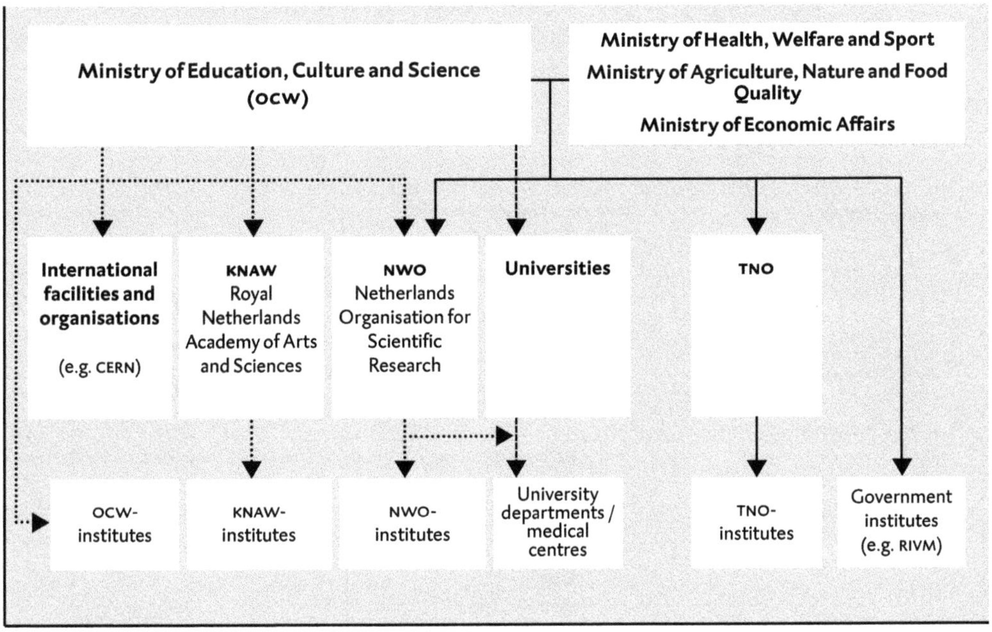

Source: NWO 2007

Wageningen for agriculture) or institute (e.g., for the environment), which is then seen as part of the exclusive purview of that governmental department. In our view, from the perspective of the importance of novel combinations and variety for innovation, argued in chapters 1 and 3, this is doubly wrong. For novel combinations, there should be connections across fields of research between disciplines, and connections across government departments in policy. For variety, areas of study should not be concentrated at a single university, and there should be room for variety of approaches and perspectives in any given field. As argued in chapter 1, there may be valid arguments of economy of scale, but these should be traded off against diseconomies of scale and arguments of variety.

4.4 RESEARCH IN THE HIGHER EDUCATION SYSTEM AND INNOVATION

4.4.1 FINANCING UNIVERSITY RESEARCH

Research at universities is mainly financed with the first money flow. For the government, criteria for the allocation of this money to individual universities are the number of students with certificates and dissertations, as well as strategic research priorities. Inevitably the history of earlier decisions and the size of the universities play a role.. The pattern of existing academic disciplines and the number of students in these disciplines will factor into the decisions made by university boards on how to spend such research money., Thus, the number of students will be an important condition for the financing of innovative academic research. This is intentional, as the academic ideal implies there is a relation between education and research at a university. However, an unintended consequence could be that important scientific research disappears if the influx of new students into different studies changes.

Figure 4.2 Funding of universities

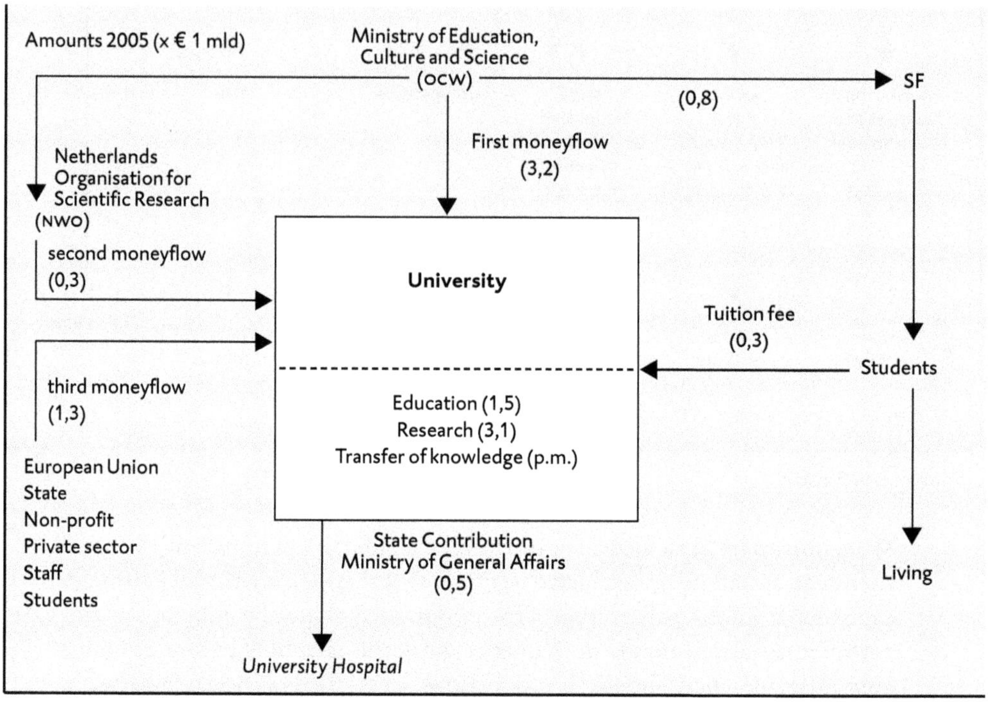

Source: Ministry of Education, Culture and Science OCW (2007)

Universities profit from large numbers of students, and therefore will be inclined to adapt available courses to the various interests of incoming students (see also *Investeren in Dynamiek* 2006: 12). Existing disciplines that are not fashionable and not successful in attracting students risk decreased funding for their courses and related research or losing funding entirely. This could have consequences for the quality of scientific research, including research that may be important for innovation. One possible partial solution to this problem is to increase the amount distributed through the second money flow, while keeping intact the relation between research and education at universities.

We will turn to a comparison of merits and drawbacks between first- and second-flow finance later. Here we note that there is a more general consequence of the relation between financing research and the academic disciplinary landscape, in both first and second stream funding. In both, existing disciplines and reputations tend to be reproduced. In the first stream, the financing of research has to duplicate the existing landscape, otherwise universities or disciplines who feel disadvantaged will complain. In the second stream, established schools of thought and scholars that dominate selection committees reproduce their lines of research. Both could lead to an unintended continuation of financing unpromising research in existing discourse coalitions. This kind of coalition consists of groups of scientist having the same theoretical views and criteria for promising research. In the research they do they limit themselves to the theories they share (Van den Boogaard 2002; Kronjee 2002). The continuation of research that was being done in the past may result in unsurprising choices of subject matters and scientific interpretations.

The amount of money in the first and second flow of funds is the result of political decisions. The third flow of funds is dependent on the benefits expected by the business community from research, but also on the responsibility this community wishes to have for good fundamental research and higher education. In section 4.4.4 we consider policy implications for the programming and funding of research.

4.4.2 EFFECT OF THE STRUCTURE OF EDUCATION

> *Dans les champs de l'observation,*
> *le hasard ne favorise que les esprits préparés.*
>
> LOUIS PASTEUR 1854

Choices in research are related to choices that are being made in education policy. The Dutch higher education system is evolving towards an American education system, including characteristics of the organisation of scientific research. Characteristic of the American system is that research takes place only later in university curricula, in the master's stage or later. This occurs next to education in professional schools (of law, business, etc.). This evolution is far from complete in the Netherlands. An obstacle is that while previously research was part of education in earlier stages (master's or even bachelor's), this is not in the interest of students going for a professional career, which exerts pressure to take out research, against the interest of careers in science. This problem is accentuated by the strong division between scientific (university) education and higher vocational education. This division follows the division in secondary education between 'vwo' (pre-university education) and 'havo' (school of higher general secondary education).

The process is as follows. In attempts to accentuate the difference between the bachelor stage of university education and training in higher vocational education (for which students also receive a bachelor's diploma), many academic disciplines begin the first phase of their education with a rather narrow, specialized training. The risk of this is that many university students received specialised training that they do not need and for which they may not even be motivated (e.g., because they only want academic training to enable them to secure a job, which is mostly not a scientific job). At the same time, those students who go for a scientific career do not get the necessary general preparatory academic education, although they need such a general education as the basis for acquiring the skills to be a researcher in their own discipline (Van Oostrom 2007).

In addition, by accentuating the difference between university education and higher vocational education the stimulus for universities to compete with other universities in offering better education is weakened. Differences between university education and vocational education might overshadow the differences *within* these broad categories of education. In this way, the belief may prevail within universities that the quality of university education is guaranteed as long as it remains sufficiently different from higher vocational education.

The preparation of students for professional practice by doing practical research during their education and by solving problems for the business community, may be an important source of innovative ideas (AWT 2005a). This kind of education could be seen as the task of higher vocational education institutions , and that is in fact how these institutions see themselves, as universities for applied science. But these institutions have limited financial and legal possibilities to function as such. In principle they are not allowed to do original research and offer a full-fledged second phase master's education. In trying to find a way out, they now cooperate with foreign higher education institutes in offering master's training. In addition these institutions try to offer students participation in practical research that is necessary for their education. The institutions try to engage in research activities through cooperation with universities, and by appointing so-called 'lectors' dedicated to this task. Despite these efforts it remains difficult for higher vocational education institutions to participate in (applied) research. The road toward the more comprehensive introduction of research to students in higher vocational education during the second phase of their education is closed (see also Kahan 2002: 47). Attempts by the Ministry of Education to loosen the strict differentiation between universities and higher vocational education institutions are met by strong opposition from universities and related institutions. For example, in a recent memorandum that accompanied the new formulation of goals for higher education, the Ministry of Education defined research at higher vocational institutions as follows: "Original investigation undertaken in order to acquire new knowledge. It is, however, directed primarily towards a specific practical aim or objective" (Memorie van Toelichting 2005). The Advisory Council for Science and Technology Policy (AWT) however responded that this should *not* be the goal of research at higher vocational education institutions. Vocational schools would then get the opportunity to do research that may lead to substantial new insights, and the AWT holds that this should not be the task of these schools (AWT 2005b). It is remarkable that maintenance of the existing educational structure is prioritised over the creation of opportunities to obtain new insight. The price of this position is that a possible research source that could lead to innovations is not being exploited. Moreover, as argued before, the training of researchers at universities may lack necessary preparatory elements, which would improve the quality of such training.

4.4.3 QUALITY MEASUREMENT OF RESEARCH AND RESEARCH POLICY

Related to the financing and programming of research are the chosen quality standards.
An important quality measure for research at universities is nowadays the success in publishing in foreign (usually American) journals, aimed at the development of often rather specific expertise in a narrow or specialised field. While we grant the merit of specialisation, when this regime is applied to everyone, negative effects arise. This practice is not conducive to a wide accessibility of research results, and also does not stimulate the practical, innovative use of research results. Apart from this, in the social sciences the usefulness of a publication sometimes decreases when it is based on data sets that are gathered abroad and not – such as in our case – in the Netherlands.

At the same time this way of measuring performance can direct investment decisions for research by government in the wrong direction, as far as these funding decisions are based on measurable criteria such as citation scores, previous investments in research, or numbers of knowledge workers. This practice may not be conducive to reaching innovation goals. It may therefore be more sensible to make choices about research investment based on content, avoiding or reducing the influence of (unclear) quantitative measures. As indicated also in chapter 7, rather than only counting publications and citations, for proper evaluation one should occasionally read publications to assess their value.

The use of a bibliometric analysis for evaluation and policy choices is elaborated in a report of the Council for the Humanities and the Social Science Council of the KNAW (Royal Netherlands Academy of Arts and Sciences). This report observes advantages of a bibliometric analysis for discovering macro-developments in research (KNAW 2005). However, the report notes that these bibliometric indicators also have limitations, when they are used to evaluate research in the humanities and social science and in research policy. It appears that the outcome may be that less attention is given to the practical use of research results. Communication with the public is necessary to get a grip on that, but this is not counted in the bibliometric statistics.

4.4.4 PROGRAMMING AND FUNDING RESEARCH IN HIGHER EDUCATION

In university research, an issue is to what extent this should be planned and programmed according to societal needs. In line with our Hayekian perspective, we are wary of bureaucrats in central offices, at ministries and national science foundations, specifying programmes or projects for which scientists or businesses may then submit tenders. Why should they know better than the totality of differentiated and dispersed knowledge? However, we do acknowledge the legitimacy for central coordination to indicate broad public priority issues, such as problems in the fields of energy, environment, water, health care, and aging, short of actually specifying projects.

First or second flow of funding?

> *The winner takes it all*
> *The loser standing small*
> *Beside the victory*
> *That's her destiny*
>
> ABBA, 1980

Some money for new research is allocated to those who have an excellent research record. They are awarded prizes or funding afterwards, to expand successful lines of research. This is understandable in view of the difficulty to predict successful research beforehand ('picking winners'), as discussed in chapter 3. Therefore, we endorse such awards to researchers that have proved their ability, in terms of carrying out research, intellectual leadership, and research management. Next to that, however, funding on the basis of prior assessment of the merits of research proposals is also needed, to allow for newcomers and outsiders. That is more problematic, in view of the difficulty to predict success and value of exploratory research. Next to the usual 'hard' measures of a researcher's past performance, in terms of publications in high 'impact-score' journals, and monographs at publishers of high repute, this requires additional, rich, context- and person-dependent knowledge, to a large extent tacit, which is best provided locally, in the research centre at the university. Furthermore, also according to the principles of Hayekian variety and experimentalism, discussed in chapters 1 and 3, local assessment allows for greater variety of judgement, based on variety of theoretical and methodological perspectives, compared to committees of a science foundation that are dominated by one school of thought or another. Thus, here we see a preference for first-flow funding. In the policy debate, it has been claimed that in that stream there is lack of competition, compared to second-flow funding. However, since the institution of research schools at universities, and an intensification of research assessments by peer review, competition has greatly increased also at the university level. Improvements and adjustments in first-flow funding are of course possible. As argued earlier, while a link between teaching and research should be maintained, first-flow funding might be coupled more to quality and novelty. In sum, we propose to maintain a two-track approach for financing innovative research: investing in promising new research on the one hand, and on the other hand at the same time backing continuation of successful research in the past, and to do so in both first- and second-flow funding. We are wary of any further shift from first to second-flow funding.

Awards should be used first of all to expand successful research, rather than increasing personal income of successful researchers. The effect of steering by rewards in terms of personal income is limited when the purpose is to encourage creative behaviour. Research shows (see also Schweizer 2004) that intrinsic rewards, such as satisfaction of scientific curiosity, is often more important for the stimulation of creative processes than economic rewards. The latter could even produce an adverse stimulus.

Research by universities and higher vocational education

The difference between exploitation and exploration may help us find new opportunities for innovative research at universities and institutions in higher vocational education. The academic world could be positioned in the sphere of exploration, with a central interest for fundamental research. Higher vocational institutions could have a role in doing research that is oriented to the exploitation of knowledge, i.e., applied research. For universities there is then no need to become too closely involved in applied research for the business world, and exploration can be protected against undesirable influence from the market system, discussed in chapter 3.

Interdisciplinarity

A further issue is the inter-disciplinarity of university research. Both radical novelty and application often arise across the frontiers of disciplines. However, universities are mostly oriented towards, and organised according to disciplines. This is one of the complications in university-industry collaboration. Indeed, one of the reasons for university-industry collaboration is to stimulate interdisciplinary exploration. Nevertheless, there are good reasons for disciplinarity. Alignment according to discipline ensures depth of specialisation, and provides a focus for teaching, scholarly societies, scientific journals, and conferences.

There may be a deeper, evolutionary argument. Perhaps disciplines may be seen in analogy to species, and in biological evolution cross-species breeding is impossible for a good reason. Inter-species breeding would yield lack of differentiation and hence lack of differential survival, by which the evolutionary process would break down.

Yet, space needs to be made for interdisciplinary research. There are large obstacles to this, from the viewpoint of the sociology of science, and over the past years those obstacles have increased rather than decreased. Junior researchers especially have to obtain scientific visibility to make an academic career. When a researcher combines A and B, he says B things to people involved only in A and A things to those involved only in B, and as a result is not fully recognisable or even identifiable to either as a contender for recognition and careers. The typical response is that one gets ignored. It is difficult for such partial outsiders to be accepted for publication in the top journals, which are mostly single-disciplinary.

Conclusions

Two central issues in investment and planning of research are the link with higher education, for diffusing scientific knowledge, and the potential contribution to scientific progress. This is to be determined within the academic world. However, the demand for innovative research creates pressures on the allocation of funds between the different flows. In that allocation one should balance different goals, such as prosperity, the solution of social problems, scientific progress and quality of education. One should also take into account relationships, for the achievement of those goals, between the development and the application of knowledge, between universities and industry and between academic institutions and institutions of higher vocational training. Earlier, it was shown that the cycle of discovery in innovation requires influences from outside the academic world and a link to applied research. Fundamental and applied research are both necessary for innovation in the cycle of discovery.

4.5 RESEARCH IN THE BUSINESS SECTOR AND INNOVATION

4.5.1 FUNDAMENTAL RESEARCH, THE BUSINESS COMMUNITY, AND INNOVATION

Much research outside universities is multidisciplinary or transdisciplinary, and aimed at application; fundamental research is done to a lesser extent (see also table 4.1). It is argued that the image of a researcher who by chance encounters an important and useful innovation while looking for fundamental insights, belongs to the past (Rathenau Instituut 2006: 9). However, this observation requires a comment.

If in reality much research is multi- or transdisciplinary and aimed at application, this should not necessarily be a norm for the organisation of all research. One can dispute whether important innovations are the result of research that aims at innovation. There are good reasons for the shift of research from the business world to universities and for the fundamental research that is being done there (Weingart 1999). In the US, increasing competition in industry has eroded its resources for exploration (Lester and Piore 2004). Yet exploitation of knowledge needs preceding exploration, in the cycle of discovery. In his plea for free and fundamental scientific research in reaction to the coalition agreement of the current Dutch cabinet, the president of the KNAW makes a similar point (Van Oostrom 2007).

As can be seen from the money flows (table 4.1), a lot of research is done by companies. A difficult question is related to the large amount of knowledge that is obtained as a result of research carried out within large companies. In the policy debate, there is much discussion of problems in the transfer of knowledge from the scientific academic world to the broader economy and society, as in the debate about the 'valorisation' of knowledge. But there are also obstacles to the accessibility of results of scientific research that is carried out in commercial environments. It would be very useful if knowledge generated there were accessible for the outside world. There is not only a problem with the protection (and thus accessibility for the rest of the world) of promising research that could lead to patents, but also with the accessibility beyond companies of research that is *not* successful. Unsuccessful research also contributes to the accumulation of knowledge. However companies fear that publication of such negative results may give clues to competitors about the direction in which the company is searching (see also De Wijkerslooth 2007).

One way of utilising knowledge from large companies lies in the phenomenon of 'spin-offs'. There, employees with innovative ideas that are generated but not exploited in the large firm, because these ideas do not fit with existing thought, competencies, or market strategy, or are considered too uncertain and risky, follow their convictions and start a new firm to further develop and commercialise the idea. This is further discussed in the chapter on entrepreneurship (chapter 5).

In the theory of innovation, a classic condition for commercialisation of research is the ability to protect profits by preventing or postponing imitation. An important policy question is to what extent such protection of intellectual property (IP) is still needed and desirable. No doubt, there is still a need in some cases, depending on the industry, but a trade-off needs to be made by firms between the possible loss of knowledge to competition and the gains from the free flow of information. Taking into account that spillover requires absorptive capacity, skills, complementary assets (including marketing assets), organisation, and the delay involved in putting these things in place, spillover may often not be a real threat (Nooteboom 1999). Several studies have shown that in fact patents and other forms of protection of IP are the instrument least used to ensure 'appropriability' of profits, compared to other instruments such as: keeping one's personnel from leaving or being poached, secrecy, speed of development, complementary assets or capabilities (needed to actually apply knowledge that spills over), and complexity of technology (OECD 2005). Several of the alternative measures for appropriability connect with the argument concerning absorptive capacity. The Community Innovation Survey (CIS) shows that only 10 percent of firms use patents to ensure profits (EUROSTAT 2007). This calls into question the use of patenting as a measure of a country's innovativeness, in 'innovation score boards'. Excessive protection of intellectual property has the traditional drawback of slowing down diffusion, thus reducing the welfare benefits of innovation. A newer insight, offered by Von Hippel is that it also obstructs the openness of innovation, with a variety of contributors, including users. This is discussed in chapter 8. Also, patents are sometimes used not to ensure profitable exploitation but for blocking access to new products by competitors.

A curious perverse effect of patents is the following. In collaborative R&D projects in the EU framework programmes, legal divisions of especially large firms sometimes keep on sending back and correcting draft agreements for projects, for improvement of IP protection of the firm. This slows down the process so much that exasperated participants force an agreement through without any regulation of IP protection, which later generates problems (private communication).

There have been increasing pressures on universities to commercialise the knowledge they developed by selling it, or by contracting applied research. This can be counterproductive. Firms complain that whereas in the past they could pass by for a quick answer to a simple question, now universities offer to set up a contract beforegiving such feedback. In view of the often substantial delay, in the view of a firm pressed for time, and other transaction costs involved, firms often decline the offer. The procedure is even unwise from a marketing perspective. A free answer to a limited problem gives parties the opportunity to test the attractiveness and mutual ability to collaborate, and to cross the 'cognitive distance' involved, in terms of the analysis of chapter 3. In other words, a substantial scope for free advice may be needed in order for contracting for more extensive research to take place. And to account for such free spending of public funds, it should be possible to measure university performance of such services, in the existing periodic research assessments of faculties, on the basis of some logging (tracking of time spent) of advisory projects and a sampling of the satisfaction of firms that were served.

4.5.2 SOURCES OF KNOWLEDGE WORK

It is claimed that there is a shortage of students in natural science or technology specialisations at Dutch universities, and that there is an untapped potential of positions for women, who until now are not sufficiently choosing those specializations (Van Langen 2005). This potential could be tapped if premature specialisation were avoided. However it is disputed whether or not there really is a shortage of personnel in those specialisations, in the Netherlands (Noailly, De Waagmeester, Jacobs, Rensman and Webbink 2005). If there was such a shortage the business world has not addressed this need by offering attractive employment conditions to such personnel. It is remarkable that companies specifically seek students who have graduated from higher vocational institutions, whom, in the Dutch setting, they can offer low wages, while it is not expected that they have done any scientific research or been on the receiving end of any scientific knowledge; indeed higher vocational institutions are prohibited from offering their students participation in scientific research.

In the higher education system, universities included, attempts are being made to prepare researchers for their future role in business. At Leiden University for example, during the training of scientific researchers attention is given to organizational and working conditions in companies (Birrer and Tobias 2003). Also worth mentioning is the Casimir program that since 2004 provides for exchanges of researchers between companies and public research institutions.

In order for research to result in innovation it must be creative, creativity is required. Creativity is not an individual characteristic but specific behaviour that results from different single personal characteristics *and* a specific environment. It is true that performance in particular fields (music, science, arts) requires specific skills or intelligence, but these are in themselves not sufficient to produce good performance. Creative behaviour is the result of several specific qualities and circumstances. Necessary individual qualities include a curious mind, perseverance, and inner directedness. For creativity in a social environment emotional intelligence is necessary, too.

Too much attention is given to brilliance as a basis for creative behaviour. Intelligence and learning capacity, as measured by a person's IQ, is just one quality amongst others that are sometimes, but not always, necessary for creative performances. IQ however does not determine on its own whether behaviour will be creative. For that to happen domain specific talents *and* the correct environment are also needed.

That attention should also be given to the environment is perhaps an unusual idea for those who think creativity is a personal characteristic that is genetically determined. But in psychology thinking in a dichotomy about the effect of personal and environmental factors (nature and nurture) is nowadays passé. Genetic make-up does not result unconditionally in specific behaviour. It is a misconception to think that creative talent provided by chance or inheritance at birth will determine whether one's life will be creative. As Smitsman (2000) argues: "(A) capability to perform a task, and a talent as the outstanding quality by which one performs a task, are *relational* properties that rest on organism and environment rather than on some hidden core inside the organism or the environment." Embodied in a person is no more than a potential. Fluctuations in performance could occur, dependent on the context, after maturing (Renzulli 1999). In chapter 9 we explore in much more detail what the organisational conditions for creativity might be.

4.6 SPECIFIC INNOVATIVE ATTEMPTS IN KNOWLEDGE POLICY

4.6.1 EUROPEAN POLICY

Considering the number of scientists that will graduate in the future in large countries with growing economies, such as India and China, it can be expected that an increasingly smaller part of total research will be done in Europe. This is one of the reasons for the increasing importance of international contacts for Dutch researchers. Related to this are the efforts in the Netherlands and Europe to fund research programs and leading technological institutes and to promote international exchange of students and researchers. An important initiative of the European Commission is the proposal to establish a European Institute of Technology (EIT). If the European Parliament and the Council approve this proposal, the institute can start in 2008 with a budget of at most 2.4 billion euros for the period 2008-2013. According to an EU press release this institute "will be a flagship for excellence in innovation, research and higher education in Europe. It will be a reference model in the way it combines the worlds of academia, research and business so as to enable Europe to face the challenges of a globalising, knowledge-based world economy more effectively."

If intuitions of the benefits of large scale form the basis for this plan, it is a dubious plan. As discussed in chapter 1, arguments of scale do not always apply, certainly not for innovation. If there are indeed economies of scale, for example in specialised support in laboratories, or large instruments and installations, economies of scale still have to be traded off against loss of diversity, in the concentration of activities. There could be a drawback if this institute were to dominate European scientific research and would take away a lot of research money, given the difficulty to predict success of research, and the bureaucratisation that could accompany such concentration of research. As argued before, fundamental and innovative research call for variety, at a certain cognitive distance and free from interference by the business community and government.

Furthermore, when programming of research is partly decided at a European level, political considerations could be included that are not in the interest of science, such as considerations of prestige and power, and the use of subsidies for research to promote the economic development of poor areas (Kahan et al. 2002: 49). There is a difference between promoting research that could lead to the solution of social-economic problems, which fits with the proposed formulation of societal goals for research, and the use of research funds in an ineffective way as an investment in a poor area, to stimulate regional development by the location of highly paid knowledge workers living in that area.

There is a tendency towards bottom-up international cooperation in research. The *Atlas of Knowledge and Innovation* shows that a large part of the knowledge flows that may lead to economic-technological innovations are related to national and international networks of companies and universities, and not primarily with other knowledge institutes and companies in the same areas (Weterings, Raspe, and Van Oort 2006). This self-organisation seeking its way in a knowledge ecology is generally better than concentration. If the EIT is set up to support rather than replace this process, it may be a good idea.

4.6.2 CREATION OF TECHNOLOGICALLY LEADING INSTITUTES (TTIS)

Probably, less rather than more policy making is necessary for innovation. Policies may have as an effect that they stimulate the development of an innovation *system* in the normative sense of designing and planning structures of education, research, and business, with the drawbacks that were discussed above. There is some doubt if existing policy initiatives are sufficiently appreciative of the danger of fixing development in this way. This could be the case with policies for the so-called strategic 'key areas' designated by the Dutch government and for technologically leading institutes (TTIs) in the Netherlands. At these institutes, partly financed with government funds, cooperation between (big) business and researchers takes place.

There are now eight TTIs in the Netherlands. These institutes are doing long-term strategic industrial research on a limited theme with international importance. An evaluation of the TTIs indicated that it is not clear what the innovative results of this approach are, although government maintains its confidence that there will be positive results in the future (Tweede Kamer 2005-2006, 30300 XII; Van der Veen, Arnold and Boekholt 2005). In our own interviews with some of the TTIs, we found that some of them focus on relations with large business, demanding large fees for participation in research and the use of its outcomes. Also, directors of TTIs are sometimes former or present part-time employees of large firms. Both conditions contribute to the exclusion of smaller firms and entrepreneurs who may need access most, in view of their limited resources of specialized expertise. In other words, there appears to be a threat of institutional 'lock-in'. For the recent creation of a new TTI on green genetics the Dutch government has promised to address the necessary conditions for wider access, under the supervision of a technological advisory committee. This committee is scheduled meet each six months and is supposed to stimulate the conversion of gathered knowledge into product innovation (Tweede Kamer 2006-2007: 30919). It is too early to evaluate this initiative.

Overall, the policy of TTIs appears to follow from the old view that innovation is mostly to be seen as a matter of technology that calls for more scientific research. However, as has been argued in various chapters in this book, innovation is not only about solving technical questions but also, or even in the first and foremost, about finding open spaces for debate, where participants discuss and deal with ambiguities, and about the exploration of opportunities (Lester and Piore 2004; Huston 2007). When exploration reaches exploitation, it requires additional, non-technological innovation in organisation, collaboration, marketing, distribution channels, technical standards, education, and training.

4.6.3 STIMULATING THE IMMIGRATION OF KNOWLEDGE WORKERS

Since 2004 it has become easier for foreigners who fit the definition of 'knowledge worker' and are sponsored by an employer to get a residence permit to live and work in the Netherlands. Now , with 4 percent foreign employees in the sector 'science and technique', the Netherlands is almost at the European average of 4.1 percent (CPB 2007). With the expectation that in the future there will be a larger demand for these workers, proposals are now being formulated to complete the demand-oriented migration with a supply-driven motivation. This means that foreign workers who meet certain criteria can migrate to the Netherlands and find themselves a job also without the invitation/sponsorship of a company in the Netherland. A complication is that a policy discussion about this same issue is taking place at a European level at the same time, with uncertain outcomes.

In a recent recommendation, the Dutch Social Economic Council (SER) found the simplification of the admission procedure to be insufficient (SER 2007). According to SER, the policy for knowledge workers should be less restrictive. In the SER proposal, which immigrants can qualify for a residence permit under the knowledge worker category would be determined via a point system based on level of education, work experience, English and/or Dutch language abilities, and age (preferably young). At the same time SER observes that there are limitations to make working in the Netherlands attractive for foreign workers. Such attractiveness includes housing, schooling, recreation, and culture.

The Netherlands has a negative immigration surplus of higher-educated people. There is a net-outflow of this category (see table 4.2 for figures from 2000; more recent comparative data was not available). The question for the Netherlands is whether a larger supply of knowledge workers will also create a larger demand for

Table 4.2

Foreign-born persons with tertiary attainment
As a percentage of all residents with tertiary attainment, circa 2000

	Immigrants from other OECD countries (A)	Emigrants to other OECD countries (B)	Immigrants less emigrants within the OECD zone (A+B)	Immigrants from the rest of the world (C)	Total 'net' foreign-born persons with tertiary attainment (A-B+C)
Australia	16,8	2,4	14,4	12,1	26,5
Austria	9,1	13,8	-4,7	5,2	0,5
Belgium	5,9	6,4	-0,5	4,2	3,7
Canada	10,3	5,4	4,9	15,5	20,4
Czech Republic	4,1	8,7	-4,5	2,2	-2,3
Denmark	4,4	7,3	-2,9	3,2	0,3
Finland	0,9	6,8	-5,9	1,3	-4,6
France	4,2	4,4	-0,2	8,2	8,0
Germany	2,8	7,3	-4,5	8,6	4,1
Greece	4,8	9,4	-4,6	7,3	2,7
Hungary	1,4	9,7	-8,3	4,5	-3,8
Ireland	14,0	26,1	-12,1	4,0	-8,1
Italy	2,8	7,3	-4,5	3,3	-1,2
Japan	0,2	1,1	-0,9	0,5	-0,4
Korea	0,2	1,4	-1,2	0,2	-1,0
Luxembourg	43,1	15,4	27,7	5,8	33,5
Mexico	0,8	6,9	-6,1	0,5	-5,6
Netherlands	3,3	8,9	-5,6	4,4	-1,2
New Zealand	14,6	24,4	-9,8	10,0	0,2
Norway	5,2	4,9	0,3	3,0	3,2
Poland	0,4	10,2	-9,8	2,3	-7,6
Portugal	4,1	11,2	-7,0	11,2	4,1
Slovak Republic	3,3	16,0	-12,8	0,9	-11,9
Spain	2,7	2,3	0,5	3,8	4,2
Sweden	6,9	5,4	1,5	7,3	8,8
Switzerland	20,0	10,8	9,1	7,3	16,4
Turkey	3,4	4,9	-1,5	2,7	1,2
United Kingdom	6,5	14,9	-8,4	9,4	1,0
United States	4,2	0,7	3,5	9,2	12,7
OECD average	6,9	8,8	-1,9	5,4	3,6

Source: OECD (2007) Factbook 2007, Economics, Environment and Social Statistics

social security for these immigrants and their family members, who might face difficulties in finding employment, either upon arrival or after they have finished a job for which they originally came to the Netherlands. In supply-driven migration the costs of recruitment and integration are in fact paid by the Dutch taxpayers and not by potential employers.

From the perspective of this book, in which innovation is primarily seen as developing bottom-up, the migration of innovative knowledge workers should be the result of an increasing demand from the business community and universities. Of course, obstacles that prevent migrant workers from satisfying that demand should be reduced. It is difficult, if not impossible, however, to know in advance which knowledge workers are attractive for an innovative economy. The risk of supply-driven migration at the European level is that the Netherlands might receive immigrants for whom this country is a second choice and in which they are less interested in integrating and staying permanently. At this time most of the highly educated immigrants into Europe go to Great Britain, France, and Germany. These countries are expected to remain most attractive; Great Britain and France especially have the advantage of being a country with a language in which many knowledge workers already are fluent. And if they are not, investing in learning English or French yields wider opportunities for moving to other countries when the need or opportunity arises. The Netherlands could wait and see what the accelerated procedures for immigration and a (hopefully) increasing demand will produce. Also, SER's advice (2007) could be followed to develop and make use of existing potential talent within the Netherlands; in the Netherlands there is still much unused talent among the ethnic minority and refugee populations.

4.7 A POLICY RECOMMENDATION: PROMOTING THE GENERATION AND UTILISATION OF KNOWLEDGE IN 'THIRD SPACES'

4.7.1 BASIC ASSUMPTIONS

Government should prevent investment in research from being fully dictated by the market value of the results of that research. Government should leave earning money to the business community that does substantially investments in (applied) research. Government has a special responsibility to solve social problems. Guiding research efforts based on their social relevance can put right some of the shortcomings of current research discussed before.

As suggested earlier, and discussed in more detail in chapter 3, exploitation and exploration require different conditions, but they should build upon each other. In the relationship between university and industry there is not only an issue of applying science to practice but also of science taking inspiration from practice. In some areas, such as engineering and medicine, close interaction between exploration and exploitation is normal, within the university. However, even there some research is more fundamental and 'blue sky', at a greater distance from exploitation. To allow for the need to combine exploitation and exploration, as well as the fact that they may require different conditions, we propose the notion of a 'third space', between university and industry. Within this space interaction between university and industry would take place for the sake of carrying exploration into exploitation as well as inspiring exploration from exploitation, while keeping institutional differentiation intact. This third space could be virtual or real, in the form of an actual location. By setting interaction apart, in a third space, industry can escape from market pressures towards exploitation and academia can protect the autonomy of its fundamental research, while being connected to each other for their mutual benefit.

Third spaces enhance the full development of the cycle of exploration and exploitation (see chapter 3). Third spaces should not be dominated by the market. They should be open places with access for researchers to the practice of exploitation, without asking these researchers to defer and limit themselves to the demands of the market. Third spaces are spaces for interpretation (Lester and Piore 2004, see also chapter 3). With these ideas in mind, we conducted a number of interviews, with both academia and industry, which yield a number of recommendations.

It appears that the industrial laboratories that formed the basis of the renowned 'Silicon Valley', such as Bell Labs, performed this function of a third space. As told by Chesbrough (2003), these arrangements have for some reason eroded. According to Lester & Piore (2004) this is due to increased competition, during the nineties, which eliminated the slack needed for exploration.

4.7.2 A RECOMMENDATION OF 'THIRD SPACES'

To test our idea of third spaces we conducted a series of interviews with universities (in Eindhoven, Groningen, and Delft), a higher vocational education school in Rotterdam, three more or less recently instituted technological top institutes connected with universities (The Centre for Food Sciences, the newly instituted top institute for water management 'Wetsus' in Leeuwarden, and the Telematics Institute), two medium-sized companies, and three large corporations (Unilever, Shell, and Philips).

The main result is that among our respondents the problem of valorisation is not perceived to be as large or urgent as it is among policymakers in the Dutch government. Many forms of collaboration have already been developed. However, as we indicated earlier, there is a tendency of technological top institutes (TTIs) to focus on relations with large firms, and the openness to outside entrepreneurs that we plead for does not always seem assured. Mostly, the respondents from university and technical institutes claimed that they already practise what we suggested, and that (something like) 'third spaces' were already a reality. We acknowledge that, but not everyone seemed aware of all options, and we believe that the people involved can still learn, mostly from each other.

The principles that emerged from our own analysis plus feedback from the informants are elaborated below.

As indicated earlier, a key principle, in our view, is that the relation between university and industry is not only a one-way issue of putting university research to use, in exploitation, but a two-way issue of also providing inspiration from practice for more fundamental university research, in exploration. Most respondents acknowledged the validity of this principle. Another key principle, to ensure exploration, is that meetings between people generate ideas and projects, around a broad theme, rather than that projects are specified beforehand for participants to subscribe to. This is related to the principle that innovation projects should be sufficiently open to allow for surprise and change of direction.

It proved useful also for other reasons to employ the distinction between exploitation and exploration. Exploitation emerges in more or less straightforward contract research, in which industry can specify and contract for desired outcomes. Exploration arises in collaboration between universities to literally explore uncertain but possible developments, in new technologies, areas of application or problems to be investigated. Here, we will focus on the latter. It was recognised that one had to carefully select staff that are able to conduct the interaction for exploration. Not all practitioners are able and willing to exchange views with academics, and not all academics are eager to obtain inspiration from industrial experience. A problem also is that universities are generally oriented towards disciplines, while exploitation is mostly interdisciplinary. For universities a side benefit may be the development of contacts for future contract research, and an increase of reputation for employment opportunities of graduates or Ph.Ds. For industry, the benefit is not only the direct acquisition of knowledge or insights into new developments, but also increased absorptive capacity to be brought back into the firm, and to develop contacts for future collaboration in both further exploration and exploitation. In the area of energy, examples are: storage of CO^2 (e.g., including the chemistry and geology of infusing CO^2 into different kinds of soil) and energy from biomass (e.g., including what to do with residual material).

In line with our earlier analysis, in chapter 7, it was recognised that in such collaboration for exploration one should take enough time to get to know and understand each other ('speak the same language') and build trust, since in such settings contractual control, e.g., of property rights, is difficult, cumbersome, unworkable, or even counter-productive. Concerning the issue of intellectual property and 'spillover', respondents noted a growing awareness that a relaxation was in order: to get knowledge one should offer and risk knowledge, in many cases exploration was sufficiently 'pre-competitive' not to yield a direct threat, contractual control would not work in early stages of exploration (there isn't yet anything well defined to appropriate), and people simply could not make progress without each others' knowledge. Among the technological top institutes that we interviewed, in one case there is an explicit policy to leave ownership of patents to firms.

For some respondents it was fine to conduct exploration more or less ad hoc, with different groups of partners on different occasions. The advantage of this is flexibility. Others were aiming at more durable, 'strategic' relations, as a basis for building understanding and trust. An element of such relations was also the detachment or exchange of personnel, for the duration of a project. In some cases there was also a desire for a certain continuity and a portfolio of activities, for the sake of synergy, spreading risks, funding less mature projects from more mature ones, and for satisfying conditions of matching for participation in EU programmes. Some respondents expressed apprehension at too close and extensive an involvement of firms in universities. That may interfere with the independence of universities and their task, generally seen as legitimate and important, to do independent fundamental research, and it may be seen by others (such as politicians or the public) as appropriation of public institutes by business.

Recently, in the US there was an initiative for a third space between BP oil company and the University of California at Berkeley and the University of Illinois at Urbana-Champaign, in the form of an Energy Biosciences Institute (EBI), with $500 million funding from BP, to conduct research into biofuels and related technologies. This caused a debate on whether this would detract too much from academic freedom and the public availability of results. What caused concern, in particular, was that BP would also have its own researchers (about 50) at the institute, in a proprietary space, who would be allowed to freely roam around in university spaces, while university staff could be excluded from BP space (http://www.berkeleydaily, planet.com). Predictably, there were also questions concerning property rights of outcomes.

A solution here may be a 'third space' in the form of an actual institute, with residential facilities, separate from both industry and university, financed jointly by business and universities, dedicated to their interaction, with the explicit task to enable and facilitate meetings and joint projects to utilise cognitive distance for the sake of exploration. A second objective of a separate institute may be to shield off and protect more fundamental research from appropriation by business interests. Participation in the activities of a third space should satisfy the principle of optimal flexibility, discussed in chapter 3: long enough to develop requisite mutual understanding and trust, but not so long as to reduce novelty and variety of ideas.

A condition for such third spaces to work is that participation by academics is seen as legitimate, and finds a place in the system of performance evaluation. On the side of industry, a condition is that participation of staff is seen as a good career move within the firm. A third space could perhaps also function as a platform for entrepreneurial spin-offs. In that case the set-up may also involve venture capitalists. However, in many cases of exploratory research exploitation is still too far off.

These arrangements are appropriate for large firms, and for small firms with adequate absorptive capacity, such as high-tech firms or spin-offs from universities or large firms. Openings should be maintained with respect to the latter, and access should not be monopolized by one or a few large firms.

For most small- and medium-sized enterprises (SME) it is more fitting to their absorptive capacity to seek collaboration with higher vocational education institutions, solely or in groups. In the Netherlands, this is done in the RAAK (Regional Attention and Action for Knowledge circulation) programme, sponsored by the Ministry of Education and executed by the Stichting Innovatie Alliantie (Innovation Alliance Foundation) (2007). To ensure diffusion of innovation, and to further enhance the interaction between exploration and exploitation, links should be forged between exploration at universities, the interaction between exploration and exploitation at third spaces, and applied, exploitative research at higher vocational institutes. As we found from one of the respondents, one way of doing this is to make use of the recent institution of 'lectors' at higher vocational institutes. Those might have part-time and/or temporary assignments at third spaces.

4.8 CONCLUSIONS

Doing excellent scientific research is one of the conditions for innovation. Other important conditions are an adequate transfer of knowledge in education and an optimal context for the applied exploitation of that knowledge in practice, in the economy, and in society.

This chapter presents a discussion of the creation and application of knowledge, in particular the role of universities and higher vocational education institutions, science policy, and relations between higher education and industry. Fundamental research and education on the one side and the exploitation of knowledge by business on the other side belong to two different institutionalised worlds. These two worlds should interact but not merge in a knowledge and innovation *system*. It is the permanent division between the two that has advantages, also because innovation is not the sole goal for acquiring knowledge and higher education does not only aim to transfer knowledge for innovation.

Among other things, an argument is made to limit the thematic programming of university research to the stimulation of areas of societal priorities, in 'second stream finance', and not to meddle in the autonomy of universities in specifying projects and approaches. A question is to what extent 'first stream finance' should remain coupled to the existing academic landscape of the universities.

In research, there is a division of labour between universities (more for exploration) and higher vocational education institutions (more for exploitation); the latter should obtain the finance to conduct corresponding applied research. For the interaction between university and large firms and high-tech small firms, ideas are proposed to further activities of exploration between them, in a 'third space' of activities, to support industry in activities of exploration and for universities to test their ideas and to gain inspiration for fundamental research with broad social applications. For the majority of small business (non-high tech) higher vocational education institutions form a more natural ally.

There are good motives for promoting research based in part on the social relevance of the research, without neglecting the other responsibilities of scientific research and higher education. The business community can be involved in a set-up of not only research that is in service of business, but also the other way round: of a business community responsible for more general societal concerns. Then the cycle of discovery will come full circle.

REFERENCES

Arundel, A. (2007) 'Innovation Survey Indicators: What Impact on Innovation Policy?' in OECD *Science, Technology and Innovation Indicators in a Changing World: Responding to Policy Needs*, Paris: OECD, 49-64.

AWT (2005a) *Ontwerp en ontwikkeling. De functie en plaats van onderzoeksactiviteiten in hogescholen*, The Hague: AWT.

AWT (2005b) *Briefadvies WHOO*, December 6, 2005, The Hague: AWT

AWT (2007) *Alfa en Gamma stralen. Valorisatiebeleid voor de Alfa- en Gammawetenschappen*, The Hague: AWT.

Birrer, F.A. and S. Tobias (2003) *Linking Science and Business: Examples of Educational Innovation*, Leiden: Leiden Institute of Advanced Computer Science.

Bogaard, A.A. van den (2002) *De verwevenheid tussen toekomstverkenning en beleid. Het ontstaan van vertoogcoalities*, Werkdocumenten W127,The Hague: WRR.

CPB (2007) 'Selectief arbeidsmigratiebeleid: Europees of nationaal? Notitie voor de Interdepartementale Commissie Beleidsplan Legale Migratie'.

De Wijkerslooth (2007) interview in *De Valorisatiekrant*, Februari .

DiJusto, P. (2006) 'Winner Takes All: Are Tech Prizes the Best Way to Cutting Edge?' www.*ScientificAmerican.com*, February.

Edgerton, D. (2007) 'The Stuff of Technofantasy', *New Scientist*. January 27: 46-47.

EUROSTAT (2007) 'Innovating Enterprises and the use of Patents and other Intellectual Property Rights', report 91/2007.

Gault, F. (2006) 'Innovation, Indicators and Policy: A Canadian Perspective', *The Future of Science, Technology and Innovation Policy: Linking Research and Practice*, SPRU 40th Anniversary Conference, 11-13 September 2006.

Hawkins, R.W. (2007) 'University Research in an "Innovation society"', in: OECD *Science, Technology and Innovation Indicators in a Changing World: Responding to Policy Needs*, OECD, 171-191

Huston, L. (2007) *Finding That Sweet Spot: A New Way to Drive Innovation*, Interview, Knowledge@Wharton, Pennsylvania University, http://knowledge.wharton.upenn.edu.

Commisie Dynamisering (Committee Dynamization) (2006) *Investeren in Dynamiek* (Investing in Dynamics), Endreport, the Hague: Ministry of Education.

Kahan, J.P. et al. (2002) *Visie op de toekomst van het wetenschappelijk onderzoek. Speerpunten voor beleid. Samenvattingen*, Cambridge UK: RAND Europe.

KNAW (Royal Netherlands Academy of Arts and Sciences) (2005) *Judging research on its merits. An advisory report by the Council for the Humanities and the Social Sciences Council*. Amsterdam: KNAW

Kronjee, G.J. (2002) *Overheid, wetenschap en toekomstverkenningen*, Werkdocumenten W130, The Hague: WRR.

Langen, A. van (2005) *Unequal Participation in Mathematics and Science Education*, dissertation Radboud Universiteit Nijmegen.

Lester, R.K. and M.J. Piore (2004) *Innovation: The Missing Dimension*, Cambridge, MA and London: Harvard University Press.

Memorie van Toelichting (2005) cited in AWT, *Briefadvies WHOO*, December 6, 2005.

Ministerie van Onderwijs, Cultuur en Wetenschap (2007) *Kerncijfers 2002-2006 Onderwijs, Cultuur en Wetenschap*, The Hague.
Noailly, J., D. Waagmeester, B. Jacobs, M. Rensman and D. Webbink (2005) *Scarcity of Science and Engineering Students in the Netherlands*, CPB Document 92, The Hague: Centraal Planbureau.
Nooteboom, B. (2007) 'De toekomst van valorisatie: third spaces', *De Valorisatiekrant*, February.
Nowotny, H., P. Scott and M. Gibbons (2003) 'Introduction. Mode 2 Revisited: The New Production of Knowledge', *Minerva*, 41: 179-194.
NWO (2007) *Fact and Figures*, www.nwo.nl.
OECD (2005) *Oslo Manual: Guidelines for Collecting and Interpreting Innovation Data*, 3rd edition, Paris: OECD/Eurostat.
OECD (2007) *Science, Technology and Innovation Indicators in a Changing World: Responding to Policy Needs*, PARIS: OECD.
Oostrom, F. van (2007) *Markt en ziel. Jaarrede voor de Verenigde Vergadering van de Koninklijke Nederlandse Akademie van Wetenschappen gehouden op 21 mei 2007*, Amsterdam, Amsterdam: KNAW
Overlegcommissie Verkenningen (1996) 'Een strategie voor de Sociale Wetenschappen', Amsterdam: Overlegcommissie Verkenningen.

Polt, W., N. Vonortas and R. Fisher (2008) Innovation Impact Study, 4th edition, Brussel: OPOCE.
Rathenau Instituut (2006) Onderzoeksprogramma 2005-2008, Science System Assessment, Webpublication Rathenau Instituut, http://www.rathenau.nl.
Renzulli, J.S. (1999) 'What is this thing called giftedness, and how do we develop it? A twenty-five year perspective', *Journal for the Education of the Gifted*, 23(1): 3-54.
Salter, A., P. d'Este, B. Martin, A. Gevno, A. Scott, K. Pavitt, P. Patel and P. Nightingale (2008) *Talent, not Technology: Publicity Funded Research and Innovation in the UK. Investing in Universities and Colleges for Global Success. Report for the CVCD in Preparation for the Comprehensive Spending Review 2000*, London: Committee of Vice-Chancellors and Principals.
Schettino, F. (2006) *Evaluating the Black Box of Innovation*, Department of Public Economics, Ph.D. Thesis, University of Rome.Webpublication, http://www. socialcapitalgateway.org/eng-phdschettino.htm
SER (2007) Arbeidsmigratiebeleid, Advies nr. 07/02, Webpublication SER, www.ser.nl.
Schweizer, T.S. (2004) 'An Individual Psychology of Novelty-Seeking, Creativity and Innovation', Ph.D. dissertation, Erasmus University.
Smitsman, A.W. (2000) 'Slumbering Talents: Where do they Reside?' p. 17-40 in: C.F.M. van Lieshout and P.G. Heymans (eds.) *Developing Talent Across the Life Span*, Philadelphia: Psychology Press.
Sociaal-Wetenschappelijke Raad (2006) *Naar een effectieve kennissamenleving. De rol van de gedrags- en maatschappijwetenschappen in de opmaat naar morgen*, Amsterdam: Koninklijke Nederlandse Akademie van Wetenschappen.
Stichting Innovatie Alliantie (2007) *Doelstellingen RAAK*, www.innovatie-alliantie.nl.
Tweede Kamer 2005-2006, 30300 XII, nr. 66, Brief van de minister van economische zaken over Technologische Topinstituten.
Tweede Kamer 2006-2007, 30919, nr. A1, Beantwoording door de minister op 13 februari

2007 van kamervragen over de oprichting van TTI GG.
Veen, G. van der, E. Arnold, P. Boekholt et al. (2005) *Evaluation Leading Technological Institutes*, Technopolis, appendix to Tweede Kamer der Staten-Generaal, Vaststelling van de begrotingsstaten van het Ministerie van Economische Zaken voor het jaar 2006, vergaderjaar 2005-2006, 30300 XIII, nr. 66.
Wetenschappelijke Raad voor het Regeringsbeleid (2002) *Van oude en nieuwe kennis. De gevolgen van ICT voor het kennisbeleid,* Rapporten aan de regering 61, Den Haag: Sdu Uitgevers.
Weingart, P. (1999) *Neue Formen der Wissensproduktion: Fakt, Fiktion und Mode, TA-Datenbank-Nachrichten,* 3-4(8): 48-57.
Weterings, A., O. Raspe and F. van Oort (2006) *De atlas van kennis en innovatie*, The Hague: RPB.

5 ENTREPRENEURSHIP AND INNOVATION

Erik Stam

5.1 ENTREPRENEURSHIP AND INNOVATION

Entrepreneurship has been recognised as a micro-driver of innovation and economic growth (Wennekers and Thurik 1999; Audretsch and Thurik 2001b; Acs 2006; Audretsch et al. 2006). What is meant by entrepreneurship, innovation, and economic growth is often not clear or is very idiosyncratic. This chapter starts with a discussion of the nature of entrepreneurship and its relation to innovation. The second section provides an overview of theory and empirical research on the relation between entrepreneurship, innovation, and economic growth. The chapter continues with a study on entrepreneurship and innovation in the Netherlands from an international and historical perspective. After these conceptual, theoretical, and empirical investigations, we turn to policy issues.

5.1.1 ENTREPRENEURSHIP DEFINED

What is meant with entrepreneurship and how does this relate to innovation? Entrepreneurship and innovation are fuzzy concepts that have been given multiple meanings. Innovation and entrepreneurship are often regarded as overlapping concepts. This can be traced back to probably the most well known definition of entrepreneurship, by Schumpeter (1934: 74), who defines entrepreneurs as individuals that carry out new combinations (i.e., innovations). Schumpeter distinguishes four roles in the process of innovation: the inventor, who invents a new idea; the entrepreneur who commercialises this new idea; the capitalist, who provides the financial resources to the entrepreneur (and bears the risk of the innovation project); the manager, who takes care of the routine day-to-day corporate management. These roles are most often executed by different persons (see for example Kenney 1986). The literature on entrepreneurship recognises a variety of entrepreneurial roles in economic change, such as:
- the person who bears uncertainty (Knight 1921);
- an innovator (Schumpeter 1934);
- a decision maker (Casson 2003);
- an industrial leader (Schumpeter 1934);
- an organiser and coordinator of economic resources (Marshall 1890);
- an arbitrageur, alert to opportunities (Kirzner 1973; 1997);
- an allocator of resources among alternative uses (Schultz 1975).

These roles all implicitly carry with them an economically positive connotation. However, if entrepreneurs are defined to be persons who are ingenious and creative in finding ways that add to their own wealth, power, and prestige (Baumol 1990), then it is to be expected that not all of their activities will deliver a productive contribution to society (cf. Murphy et al. 1991). For other reasons, many entrepreneurs do not directly contribute to an increase in for example national income: some entrepreneurship is more adequately characterised as a non-profit-seeking activity (cf. Benz 2006). Greater independence and self-fulfilment are more often mentioned as important motivations to become self-employed than increasing earning power (EOS Gallup 2004). Empirical studies have even shown that (on average) entry into self-employment has a negative effect on the monetary income of individuals (Hamilton 2000; Parker 2004). Being an entrepreneur may be rewarding because it entails substantial non-monetary benefits, like greater autonomy, broader skill utilisation, and the possibility to pursue one's own ideas; i.e., more freedom (cf. Sen 1999). These wide ranging effects of entrepreneurship are reflected in entrepreneurship policy.[1]

There have been dozens of definitions of entrepreneurship (see for example Hebert and Link 1989; Thurik and Van Dijk 1998). There is certainly not one answer to the question of what the phenomenon entrepreneurship 'truly' is. Taking all entrepreneurship definitions together, they broadly reflect two relatively distinct social realities (Davidsson 2004). The first of those is the phenomenon that some people, rather than working for somebody else under an employment contract, strike out on their own and become self-employed.[2] These economic entities involve some element of innovation at start-up, and some degree of innovativeness is needed to survive over time. However, innovation is not central to this phenomenon. It is to the second social reality. This reality involves the development and renewal of any society, economy or organisation, which is based on micro-level actors who have the initiative and persistence to make change happen. In this reality, 'entrepreneurship' means the creation of new economic activities and organisations ('Schumpeterian entrepreneurship') as well as the transformation of existing ones ('corporate entrepreneurship').

In the context of this chapter we are especially interested in this second social reality ('entrepreneurship') and less so in the first. In order to narrow down the discussion we would like to propose a working definition of entrepreneurship as "the introduction of new economic activity by an individual that leads to change in the marketplace" (cf. Sarasvathy 2000; Davidsson 2004). This means that we exclude some other interpretations of entrepreneurship (as non-innovative self-employment) and parts of the innovation phenomenon (see figure 1). For example, we exclude non-market activities such as not-for-profit endeavours, changes in contract (e.g., from employee to self-employed) and internal, organisational innovations. We also exclude mere contemplation over new ideas or introduction of fatally flawed ones that do not change the market (directly or indirectly, via learning mechanisms). We thus do not include novelty and creativity in *any* domain of human behaviour in our concept of entrepreneurship (see also chapter 1). Inclusion of all this novelty and creativity would make the events of September 11, 2001, an entrepreneurship masterpiece: "To conceive of a fully fuelled passenger jet as a missile and to combine the idea of hijacking with that of kamikaze attacks is certainly innovative, and in terms of impact – economic and otherwise – it has few parallels. However, regarding these attacks as driving market processes is far-fetched" (Davidsson 2004: 7). This example also shows that innovations can have devastating effects on society.

Figure 5.1 Entrepreneurship, innovation and self-employment

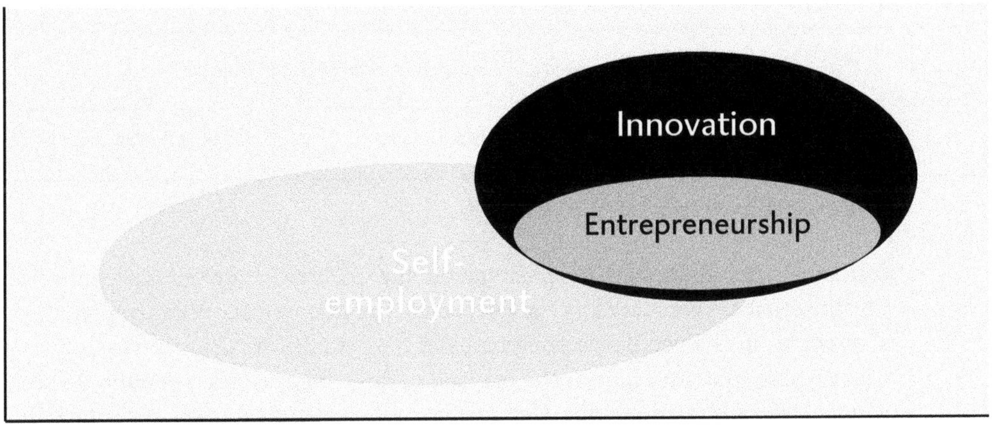

In line with our definition of entrepreneurship as the introduction of new economic activity by an individual that leads to change in the marketplace, we can formulate several necessary conditions for entrepreneurship (cf. Shane 2004: 6):

- existence of entrepreneurial opportunities (environmental changes: technological, political/regulatory, social/demographic);
- difference between people (in their willingness and ability to act upon an opportunity);
- risk bearing; uncertainty until the entrepreneur pursues the opportunity (does demand exist? can the entrepreneur compete with others? can the value chain be created? etc.);
- organising (exploiting the opportunity); either creating a firm, or using the market mechanism (for example, licensing);
- innovation: recombination of resources into a new form that is by implication not a perfect imitation of what has been done before, and thus involves a change in the marketplace.

These are necessary conditions for entrepreneurship. It is however contingent upon whether the individuals discovering an opportunity are employees or independent individuals, and whether new firms or incumbent firms are used for the exploitation of the opportunity. See figure 5.2 for a typology of entrepreneurial efforts as a function of the locus of discovery and exploitation.

5.1.2 ENTREPRENEURIAL OPPORTUNITIES

Because the range of options and the consequences of exploiting new things are unknown, entrepreneurial decisions cannot be made through an optimisation process in which mechanical calculations are made in response to a given set of

Figure 5.2 Typology of entrepreneurial efforts

		Discovery	
		Independent individual	Employee
Exploitation	Independent individual	Independent start-up	Spin-off
	Incumbent firm	Acquisitions	Corporate venturing

Source: Adapted from Shane and Eckhardt 2003: 186

alternatives (Baumol 1993). People must be able to identify new means-ends relationships that are generated by a given change in order to discover entrepreneurial opportunities. Even if a person possesses the prior information necessary to discover an opportunity, he or she may fail to do so because of an inability to see new means-ends relationships. Unfortunately, visualising these relationships is difficult. History is rife with examples in which inventors failed to see commercial opportunities (new means-ends relationships) that resulted from the invention of important technologies – from the telegraph to the laser.

Every entrepreneur who starts a new business has ideas. The real challenge is to discover an opportunity that is more than just a good idea. These opportunities can have a radical nature (Schumpeterian) or be relatively incremental (Kirznerian). Schumpeterian opportunities originate from changes in the environment (Shane 2003). These can be technological, social /demographic, and political/regulatory changes. First, technological change, often based on progress in the research base of society, is a prime source of entrepreneurial opportunities for new technology-based firms (for example in the ICT and biotech industries). See section 5.2 for the implications of this on economic growth. Second, social and demographic changes can be quantitative changes like an ageing population that offers new opportunities for entrepreneurs. It may also involve more qualitative changes: changing preferences or wants, for example reflected in the increase in the creative industries that satisfy new wants (e.g., Reality TV production companies), or in the trend toward health and nutrition and the supply of diet and organic food. In that sense people's necessities are few but their wants are endless. Third, deregulation, privatisation, and liberalisation have opened up many opportunities for entrepreneurship.[3] An example of deregulation is labour market flexibility policy. Flexibilisation of the labour market has opened up several opportunities for entrepreneurship. One the one hand many employees have become self-employed, partly lured by the lower tax rates in comparison to wage-labour. On the other hand, there have been high-growth start-ups that have used this new trend of labour flexibility to specialise in temporary staffing. Other examples of privatisation as sources of entrepreneurial opportunities are the downsizing of municipal services and the privatisation of the care market, which have provided opportunities for high-growth start-ups.

Some of the examples just given about deregulation as a source of opportunities can also be characterised as Kirznerian. Related to that, the fourth – Kirznerian – source of opportunities can be formulated, namely opportunities proven elsewhere that can be pursued in a new context ('filling a gap in the market'). An example of such a Kirznerian opportunity is the imitation of the Italian coffee bar by Starbucks: the founder of Starbucks was alert enough to see that the coffee bar culture in Italy and its social role might translate to the US. These Kirznerian opportunities most often do not involve straightforward replications, as the business idea has to be adapted to the new local context (see 'differentiation' in chapter 3). Thus, although conceptually it is an imitation, the implementation can be seen as an example of (innovative) entrepreneurship.

Finally, customers can themselves be a source of entrepreneurial opportunities, involving so-called user-entrepreneurship (see also user-innovation in chapter 8).

"The mountain bike industry emerged out of a group of hippies that constructed mountain bikes out of existing bike parts, in order to fulfil their want to ride the bike on off-the-road single track downhills. Gary Fisher – one of those hippies – started to produce these mountain bikes in 1979,[4] and stood at the cradle of what turned out to be a huge industry (Buenstorf 2001). Another example of user-entrepreneurship is the online communication platform Hyves, started out of a personal need to have a computer-mediated social network that connects people. Initially (in 2004) only friends and acquaintances of the founders joined their website; in 2007 Hyves has millions of members, has grown into an enterprise with 30 employees, and has an estimated market value of 50 million euros" (*Intermediair* 2007: 26).

5.1.3 ENTREPRENEURSHIP AS AN ORGANISATIONAL PRODUCT

Figure 5.2 showed that the discovery of an entrepreneurial opportunity can be made by an employee (i.e., a paid organisation member) or an independent individual. The latter situation is reflected in so-called user-entrepreneurship: i.e., a personal need as a consumer is the source of the opportunity. Empirical research has shown that the prior situation occurs much more often, as most founders start a new business in an industry that is similar or related to their prior experience (Klepper 2001). 'Producer-entrepreneurship' is thus a much more widespread phenomenon than user-entrepreneurship. In organisational terms the most important question is whether this opportunity is pursued and exploited within or outside the organisation of origin: i.e., in the form of a spin-off or of (internal) corporate venturing. Spin-offs involve the exploitation of an opportunity by an employee who leaves an organisation to start a firm of her own that is independent of the parent organisation. Corporate venturing or corporate entrepreneurship has been defined as "the process whereby an individual or a group of individuals in association with an existing organisation, create a new organisation or instigate renewal or innovation within that organisation" (Sharma and Chrisman 1999: 18). Two sub-types of corporate venturing are typically distinguished: 'internal corporate venturing' which focuses on opportunities identified within the company (also called intrapreneurship, sometimes leading to a spin-off firm that commercialises this opportunity outside the mother firm); and 'external corporate venturing' or 'corporate venture capital' which focuses on opportunities external to the company, in the form of investments in independent start-ups. Frequently, corporate venture units pursue some combination of internal and external opportunities. Sometimes a third type – alliances – is also included. Alliances offer the advantage of combining the assets of the larger company (brand strength, market channels, investment capital, and other scale-related advantages) with the more focused and flexible characteristics of the smaller, younger partner.

Why would an opportunity be pursued outside the organisation in which it was discovered? When the opportunity depends more on firm specific (e.g., physical or intellectual) assets than on human capital, spin-offs are less common, because entrepreneurs cannot move these proprietary assets with them when they exit a firm. This explains the high number of spin-offs in business services, because the most important asset in this industry is human capital. When innovations are architectural and therefore reconfigure the way in which products are developed, spin-offs will also be more common because established firms have a hard time changing their organisation in order to exploit such innovations (Henderson and Clark 1990). Spin-offs are also more likely when established firms are incapable of responding to radical technological changes that upset the established ways of organising their businesses, i.e., their business model (see Chesbrough and Rosenbloom 2002). High-level managers may be incapable of evaluating the new entrepreneurial opportunities or they choose to focus on their company's core line of business. Likewise, when a new good or service only serves a small market niche, spin-offs are more common because an existing customer base will restrict an incumbent firm from focusing attention on the new niche (Christensen 1997). The risk averseness of the discovering person and the organisation in which she is employed will be negatively and positively related to the likelihood of spin-off: risk averse persons will not be eager to leave a secure job, while on the other hand risk averse organisations will not be open to accommodate risky ventures.[5] Taking this latter mechanism into account, an increase in the number of spin-offs (and thus new firms in general) could also be an effect of the increased risk averseness of incumbent organisations.[6]

5.2 ENTREPRENEURSHIP, INNOVATION, AND ECONOMIC GROWTH

How can we explain the relationship between entrepreneurship and economic growth? Several mechanisms may be at work here, which explain why new and small firms in combination with large organisations may drive innovation and ultimately economic growth. These mechanisms are knowledge spillovers, decentralisation, experimentation, and competition. We will first discuss these mechanisms. Next, we will provide an overview of empirical studies that have tested the effect of (different types of) entrepreneurship on (different types of) economic growth.

5.2.1 KNOWLEDGE SPILLOVERS

First, as has been mentioned before, new scientific and technological knowledge is an important source of entrepreneurial opportunities. Organisations investing in research or technology development often end up facilitating other agents' innovation efforts, either unintentionally, as when inventions can be imitated, or intentionally as where scientists report on their research. Economists have termed this knowledge spillovers: 'any original, valuable knowledge generated somewhere that becomes accessible to external agents, whether it be knowledge fully characterising an innovation or knowledge of a more intermediate sort. This knowledge is absorbed by an individual or group other than the originator' (Foray 2004: 91). There has been much empirical research showing that firms located near knowledge sources introduce innovations at a faster rate than rival firms located elsewhere (Audretsch et al. 2006). These can be incumbent firms, but more likely involve firms that have been set-up by prior employees of the knowledge producing organisations. They are the Schumpeterian entrepreneurs that commercialise inventions. Many major inventions have been reshaped, speeded, and expanded by (individuals and their) new firms with different objectives, interests, and ideas from those of the original inventor (cf. Shane 2000) or originating organisation. These innovative new firms are started because their innovations would have been turned down or severely delayed in the organisations in which the initial idea was developed.

Several case studies throw some light on how potential entrepreneurs may recognise opportunities that are not recognised as valuable by the knowledge-originating organisation. Well-known examples of companies developing resources that they failed to exploit are Bell and Xerox, private companies that incubated emerging technologies.

During the emergence of the semiconductor industry, the growth of knowledge developed at the Bell Labs and the Bell System provided more opportunities for new semiconductor firms than the Bells could exploit (Holbrooke et al. (2000: 1037); cf. Moore and Davis (2004) for a similar situation at Fairchild Semiconductors). In the early semi-conductor industry, a diversity of new companies were started, based on newly developed knowledge, which ensured that a wide opportunity space presented by the transistor's invention was explored and exploited. The use of semiconductors was appreciably accelerated and broadened as a result of the ready formation of firms (like the multiple generations of spin-offs Shockley Laboratories, Fairchild, and Intel) with different development criteria than Bell's (cf. Rosenberg and Birdzell 1986; Holbrooke et al. 2000). It has also been claimed that roughly half the population of Silicon Valley semiconductor manufacturers can be traced back to the Bell Labs (Rogers and Larsen 1984: 43-45).

Another well-known source of entrepreneurial opportunities was Xerox Corporation. In the 1960s and 1970s managers at Xerox who understood the potential of digital electronics and computing set up Xerox PARC near Stanford University. PARC (its employees aided by Pentagon funding) created many of the key technologies of the PC industry, but failed to take advantage of their opportunities

(Smith and Alexander, 2003). Xerox's innovations in computing were largely underexploited because its business model was based on developing copier systems in-house with proprietary standards. PARC employees were alert to business opportunities neglected by Xerox and chose to leave to found new companies based on novel business models (Chesbrough and Rosenbloom 2002).

Large research organisations are often repositories of unused ideas: big firms have natural diseconomies of scope that a cluster of start-ups does not have (Moore and Davis 2004; cf. Nooteboom 2000), and public research organisations often do not have incentives to commercialise ideas. One of the arguments behind the so-called open innovation strategies of large firms like Philips Electronics is exactly this: the intellectual property developed in these firms could be exploited much more widely by firms outside its organisational boundaries than by divisions from within. Technology transfer and 'valorisation' has also become an important function of public research organisations. University based spin-offs commercialising knowledge have become more common world wide (Shahid and Kaora 2007). These companies explore applications of knowledge beyond the academic remit which established firms find commercially uncertain or which conflict with their current activities. The pioneer in Europe among centres of high tech activity was the University of Cambridge. The first spin-out company from the university was the Cambridge Scientific Instrument Company, founded in 1881 by Horace Darwin, son of Charles Darwin. The current cluster of high-tech activities resulted from multi-generational spin-out from the university (Garnsey and Heffernan 2005).

Regions without larger research organisations (at the scientific or technological frontier), will probably have fewer spin-off firms, both because a lack of technically trained people and a shortage of ideas (Moore and Davis 2004). A mix of large and small knowledge-based organisations is thus a better starting point for the exploration and exploitation of new ideas than a concentration of small entrepreneurial firms only (Baumol 2002; Moore and Davis 2004; Rothwell and Dodgson 1994; Nooteboom 1994). The combination of high investments in new knowledge (exploration) and high levels of entrepreneurship exploiting this knowledge is a key driver of growth in advanced capitalist economies (Acs et al. 2005; Audretsch et al. 2006). Large firms are not only important as sources of entrepreneurial opportunities. They are also important for more downstream functions. The most useful innovations are likely to produce one or more giant firms, simply because useful often means 'widely used' and widely used may well mean 'mass-produced' (Rosenberg and Birdzell 1986: 271). See for example the dominance of a few large firms in chemistry, electronics, drugs, and automobiles. Some of these large firms owe their size to an innovation that occurred while the firm was still small, while others had to shift their production and marketing to a new field in order to become large corporations. Large and small firms have dynamic complementarities in technological development (cf. Nooteboom 1994).

"Large corporations invented neither the airplane nor the automobile, but they contributed both technologically and commercially to filling the gap between the horseless carriage and the everyday family car, and between the plane of Kitty Hawk and the commercial airliner" (Rosenberg and Birdzell 1986: 288).

Diversity of enterprise is a necessary condition for economic growth and prosperity. History has shown that long-term economic growth and prosperity depends on a mix of large and especially small enterprises (Rosenberg and Birdzell 1986; Landes 1969). Many types and sizes of enterprise are useful under the right conditions and circumstances, but what matters is the diversity of economic organisation in economic systems – the variety of the system's organisational repertoire rather than the size of particular enterprises (Rosenberg and Birdzell 1986: 270). The role of diversity of enterprise in economic growth and prosperity has two key elements (Rosenberg and Birdzell 1986: 296-297): experimentation and decentralisation.

5.2.2 EXPERIMENTATION

First, experimentation is almost always best conducted on the smallest scale necessary to prove or disprove a point. Since experimentation is so important to economic change, a great part of the activity in progressive economies will be conducted on a small scale. Economic growth implies change and adaptation, and much of this adaptation takes place through the formation of firms that are, at least initially, small. New firms are useful devices for experimenting with innovation, because they can be established at a small, experimental scale at relatively low cost and therefore in large numbers, and their efforts can be intensively focused on a single innovation. The experimental aspect of new firms is reflected in the facts that they usually start small, their number is large, and as with other kinds of experimentation, most of them fail. High rates of firm entry and exit (so-called churning or turbulence) can even be regarded as a necessary price to pay in order to allow 'exploration' of new technological and market possibilities: failures at the micro level may be consistent with social benefit at the aggregate level (see March 1991; Saxenian 1994; Dosi and Lovallo 1997). A high level of new variety is needed to produce a few very successful new innovative industry leaders, like Microsoft, Google, and Ebay. The experimental approach to the organisation of economic activity is a key mechanism for economic progress. New firms often provide the seedbed for the emergence of new industries.[7] They have been instrumental in the introduction of electricity, the internal-combustion engine, automobiles, aircraft, electronics, aluminium, petroleum, plastic materials, and many other advances (Rosenberg and Birdzell 1986; Baumol 2002; Audretsch 1995).

5.2.3 DECENTRALISATION

Second, a fundamental characteristic of organisation in highly developed economies is decentralisation – a diffusion of authority and responsibility and a limitation of the pyramiding of managerial hierarchies. The resistance to agency costs and the complexities of controlling those costs are not limited to that part of the pyramid that extends from a government board of planning and control down to individual enterprises; they are reflected in the organisation of economic activity at all levels. The organising principle is that the costs and benefits of hierarchy must be balanced out (including the static and dynamic transaction costs; see Nooteboom 1992; Langlois and Robertson 1995).

"That the benefits [of hierarchy] outweigh the costs in comparatively few situations is a fact of social life, as evidenced by the predominance of relatively small hierarchies in Western economies. The strength of the tendency to decentralization in Western economies is chronically underestimated, if one may judge from the many prophecies that capitalism would end in the centralization of Western economies in the hands of a few capitalists – prophecies repeated by now for more than a hundred years and still unfulfilled" (Rosenberg and Birdzell 1986: 297).

Although a large part of economic change is brought about by the expansion and conversion of old firms, innovative change is to a large degree brought about by new firms (see Rosenberg and Birdzell 1986; Acs and Audretsch 2003). That small firms have played a large part in economic growth is not accidental; it can be explained, at least in part, by their smaller agency costs (in addition to their special suitability to the experimental stage of innovation). Innovation is more likely to occur in a society that is open to the formation of new enterprises than in a society that relies on its existing organisations for innovation (Rosenberg and Birdzell 1986: 258).[8] New, usually small, firms have an important role in bringing about change – a role that may well depend on the degree of inertia accumulated in older bureaucracies.

5.2.4 COMPETITION

Competition has been the principal source of diversity in enterprise organisation: differentiation via the development of unique products, methods of production and distribution, and forms of organisation is central to the strategy of competition. Diversity of enterprise is closely related, both as cause and consequence, to diversity of products and services available to customers.[9] See Porter (1980) on the micro-economic, and Helpman (2004) on the macro-economic relevance[10] of product differentiation (see also chapter 1).

New firms played a direct role in economic growth, with the introduction of new products, but also an important indirect role in triggering old firms to improve or restructure their activities (or to exit the market if they fail to do so). The easy formation of new firms acts as a disciplinary device for existing firms (cf. Aghion et al. 2006). New innovative firms circumvent bureaucratic rigidity and supply older firms with an incentive – self-preservation – for taking internal measures to avoid the habits and practices that eventually lead to rigidity. This is for example reflected in the rise of corporate venturing, as a means for corporate renewal.

5.2.5 ENTREPRENEURSHIP AND ECONOMIC GROWTH: EMPIRICAL EVIDENCE

Already at the start of the twentieth century the economist Schumpeter made a plea for the entrepreneur as the person who brings new ideas to the market and in that way causes economic renewal and progress. A necessary condition is that these innovations have to offer more (or the same for a lower price) than the pre-existing supply. If this condition is fulfilled there might even be creative destruction: innovations that make the 'old economy' superfluous. A recent example in the Dutch economy is the success of the digital TomTom route planner that has substituted a large part of the production of roadmaps. An indirect effect of the introduction of these innovations by new firms is that incumbents are triggered to upgrade their product offerings in order to remain competitive. How and to what extent does entrepreneurship lead to innovation and economic growth?

Why should entrepreneurs start with an uncertain innovation process at all? A recent study (CBS 2006: 153) shows that entrepreneurs innovate because they want to improve the quality of goods and services (cf. Aghion and Howitt 1992), to offer a broader range of goods and services (cf. Romer 1990), and in the end they want to access new markets or a larger market share. A recent review of empirical studies by Van Praag and Versloot (2007) shows mixed evidence on the assumption of the relatively high innovativeness of small and new firms. They conclude that 'entrepreneurs and their counterparts [large incumbent firms] contribute equally importantly to the innovativeness of societies. However, they serve different goals in terms of quality, quantity and efficiency, as well as in terms of producing (and adopting) more radical (and higher cost) innovations' (Van Praag and Versloot 2007: 18). They show that new and small firms have relatively high levels of innovative sales, and are relatively less likely to adopt high-cost innovations.

A key question is whether and how entrepreneurship causes economic growth. Before we can answer this question with empirical research, we have to choose empirical indicators for entrepreneurship and economic growth. Traditionally, economic growth has been referred to as the growth of employment or national income, while recently productivity growth is seen as a more relevant indicator (see chapter 2). The two dominant empirical definitions of entrepreneurship are the creation of new organisations (a new legal entity; including both independent start-ups and spin-offs) and self-employment (performing work for personal profit rather than for wages paid by others). Some studies also take into account people with a preference for entrepreneurship ('latent entrepreneurship'), or people who take steps to start a new business ('nascent entrepreneurship'). The latter two indicators can be seen as potential entrepreneurship. Corporate entrepreneurship is not easily identified, and is unfortunately largely an invisible aspect of entrepreneurship in empirical research. In addition to these operational definitions of entrepreneurship, there are several measures of performance, like survival, growth, profitability, and realising an initial public offering (IPO) of the business. These performance measures are indicators of entrepreneurship to a lesser or greater degree. Take for example survival: new firms that survive in the long term but remain relatively small often become more conservative (i.e., less innovative) while new firms that grow into substantial corporations often revolutionise the economic structure (cf. Schumpeter 1942: 83). In addition, there are habitual entrepreneurs that 'specialise' in setting up new firms and often leave the newly created firms (either successfully, for example via an IPO, or less successfully with a liquidation) to set up other ones.

Empirical research shows an ambiguous relationship between entrepreneurship and employment growth: the relationship is often positive (Audretsch and Thurik 2001a; Audretsch and Fritsch 2002; Bosma et al. 2006; Acs and Mueller 2008; Acs and Armington 2004; Carree and Thurik 2007; Van Stel and Suddle 2008; Thurik et al. 2008),[11] sometimes non-existent (Audretsch and Fritsch 2002; Acs and Mueller 2008),[12] or even negative (Van Stel and Storey 2004; Mueller et al. 2008). Growth in national income is unambiguously related to high levels of new firm formation and high-growth start-ups (Stam et al. 2006; Wong et al. 2005). Research on the effects of entrepreneurship on productivity growth is less abundant, and only shows an ambiguous positive effect of new firm formation (Callejon and Segarra 1999; Audretsch and Keilbach 2004; 2005[13]; Bosma et al. 2006) or no effect of changes in self-employment (Carree and Thurik 2007).

The review of recent research on entrepreneurship and economic growth reveals that high levels of new, growing firms are strongly related to economic growth. There is no consistently positive relationship between new firms in general or the level of self-employment with economic growth (see table 1). The latter outcome is not that remarkable: many new firms are a continuation of the activities that were previously done as by employees before – so these involve no new economic activities (for example the construction worker who becomes an independent handyman, and the graphic designer who is made redundant with a round of restructurings, but still supplies the same services to her previous employer). The decision to enter into self-employment is seldom driven by innovation, and relatively often by lifestyle reasons, like the combination of labour and care tasks and a focus on a particular craft (Dirks et al. 2003).[14]

Table 5.1 Entrepreneurship and economic growth (in OECD countries)*

	Employment	Income	Productivity
Self-employed	+ / 0	+	0
New firm formation	+ / 0 / -	+	+ / 0
High-growth start-ups	+ / 0	+	x

* + = statistically significant positive relation; 0 = no statistically significant relation;
- = statistically significant negative relation; x = no empirical research

A critical interpretation of the overview of empirical research could be that entrepreneurship as measured in these studies does not have much to do with innovation at all: productivity growth is probably the best output indicator of innovation, and the studies reviewed showed that entrepreneurship has hardly any effect on this. The positive effects on income and employment do not have to be explained with innovation: consider the situation in which increased labour market participation via self-employment is registered both as an increase in new firm formation and in self-employment, this is likely to lead to an increase in employment and income, as members of society that were not involved in paid labour, now contribute both to total employment and to total income. In this situation both employment and income are growing, but innovation is not a necessary ingredient in this.

In addition to economic growth, a more relevant indicator may be welfare (Layard 2005). Unfortunately, the relation between entrepreneurship and welfare has not been researched to a large extent. However, there are several indications that entrepreneurs are on average more satisfied with their occupation than employees are (Blanchflower and Oswald 1998; Benz and Frey 2003; Frey and Benz 2003).

5.3 ENTREPRENEURSHIP IN THE NETHERLANDS

A record number of new firms was started in the Netherlands in 2007: 85,000. At that time the number of new firms had almost tripled since the end of the 1980s. On the basis of these numbers one would tend to conclude that the Netherlands has become more entrepreneurial in the last decennia. If ones assumes that these new firms also supply something that is sufficiently new and different from the existing supply of goods and services, and even make a profit, than it is not such a strange idea to regard new firms as the driving force of an innovative economy (see Coalitieakkoord 2007, especially "pijler II").
In spite of this record number of new firms, there is still a common opinion that entrepreneurship is in short supply in the Netherlands, especially in comparison with Anglo-Saxon economies like the UK, Ireland, and the US. In order to understand whether there is 'sufficient' entrepreneurship in a particular economy or innovation system, one should compare systems in time or over space. Only such a comparison makes it possible to identify a 'problem'. We will attempt to get a better insight into the relative position of entrepreneurship in the Netherlands internationally (in comparison to other relevant 'benchmark' countries) and over time.

5.3.1 INTERNATIONAL COMPARISON

We start with an international comparison of three key indicators of entrepreneurship: self-employment, new firms, and ambitious entrepreneurs. With respect to the percentage of self-employed in the labour force – the most static measure for entrepreneurship – the Netherlands leads all benchmark countries (see figure 5.3). When we look at more dynamic measures of entrepreneurship,[15] the picture looks a bit less rosy: the Netherlands is now behind almost all benchmark countries, and is – just like all other European countries – miles behind the US (see figure 5.4). Especially with regard to ambitious entrepreneurship – one of the most important types of entrepreneurship for economic growth – the Netherlands is at the back of the pack (see also Hoffmann 2007).[16]

Figure 5.3 International comparison of self-employment (2004)

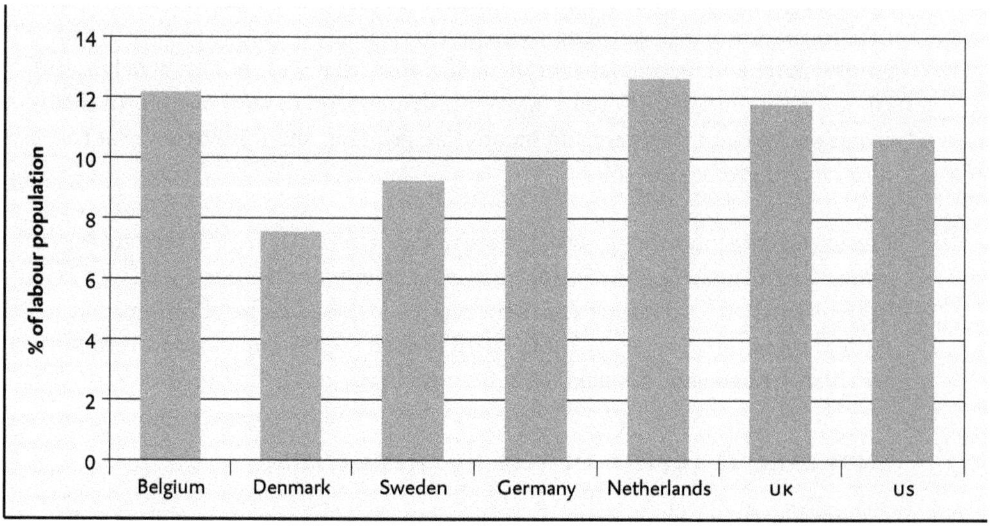

Source: CBS

In spite of the record number of new firms in the Netherlands and the high percentage of self-employed people,[17] the Netherlands can still not be seen as a leading entrepreneurial economy. Especially with respect to ambitious entrepreneurship, the Netherlands is behind other relevant countries. This is confirmed in international comparisons on the share of high-growth firms (see figure 5.5). This form of entrepreneurship is very important for economic growth. The Netherlands is thus a country with many self-employed people but with few ambitious entrepreneurs and high-growth firms.

The European Trend Chart on Innovation (European Commission 2006a) includes one set of indicators reflecting innovation and entrepreneurship (based on the Community Innovation Survey and Eurostat data), which measure the efforts towards innovation at firm level. The six indicators are 'SMEs innovating in-house (percentage of all SMEs)', 'innovative SMEs co-operating with others (percentage of all SMEs)', 'innovation expenditures (percentage of total turnover)', 'early-stage venture capital (percentage of GDP)', 'ICT expenditures (percentage of GDP)', and 'SMEs using organisational innovation (percentage of all SMEs)'. The Netherlands performs particularly well with respect to ICT expenditures, but performs particularly badly with respect to innovation expenditures, cooperating with others for innovation, and innovating in-house. Taken together, the Netherlands performs less than average on the innovation and entrepreneurship indicators in the European context (see figure 5.6).

Figure 5.4 International comparison of TEA and ambitious entrepreneurship (2006)

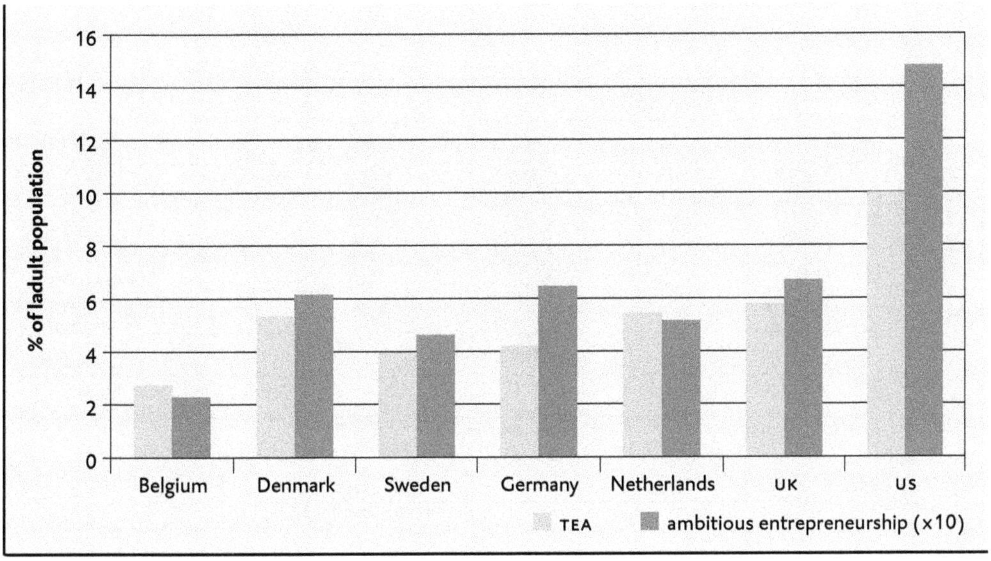

Source: GEM

Figure 5.5 International comparisons of high-growth firms

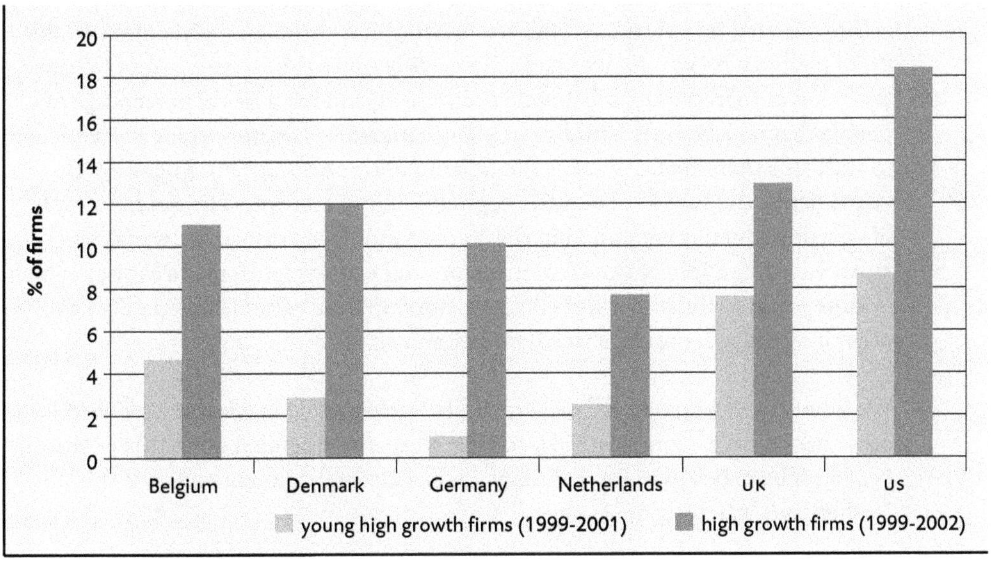

Sources: Hoffmann and Junge 2006; EIM 2006

What drives entrepreneurship? The literature on entrepreneurship has shown that two of the necessary conditions of entrepreneurship are the (perceived) skills and knowledge to start a business and the (perceived) opportunities to start a business. With respect to entrepreneurial skills, the Netherlands adult population is rather average within Europe (Bosma and Schutjens 2008), while with

Figure 5.6 Innovation and entrepreneurship indicators, NL relative to the EU 2003

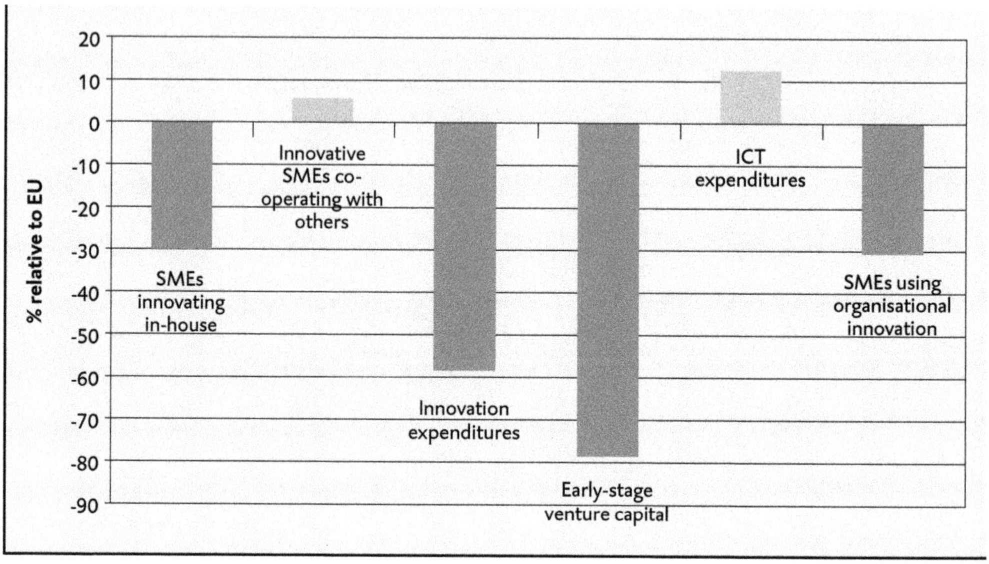

Source: European Commission 2006

respect to opportunities, the Netherlands seems to be relatively abundant (Bosma and Schutjens 2008). Another aspect of national culture that is often seen as a major constraint to entrepreneurship in Europe in comparison to the US, is the so-called fear of failure. The Netherlands has the lowest percentage in the EU of people who would not start a business because they have a fear of failure (even lower than the US; see Bosma et al. 2008), and has a lower percentage of people that regard the possibility of going bankrupt as an important risk attached to start-ups than the EU average (EOS Gallup Europe 2004). For high growth (new) firms, the supply of venture capital is highly relevant. The overall supply of venture capital is very good in the Netherlands (see Porter et al. 2007). However, the supply of informal investors and early-stage venture capital, which is more relevant for successful entrepreneurship, lags behind considerably to other benchmark countries (see figures 6 and 7).

We now have a 'worrisome' picture of the Netherlands, in comparison with other benchmark countries. How has the Netherlands ended up in this position, i.e., has it been better in former times, or has it even been worse, and is the Netherlands 'catching up'?

5.3.2 ENTREPRENEURSHIP IN THE NETHERLANDS OVER TIME

How has entrepreneurship developed in the Netherlands in the last two decennia? If we look at the annual number of new firms, there has been a huge increase, with only a hesitation after the technology-bubble in the early 2000s (see figure 5.8). A similar, but more moderate, trend can be observed if we look at the number of self-employed over time (see figure 5.8).

Figure 5.7 International comparison of informal investors and early-stage venture capital

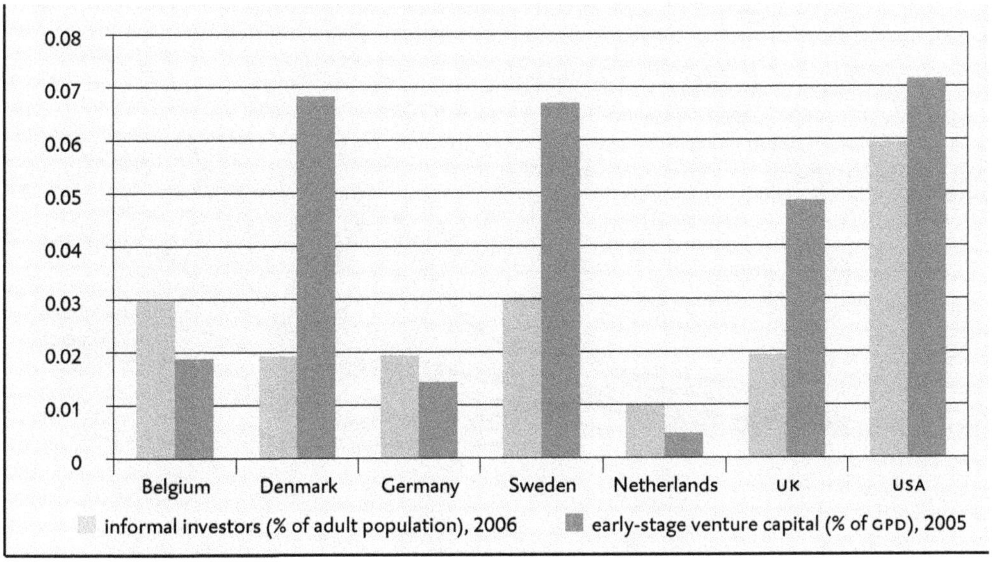

Can we trace back the increase in the number of new firms to particular sectors, or have all sectors contributed evenly to the increase in new firm formation? Figure 9 shows that the construction and (business and personal) service sectors have grown disproportionately. These two sectors are certainly not a marginal group as they also include more than half the population of new firms in

Figure 5.8 Total number of new firms and self-employed in the Netherlands, 1987-2006

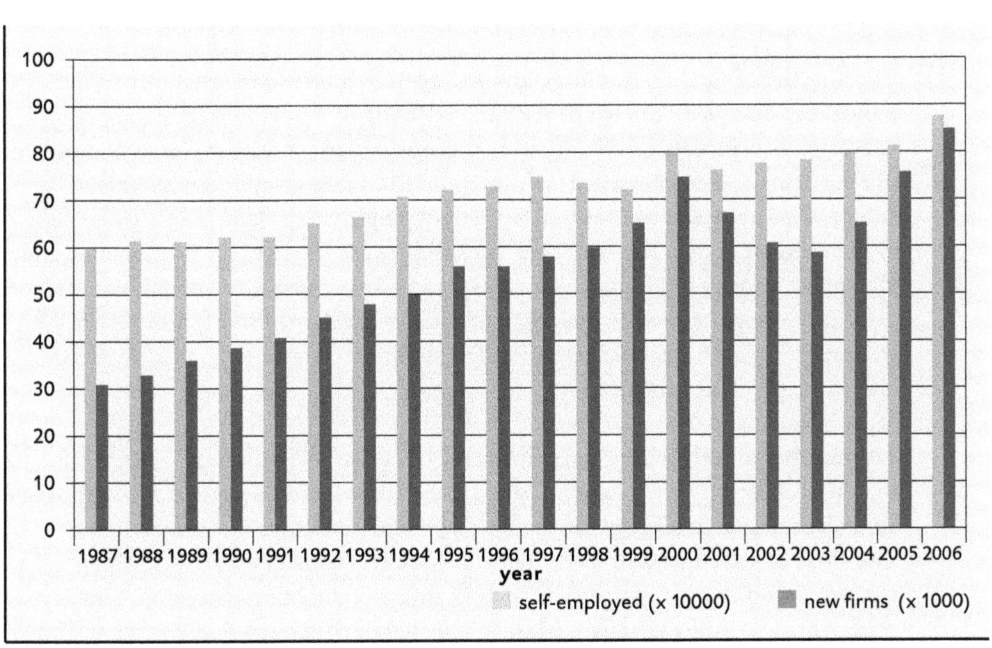

Source: CBS

Figure 5.9 New independent start-ups per sector, 1987-2006 (index 1987 = 100)

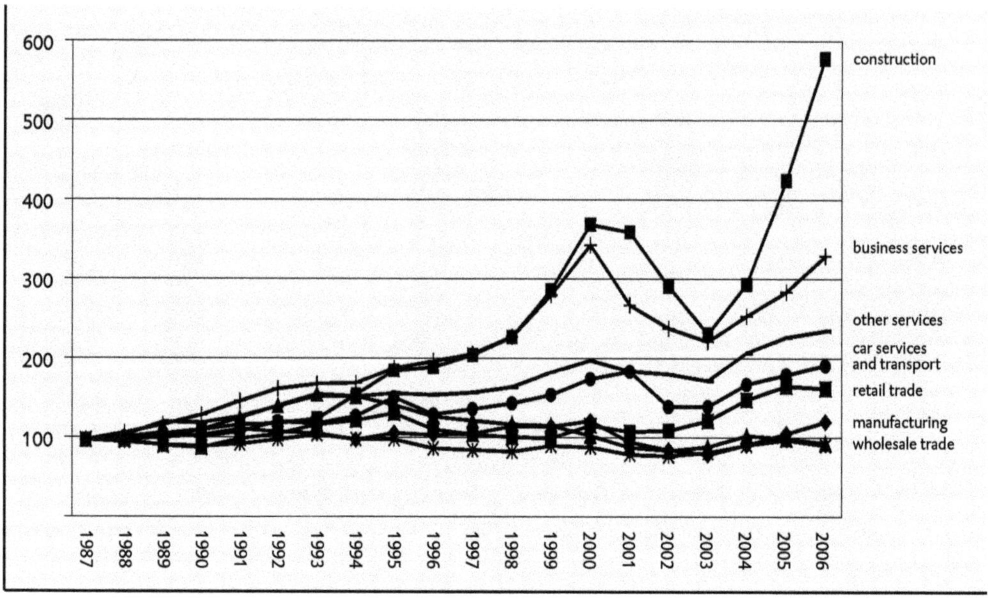

Source: EIM, based on data from the Dutch Chamber of Commerce (KVK)

2006. Two increasingly important contributing groups of entrepreneurs are female and foreign entrepreneurs. Female entrepreneurs are over-represented in personal services, and relatively often start part-time businesses, while – especially Eastern European – foreign entrepreneurs have provided a substantial impetus to the growth of new construction firms (Braaksma and Meijaard 2007).

The number of subsidiaries has increased even stronger than the number of independent start-ups. In the 1980s the number of firms and employment share of subsidiaries was respectively about 15 percent and 30 percent of the total number of new firms and employment, while in the 2000s the employment share of subsidiaries has risen to about 40 percent of employment (and about 30 percent of the number of new firms). These subsidiaries are started out of 'offensive' and 'defensive' motives. Offensive motives lead to distributing different business activities in separate profit centres, and (thus) stimulating 'intrapreneurship'. An important defensive motive is reducing risk for corporations by creating financially independent subsidiaries.[18]

The number of technology start-ups, a relatively important category for technological innovation, has also increased over the last decade, however less strongly so than for the general population of new firms (Braaksma and Meijaard 2007). The share of start-ups with R&D has increased to a considerable degree from 10.4 percent in 1994 to 19.2 percent in 2003 (Braaksma and Meijaard 2007). This is in contrast to the innovativeness of SMEs in general, which has decreased substantially over the last eight years (see figures 10 and 11).[19] The latter trend resembles

Figure 5.10 Innovation inputs of SMEs in the Netherlands, 1999-2007

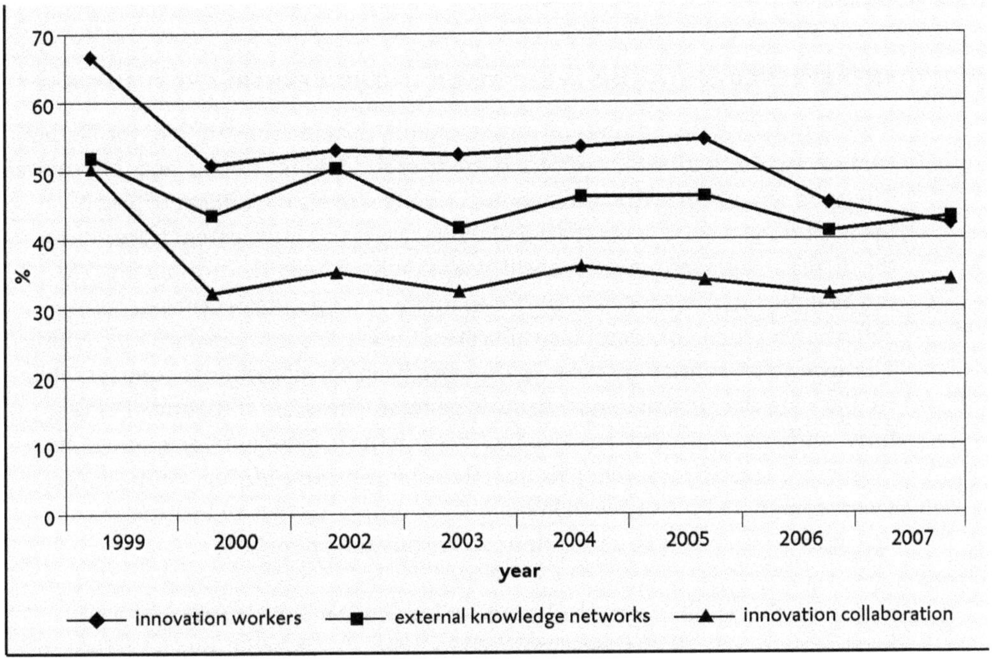

Source: De Jong and Jansen 2007

Figure 5.11 Innovation outputs of SMEs in the Netherlands, 1999-2007

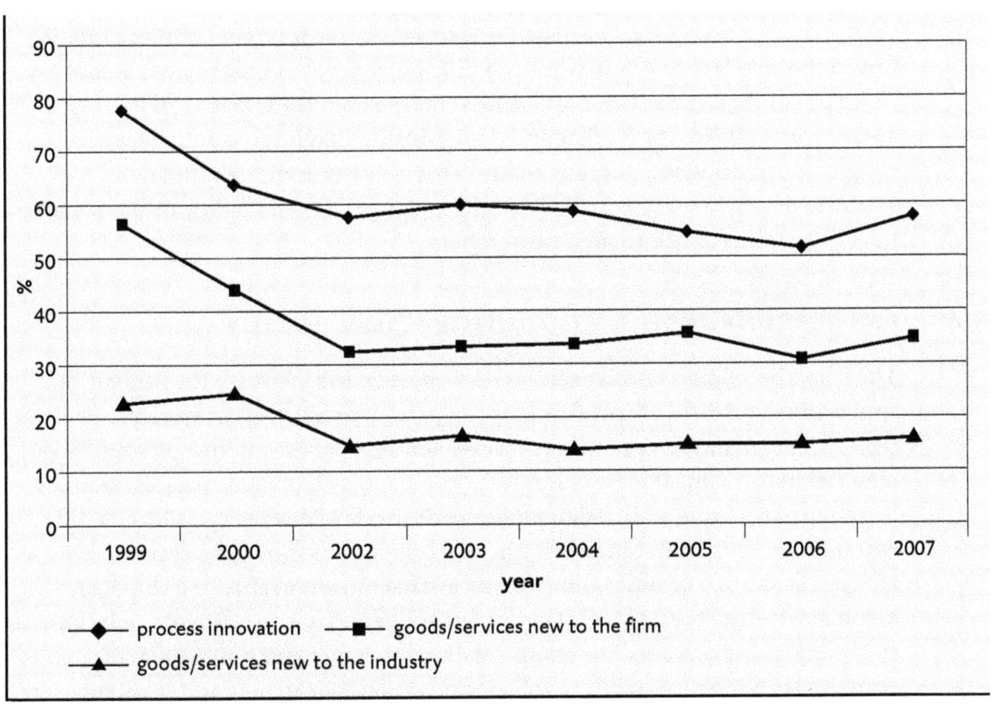

Source: De Jong and Jansen 2007

the general decline in the innovative output of the Dutch business population (firms with more than 19 employees) over the last decade (CBS 2007).[20]

5.3.3 ENTREPRENEURSHIP AND INNOVATION IN THE NETHERLANDS: SUMMARY

If we summarise the findings on entrepreneurship and innovation in the Netherlands in an international perspective and over time there is 'good' and 'bad' news. The good news is that the annual number of new firms has almost tripled over the period 1987-2006. This spectacular increase is to a large extent traceable to the increase in the number of subsidiaries that have been set up annually. This can be done for offensive reasons, for example as corporate venturing, or for defensive reasons, for example to allocate risky activities to separate legal entities. The growth of technology start-ups has also been considerable, though less spectacular than the growth of new firms in general. The share and number of new firms with R&D activities has also increased in the last decade (1994-2003), indicating an increasing number of innovative start-ups. In addition, young SMEs seem to be more innovative than older SMEs, implying that an increase in start-ups will probably lead to a more innovative business population.

However, there is also some bad news for innovation. First, many of the new firms seem to actually be self-employed individuals who continue with the same activities (mainly in the construction and services sectors) that they were previously engaged in as employees. This is not likely to improve the (product) innovativeness of the economy at all. At most this means a more flexible economy, and an improved allocation of resources (static efficiency), and in the end a higher productivity.[21] Second, on average SMEs have become less and not more innovative in the last decade (1999-2007), and the percentage of innovative SMEs is much lower than the EU average. Most SMEs are not innovative at all (see also Parker 2001). Third, the Netherlands is lagging behind internationally with respect to ambitious entrepreneurship. This is not a recent phenomenon, but seems to be consistent in the last decade, in spite of several policy measures that have been taken to improve these numbers.

5.4 ENTREPRENEURSHIP AND INNOVATION POLICY

The objective of innovation policy is to encourage and facilitate the generation, application, and diffusion of new ideas, in particular innovation systems. The systemic nature of innovation means that we should not base our insights and recommendations on a linear view of innovation. Chapter 3 has shown that entrepreneurship plays a role in different phases of the cycle of discovery. We know that entrepreneurship fulfils an important function in the innovation system (cf. Hekkert et al. 2007), but this does not mean that more entrepreneurship is always better for the functioning of the innovation system. Two disclaimers apply: an empirical and a systemic. The empirical disclaimer is that not all that is counted as entrepreneurship necessarily involves innovation (not all new and small firms are innovative) and not all entrepreneurship is counted (corporate entrepreneurship is hard to trace with empirical research). Aiming at higher numbers of new firms, small firms, or self-employed will thus not automatically imply an increase in innovation. The systemic disclaimer has two aspects. First, the elements of an innovation system should be aligned in some way in order to function properly: the elements should be sufficiently present and well connected to the other elements. This means for example that if the number of innovative new firms increases, but without sufficient venture capital supply or advanced customers, this will not lead to higher levels of economic growth: the newly introduced products are not commercialised to a large extent due to a lack of investment means or a too small market base. Second, not every system needs the same level of entrepreneurship. Mowery and Rosenberg (1993: 29) showed that new firms in the US have played a significant role in the commercialisation of new technologies; a role which the new small firm does not necessarily play in the innovation system of other countries (e.g., Japan and Sweden).[22] Several configurations could explain this. First, internal corporate venturing could act as a substitute of new firm formation. This means that if large firms in a country are relatively less inert and bureaucratic (and risk averse) than in other countries – for example due to different formal and informal institutions – there is less need for employees to start a new firm in order to exploit a new idea. Second, organisational innovations, for example in the form of new forms of network organisations, could also enable the recombination of resources without installing new legal entities. Third, certain – especially large-scale, and incremental (Winter 1984) – innovations are better realised in large firms than in small firms.

Does all this mean that nothing remains to be said about entrepreneurship in innovation policy? No, but first we should be careful not to start with sweeping statements about the supposed backwardness of the Netherlands and to take the US as the role model entrepreneurial economy. Entrepreneurship policy in the Netherlands can take inspiration from top-performing countries in this respect, but the initiatives have to be tailored to the Netherlands' context. Second, we should use indicators of entrepreneurship that are as close as possible to innovation, including both exploration and exploitation. This means that for example spin-offs and new technology based firms are probably better indicators of exploitation of unused ideas than the general population of new firms. High-growth start-ups are even stronger indicators of successful exploitation on a relatively large scale. However, we should not neglect exploration. Exploration could be involved in many 'experimental' new firms that just try out new ideas, and which might act as a source of learning for the firms in related activities, even if these start-ups fail in the short run.[23] In this sense, it is important to 'let a thousand flowers bloom'. An entrepreneurial climate – with little stigma for failure – is likely to facilitate these experimental firms. With these remarks in mind, the next subsections deal with how institutions might be changed in order to stimulate new firm formation in general, and technology start-ups, spin-offs, and corporate venturing, and finally high-growth start-ups in particular.

5.4.1 NEW FIRM FORMATION

The total supply of entrepreneurs varies among societies due to different prevailing values and beliefs related to entrepreneurship, i.e., its entrepreneurial culture. Economists generally share the opinion that it is not the role of government to change the attitude of its people, perhaps even leading to "social engineering" (Storey 2002), or that public policy cannot change the culture of a country in order to stimulate the supply of entrepreneurship, on the short term (Baumol 1990). Some economists have argued that entrepreneurship is an omnipresent aspect of human action, and that for economic development to take place, certain institutions must be present in order for the entrepreneurial aspect of human action to flourish (Boettke and Coyne 2003). This omnipresence also means entrepreneurship cannot be the "cause" of economic development: it is caused by proper institutions that channel entrepreneurship in a direction that spurs economic growth (cf. Baumol 1990). Entrepreneurship policy in the Netherlands also includes integrating entrepreneurship in the education system in order to develop entrepreneurial skills and promote an entrepreneurship culture in the long run. The latter aspect is a clear example of the convergence of entrepreneurship and innovation policy, in that a strong entrepreneurial culture is an enabling context for the emergence of innovations.

The other more direct role for public policy involves changing the formal institutions in order to stimulate productive entrepreneurship. Examples of these formal institutions relevant for entrepreneurship are taxation rules, bankruptcy regulations, social security rules, and immigration laws. Enough has been said and done about the first three of these. The latter is less straightforward.

Innovation systems are not closed systems: exploration taking place in one country, might lead to exploitation in another country, and the other way around. This also means transnational entrepreneurs might function as important linksbetween different national innovation systems. This legitimises accelerating changes in the immigration system in order to attract (potential) entrepreneurs from elsewhere (cf. Saxenian 2006) as part of – person-level – entrepreneurship policy and innovation policy. This could also involve targeting ambitious expatriates to relocate back to the Netherlands (or to become transnational; with firms registered in the Netherlands and elsewhere) and start an internationally oriented new firm.[24]

5.4.2 TECHNOLOGY START-UPS

Technology start-ups are a particular kind of new firm, which are very important for the diffusion of new technologies, possibly in new applications. The stimulation of new technology-based firms is an example of the convergence of innovation and entrepreneurship policy, as this improves both the creation and diffusion of new technologies and the levels of entrepreneurship (cf. Garnsey 2001).[25] Indirectly, public R&D spending (as part of innovation policy) could increase opportunities for new firms (part of entrepreneurship policy).

As a form of economic organisation, new technology-based firms are ideally suited to minimising the social risk of experimentation in developing applications of new technologies. They have every incentive to find such applications, but at the same time the cost of the inevitable failed experiments is not multiplied by the bureaucrat's temptation to pour good money after bad rather than admit failure (Rosenberg and Birdzell 1986: 312).

There has been much debate in economics as to whether R&D divisions of large firms provide better incentives for innovation than entrepreneurial firms. A study by Kortum and Lerner (2000) showed that the ratio of venture capital (invested in young technology-based firms) to corporate R&D investments in manufacturing is about 3 percent, while venture capital accounts for about 15 percent of industrial innovations (measured by patents). The apparently greater efficiency of venture funding in spurring innovation raises questions about whether industrial R&D spending has been optimally directed or exploited. Jensen (1993) has argued that agency problems have hampered the effective management of major corporate industrial research facilities. Indeed, it appears that many major corporate research facilities are today in the process of being restructured. One striking change is an emphasis by many large firms on adopting (open innovation) programs, such as joint ventures with smaller firms and strategic investment programs, whose structures resemble that of venture capital investment (for an overview, see Rosenbloom and Spencer 1996).

One of the key market failure arguments for innovation policy is that the private return of developing and commercialising a new technology is below the social rate of return (see chapter 1). This situation might be explained if (new technology-based) firms are unable to fully appropriate the rents on their innovations. In such a situation entrepreneurial activities could be lower than socially optimal. Three key problems in this respect are:
- problems of appropriating the rents of the innovation (spillovers);
- financial constraints to invest in the development of new technology (asymmetric information between potential investors and the entrepreneur);
- uncertainty about demand for the new technology/product in the market place.

Government interventions to tackle these problems are:[26]
- securing intellectual property rights;
- stimulating seed capital and early stage capital (with subsidies, tax reductions, or favourable interest rates);
- government as leading customer (or setting standards/certifications).

Intellectual property right policy has become more of a European policy issue (European Patent Office) than a strictly national one. Examples of the second type of interventions are the Small Business Investment Company (SBIC) Program in the US (providing long-term funds with favourable interest rates to private companies that invest in small firms), the Carbon Trust in the UK (co-investor in the low carbon technology field), and the Technopartner program in the Netherlands.[27] The goal of Technopartner is to realise an increase in and improvement of the quality of technology start-ups by mobilising the risk capital market for technology start-ups through the Seed Facility (comparable to the SBIC Program in the US). This facility accommodates loans to private investment funds. Technopartner also offers direct support and financial scope to new technology-based firms and stimulates knowledge institutions to professionalise their patent policy through the Subsidy programme Knowledge Exploitations (SKE).

An example of the third type of intervention is the Small Business Innovation Research (SBIR) program in the US.

The US Congress enacted the SBIR program in the early 1980s as a response to the loss of American competitiveness in global markets, especially in the face of the 'Japanese threat'. The US regulation underpinning the SBIR programme requires that 2.5% of all federal government agencies' external R&D budgets are distributed to innovative small firms. Each year the SBIR program makes over 4,000 awards to US small firms, amounting to over $2 billion in value (Connell 2006). The SBIR consists of three phases. Phase I is oriented towards determining the scientific and technical merit (technological creativity) along with the feasibility (economic creativity) of a proposed research idea. A Phase I award (typically around $100 k) provides an opportunity for a small business to establish the feasibility and technical merit of a proposed innovation. This is a venture capital segment – seed capital – that is generally ignored by private venture capital. Phase II awards are more selectively aimed at developing new technologies and products, which involves about 50% of the phase I award winners, and delivers up to $750 k. Phase III awards are funded from mainstream (i.e., non SBIR budgets), and add probably again as much as Phase I and II in total to overall R&D expenditure on SBIR projects (Connell 2006). These Phase III projects also bring small firms the opportunity to win valuable sole supplier contracts with federal agencies. Some of the most innovative American companies, like Genzyme, Amgen, Genentech, and Qualcomm received early stage SBIR finance. Lerner (1999) showed that SBIR funded firms enjoyed substantially greater employment and sales growth than other similar firms. It is not just the size of the subsidy that is important for the recipients: these awards also play an important certification function, increasing the trustworthiness of the recipients.

Programs like the SBIR are highly valuable in making governments lead users/customers for innovative new and small firms in 'public sectors'. There are indications that the Netherlands is lagging behind with respect to this kind of government procurement of advanced technological products (Porter et al. 2007).

5.4.3 SPIN-OFFS AND CORPORATE VENTURING

We discussed the conditions under which spin-offs are more likely than corporate venturing as a means of exploitation. Both types of entrepreneurial effort involve an entrepreneurial opportunity that is based on the existing knowledge base of the parent firm. Spin-offs and corporate venturing are thus likely to be the vehicles of innovation, and sources of so-called related variety: exploring new options based on existing capabilities (see chapter 11). In addition spin-offs provide new nodes in an inter-organisational network (see chapter 10). There has been much public policy and media attention for university spin-offs, while corporate spin-offs and corporate venturing have largely been neglected. Corporate spin-offs are both more numerous and more likely to be successful (due to better capabilities and market orientation: exploration based on exploitation), while corporate venturing has a direct positive effect on the innovativeness of large firms, which is likely to lead to an increase in macroeconomic growth. Should public policy pay more attention to spin-offs and corporate venturing; if yes, how?[28]

There are several problematic issues in the practice of corporate venturing, which explain why corporate venturing units often fail (Birkinshaw and Hill 2005). For example, it has been argued that mixing both corporate-trained people and people from the venture capital world is one of the key factors behind corporate venturing (Birkinshaw and Hill 2005; Ernst et al. 2005); however, many companies are not willing to have venture capitalists as 'intruders' in their business. This might constrain the development of high-potential new businesses. Another delicate issue is the appropriation of the rents from innovation: how are these distributed over the top-management, corporate venturing unit staff, and the employees that started the initiative? In addition, how and when will the new venture be integrated into the incubating organisation? If the venture is integrated too early, the innovation might die because of the parent's bureaucracy, while if it is integrated too late, the venture has developed an identity and routines of its own, which constrain re-integration.

"Corporate venturing is never easy, but it is beginning to be recognised as something that far-sighted companies cannot do without" (Birkinshaw et al 2002: 43).

Stimulating an entrepreneurial culture is not only a mission of the government: large firms also play a role here, for example with their corporate venturing practices. Corporate venturing could lower the risk averseness in large firms, but this should also include the alignment of incentives and the attitude towards failure. Large firms' corporate venturing practices will only create value in the long term if the entire organisation has been made more 'entrepreneurial'. In addition, corporate venturing is a means to stimulate investment behaviour and in the end the innovativeness of large firms. This should not be a one-off action, but should be a structural element of corporate strategy ('serial corporate ventures', like Intel (external VC) and 3M (internal VC)). Corporate venturing can also create a platform where entrepreneurs, managers, and investors can find each other. Entrepreneurs may acquire the (technical and commercial) expertise of large firms and the financial means of (corporate or private) venture capitalists.

Stimulating entrepreneurship in existing firms (corporate entrepreneurship) is probably the best policy to start with: stimulating entrepreneurial initiatives that have access to a large range of resources within or via the mother firm. However, some characteristics of large firms (inertia, incentive structures) make this an unlikely success. That is why stimulating spin-offs might be a second-best policy option. To some extent this is also easier to affect by government policy as it is very much related to labour-market and competition regulation. One important entry barrier especially for growth-oriented spin-offs is anti-competition clauses (also called 'non-compete agreements'). Government regulation might reduce the use of these clauses. However, this might have negative effects on investments of firms in their employees and of venture capitalists in their portfolio firms, as the (potential) entrepreneur can leave to join competitors or set up another firm with knowledge from the proprietary organisation (cf. Gilson 1999; Fallick et al. 2006).

Corporate venturing might be connected to universities in order to tap their knowledge and look for commercial applications. A related trend can be identified in the UK where university incubators have become private organisations (like ISIS Oxford, Imperial Innovations; see Library House 2007). This privatisation of university incubators both provides these organisations with more competent staff and 'purifies' the university as producers of public knowledge (cf. Lester and Piore 2004).

5.4.4 HIGH GROWTH START-UPS[29]

High growth start-ups are the economic entities that are successful in commercialising new ideas on a large scale in a short term. These firms are serious candidates for the industrial leadership of tomorrow. However, the contradiction for public policy is that policymakers grant themselves an important role in stimulating these (potentially) successful firms (Smallbone et al. 2002; MinEZ 2006), but that these firms themselves regard government intervention as a very marginal influence on their success (see Fischer and Reuber 2003; Perren and Jennings 2005; Te Peele and Brummelkamp 2007).

This brings us back to the discussion of picking the winners or backing the challengers.[30] Challengers are ambitious new firms that have already revealed some of their potential to become leaders in their (sector or geographically unique) niche.[31] Leadership in a sector unique niche involves the creation of a new technology or product that might be diffused globally (e.g., the development of new ERP software by BAAN, or digital route planners by TomTom), while leadership in a geographically unique niche (i.e., the Netherlands as a whole or a specific region initially) is likely to involve the application or adoption of technology or product that has been developed in another country (e.g., the early leaders in the Dutch Internet Service Providers industry: XS4All and Planet Internet). The latter type of leadership is not based on pushing the technological frontier (technology push), but on advanced users that spur the development and diffusion of new applications.

One measure to stimulate high risk and potentially high-growth start-ups is to lower the barriers to incorporation. Incorporation provides a means to safeguard ambitious entrepreneurs from the high risks of growing a new business, by making the incorporated firm and not the entrepreneur responsible for the eventual financial losses. Lowering the barriers to incorporation is likely to stimulate firm growth (Storey 1997), but also leads to higher bankruptcy rates (Harhoff et al. 1998). The net effect of these two outcomes – increase of the number of high-growth start-ups and of the number of bankruptcies – on the societal level is still unknown.

Research has shown that probably the best that entrepreneurship policy could do for young high-growth firms is to stimulate communities of practice for entrepreneurs leading (potentially) high-growth firms (Smallbone et al. 2002; Fischer and Reuber 2003). Such peer networks have already been initiated in many ways, and are also facilitated by the Dutch Ministry of Economic Affairs (see Waasdorp and Bakkenes 2006).

5.5 SUMMARY

This chapter has provided insights into the nature of entrepreneurship and its role in innovation and economic growth. We have defined entrepreneurship as the introduction of new economic activity by an individual that leads to change in the marketplace. Innovation is a necessary condition of entrepreneurship, just like the existence of entrepreneurial opportunities and heterogeneous risk-taking individuals that organise the exploitation of these opportunities. Entrepreneurial opportunities can emerge from major scientific breakthroughs, but also from more mundane applications of existing solutions in new contexts. The particular characteristics of the opportunities, the organisations in which they were discovered and the persons that perceived them affect the trade-off between exploiting these opportunities in the incumbent firm (corporate venturing) or in a spin-off firm. What matters for macroeconomic growth is the performance of these entrepreneurial efforts. Both new firm formation in general and high-growth start-ups in particular play their role in economic growth. New firms are useful devices for exploring the viability of innovations. They start small, with relatively low costs and in large numbers. The provision of requisite variety comes at a cost, because most of these new firms fail. However, without these large numbers of experiments, breakthrough innovations and new industries and industry leaders are less likely to emerge. Still, if this pool of experiments does not lead to a sufficient number of high-growth start-ups, they are less likely to lead to economic growth. They might have other indirect effects on economic growth by providing knowledge about 'failed experiments' to incumbent firms, and acting as a potential threat to incumbents that will spur corporate renewal. Investments in the knowledge base of existing organisations also provide a source of entrepreneurial opportunities, to be commercialised by knowledge-based new firms. In this perspective new firms provide the missing link between investments in science and technology and economic growth.

We have zoomed in on the development and presence of diverse forms of entrepreneurship in the Netherlands. Even though the annual number of new firms has increased spectacularly (almost tripled) during the period 1987-2006, there are more weaknesses than strengths with regards to entrepreneurship in the Netherlands. First, a large part of the population of new firms seems to involve self-employed individuals who continue with the same activities (mainly in the construction and services sectors), which they had carried out as employees before. This is not likely to improve the (product) innovativeness of the economy at all. Second, on average SMEs have become less and not more innovative in the last decade, and the percentage of innovative SMEs is much lower than the EU average. Third, the Netherlands is lagging behind internationally with respect to ambitious entrepreneurship. This is not a recent phenomenon, but seems to be consistent in the last decade, in spite of several policy measures that have been taken to improve these numbers. The Netherlands seems to have become a good place for self-employed people with low ambitions, but this has not been accompanied by a rise in innovativeness and high-growth firms. Especially the low number of ambitious entrepreneurship seems to be worrisome, as this is a strong driver of national economic growth.

What is the role of entrepreneurship in innovation policy? We have stated two important disclaimers concerning the measurement (not everything that is counted as entrepreneurship concerns innovation) and systemic effects (more entrepreneurship does not always mean more economic growth) of entrepreneurship. A small set of specific types of entrepreneurship – technology start-ups, spin-offs, and corporate venturing, and high-growth start-ups – seems to be more relevant for innovation than other types. The capital provision for technology start-ups is already stimulated by a set of policy measures within the Technopartner program in the Netherlands. Implementation of a SBIR-like program would probably lower the uncertainty about demand for new technologies and further stimulate the emergence and growth of new technology-based firms. Corporate venturing is a highly important means to stimulate the innovativeness of incumbent firms, but unfortunately corporate venturing strategies seem to be less often successful than is needed for a dynamic innovation system. A second best solution in this respect involves – university and especially corporate – spin-offs. The last but not least important type of entrepreneurship that is discussed in the context of innovation policy is the high-growth start-up. These firms probably need the least government support, but policy could play a role here in facilitating communities of (high-growth) practice that stimulate peer-to-peer learning of the entrepreneurs involved.

NOTES

1 The aims of entrepreneurship policy in the Netherlands are to increase employment, the flexibility and innovativeness of the economy, individual development, emancipation, and integration (Rekenkamer 2002).

2 In a similar way, entrepreneurship is often equated with self-employment and SMEs in other EU documents (EOS Gallup Europe 2004; European Commission 2006b)

3 See Phillips (1985) for evidence on the positive effects of deregulation on new firm formation in the US and Berkowitz and Holland (2001) on the positive effects of privatisation on new firm formation in Russia.

4 This was the first mountain bike producing firm ever (called MountainBikes). This firm dissolved in 1983, the year in which Fisher founded his better-known company Fisher MountainBikes, which was acquired by Trek in 1993.

5 In addition, a risk averse firm is unlikely to offer return options for the entrepreneur if the new venture fails.

6 This is in line with a study by Wennekers et al. (2007) which shows that uncertainty avoidance is positively correlated with the prevalence of business ownership: a restrictive climate of large organizations in high uncertainty avoidance countries pushes individuals striving for autonomy towards self-employment.

7 According to Pasinetti (1993) an economy that does not increase the variety of industries over time will suffer from structural unemployment, and will ultimately stagnate. In this view, the development of new industries in an economy is required to absorb labour that has become redundant in pre-existing industries. This labour has become redundant due to a combination of productivity increases and demand saturation in pre-existing industries, characterising the product lifecycle dynamics in each sector.

8 In comparison to other small economies like Belgium, Denmark, and Ireland, the Netherlands has a 'water head': relatively many large dominant firms (the Netherlands had the most Fortune 500 firms per capita in 2006, after Luxembourg and Switzerland), which have a more than proportional influence on public policy, and have received a more than proportional part of government spending. See Banning and Meeus (1998) for the role of Philips in the Netherlands. Studies in the US have shown the negative effect of large firm power on small firm innovation (Christopherson and Clark 2007) and on new firm formation (Choi and Phan 2006). This overrepresentation of large dominant firms is likely to constrain the experimental nature of the Dutch economy. More research is needed to confirm (or reject) this hypothesis.

9 The very limited variety of products that was available in communist economies seems to confirm this generalisation. Wealth can even be defined as the range of choice people have, not just the quantity of supply.

10 Funke and Ruhwedel (2001) found that a greater degree of product variety is highly correlated to per capita GDP levels and TFP growth in OECD countries.

11 Even when controlled for recent macroeconomic growth and time lags of the effect on economic growth. Prior economic growth has positive and negative

relations with entrepreneurship rates: positive because of growth opportunities ('prosperity-pull'), and negative because unemployed workers are encouraged to become self-employed because the opportunity costs of self-employment have decreased ('recession-push') (see Thurik et al. 2008).

12 In what they call 'revolving door' regimes: inefficient entrants, which exit soon after entry because they cannot make a valuable contribution to the economy.

13 The studies of Audretsch and Keilbach find no (2005), or only very weak (2004) associations of new firm formation in general- and labour-productivity growth. Only specific forms of entrepreneurship, like new firm formation in high-tech or ICT industries (i.e., technology start-ups) have strong positive associations with labour productivity growth.

14 The Global Entrepreneurship Monitor makes a distinction between 'necessity entrepreneurship', which is having to become an entrepreneur (often 'self-employed') because you have no better option, and 'opportunity entrepreneurship', which is an active choice to start a new enterprise based on the perception that an unexploited or underexploited business opportunity exists. Analyzing data in 11 countries, Acs and Varga (2005) found that effects on economic growth and development of necessity and opportunity entrepreneurship vary greatly: necessity entrepreneurship has no effect on economic development while opportunity entrepreneurship has a positive and significant effect. They also found that the ratio of opportunity to necessity entrepreneurship in a country is positively related to GDP per capita.

15 'TEA' reflects the percentage of the adult population that is actively involved in setting up a business or owns a young (<42 months old) firm; 'percentage ambitious entrepreneurs' reflects the percentage of the adult population that is actively involved in or owns a young firm and have the ambition to expand to a size of more than 20 employees (Autio 2007).

16 This might be less bad for the economy if this is caused by new forms of governance, like network organisations: entrepreneurs may have low ambitions for employment growth, but high ambitions in sales growth and growth in value added, realised as a self-employed individual in a network organisation. However, it is far from clear that this phenomenon is over-represented in the Netherlands, and that this causes the relatively low number of entrepreneurs with employment growth ambitions in the Netherlands.

17 The high percentage of self-employed individuals in the Netherlands can partly be explained by the relatively high survival rates of new firms in the Netherlands, in comparison with for example the US (Bartelsman et al. 2005).

18 A low-performing subsidiary may eventually go bankrupt without dragging along the corporation as a whole.

19 Unfortunately the data for 2001 was not available for figures 10 and 11.

20 A business cycle explanation of these dynamics does not seem to match the data, as there had already been a strong downward trend before the post-2000 recession started. In addition, research has shown that in general innovation behaviour is pro-cyclical, but that innovation behaviour of SMEs is countercyclical (Heger 2004). This latter phenomenon is explained by the lack of access to qualified labour for SMEs in boom periods.

21 More research is needed to disentangle different types of self-employment and their direct and indirect effects on innovation and economic growth. An interesting question in the Dutch context is whether the institutions that have fostered flexibility and modest wage increases have improved self-employment (static efficiency) but not innovation (dynamic efficiency) (cf. Engelen 2002).

22 This means that it is unlikely that there will be institution-free descriptions of best practices in entrepreneurship and innovation policy (cf. Chesbrough 1999, 2003). Sweden, for example, is renowned for the economic dominance of large firms and is often cited in comparative entrepreneurship studies as a country whose institutional environment discourages entrepreneurship (see also figures 3 and 4 in this chapter), however, it is a leader with respect to innovative entrepreneurship indicators (European Commission 2006a; Parker 2006).

23 In addition to this, empirical research has shown that entrepreneurs that have started multiple firms ('habitual entrepreneurs') are more often innovative than one-off entrepreneurs (Ucbasaran et al. 2008).

24 More research is needed to gain insight into the effects of migration of highly skilled labour – and the subsequent transnational entrepreneurship – on innovation and economic growth.

25 New technology-based firms are more likely to be innovative (due to the technologically dynamic nature of the industries in which they are active), and are likely to involve viable businesses due to the high-opportunity costs of the founders (leaving behind secure well-paid jobs).

26 These are very general policy implications: they do not indicate how large the subsidies should be or within which specific area one should intervene, and they also do not say which policy instruments should be given priority. Unfortunately, the systems of innovation approach does not do better than the market failure theory in providing specific policy implications (see Edquist 2001).

27 See Boot and Schmeits (2004) for a useful review of imperfections in the capital market for innovative firms, and public policy in the Netherlands.

28 For example, the UK government introduced tax relief for corporate venturing in 1999. However, one could doubt the effectiveness of such a measure, given its low level of additionality: i.e., if corporate venturing is important to (especially large) firms, they will do it anyway, and only the accountant will notice the advantage of this tax relief.

29 See Stam et al. (2007) for an overview of the literature on the economic importance of high-growth entrepreneurs and the rationales for public policy.

30 The so-called *uitdagerskrediet* (challengers credit) partly covers this field. It is a public credit facility for financing challenging innovation projects by SMEs, in which new goods, processes, or services will be developed, which will lead to fast and substantial growth of the firm.

31 Focus on post-entry policies minimises the risk of waste and the possible substitution effect of government subsidies. A possible deadweight effect should be avoided by the identification of those firms that are good enough to survive but not strong enough to grow (i.e., that do not have the financial means to invest substantially) (cf. Santarelli and Vivarelli 2007).

REFERENCES

Acs, Z. (2006) 'How Is Entrepreneurship Good for Economic Growth?', *Innovations: Technology, Governance, Globalization*, 1(1): 97-107.

Acs, Z. J. and C. Armington (2004) 'Employment Growth and Entrepreneurial Activity in Cities', *Regional Studies*, 38: 911-927.

Acs, Z., D. Audretsch, P. Braunerhjelm, and B. Carlsson (2005) 'The Knowledge Spillover Theory of Entrepreneurship', CEPR Discussion Paper No. 5326, CEPR: London.

Acs, Z.J. and P. Mueller (2008) 'Employment Effects of Business Dynamics: Mice, Gazelles and Elephants', *Small Business Economics*, 30(1): 85-100.

Acs, A. J. and A. Varga (2005) 'Entrepreneurship, Agglomeration and Technological Change', *Small Business Economics*, 24(3): 323-334.

Aghion, Ph., R. Blundell, R Griffith, P. Howitt, and S. Prantl (2006) 'The Effects of Entry on Incumbent Innovation and Productivity'. NBER Working Paper 12027, Cambridge, MA: NBER.

Aghion, Ph. and P. Howitt (1992) 'A Model of Growth Through Creative Destruction', *Econometrica*, 60(2): 323-351.

Audretsch, D.B. (1995) *Innovation and Industry Evolution*, Cambridge, MA: MIT Press.

Audretsch, D.B. and R. Thurik (2001a) *Linking Entrepreneurship to Growth*, OECD Science, Technology and Industry Working Papers 2001/2, OECD, Paris.

Audretsch, D.B. and R. Thurik (2001b) 'What is New About the New Economy: Sources of Growth in the Managed and Entrepreneurial Economies', *Industrial and Corporate Change*, 10(1): 25-48.

Audretsch, D.B. and M. Fritsch (2002), 'Growth Regimes over Time and Space', *Regional Studies*, 36: 113-124.

Audretsch, D.B. and M. Keilbach (2004) 'Entrepreneurship and Regional Growth: An Evolutionary Interpretation', *Journal of Evolutionary Economics*, 14(5): 605-616

Audretsch, D.B. and M. Keilbach (2005) 'Entrepreneurship Capital and Regional Growth', *Annals of Regional Science*, 39: 457-469.

Audretsch, D.B., M.C. Keilbach, and E.E. Lehmann (2006) *Entrepreneurship and Economic Growth*, Oxford University Press: Oxford.

Autio, E. (2007) *GEM 2007 High-Growth Entrepreneurship Report*, Global Entrepreneurship Consortium: London.

Banning, C. and T.J. Meeus (1998) *De onzichtbare hand van de politiek. De achterkamers van Paars en de heimelijke steun aan Philips*. Amsterdam: Balans.

Bartelsman, E., S. Scarpetta, F. Schivardi (2005) 'Comparative Analysis of Firm Demographics and Survival: Evidence from Micro-level Sources in OECD Countries', *Industrial and Corporate Change*, 14(3): 365-391.

Baumol, W.J. (1990) 'Entrepreneurship: Productive, Unproductive, and Destructive', *Journal of Political Economy*, 98(5): 893-921.

Baumol, W. (1993) 'Formal Entrepreneurship Theory in Economics: Existence and Bounds', *Journal of Business Venturing*, 8: 197-210.

Baumol, W.J. (2002) *The Free-Market Innovation Machine – Analyzing the Growth Miracle of Capitalism*, Princeton and Oxford: Princeton University Press.

Benz, M. (2006) 'Entrepreneurship as a Non-profit-seeking Activity', Institute for Empirical Research in Economics Working Paper #243, University of Zurich.
Benz, M. and B.S. Frey (2003) 'The Value of Autonomy: Evidence from the Self-Employed in 23 Countries', Institute for Empirical Research in Economics Working Paper #173, University of Zurich.
Berkowitz, D. and J. Holland (2001) 'Does Privatisation Enhance or Deter Small Enterprise Formation?' *Economics Letters* 74: 53-60.
Birkinshaw, J., R. Van Basten Batenburg, and G.C. Murray (2002) *Corporate Venturing: The State of the Art and the Prospects for the Future*, London: London Business School.
Birkinshaw, J. and S.A. Hill (2005) 'Corporate Venturing Units: Vehicles for Strategic Success in the New Europe', *Organizational Dynamics,* 34(3): 247-257.
Blanchflower, D.G. and A.J. Oswald (1998) 'What Makes an Entrepreneur?', *Journal of Labor Economics*, 16(1): 26-60.
Boettke, P. and C.J. Coyne (2003) 'Entrepreneurship and Development: Cause or Consequence?', *Advances in Austrian Economics,* 6: 67-88.
Boot, W.A. and A. Schmeits (2004) 'Imperfecties in de vermogensmarkt en overheidsbeleid' in: Jacobs, B. and J.J.M. Theeuwes (eds.) (2004) *Innovatie in Nederland: De markt draalt en de overheid faalt.* Amsterdam: Koninklijke Vereniging voor de Staatshuishoudkunde, 95-136.

Bosma, N., K. Jones, E. Autio, and J. Levie (2008) *Global Entrepreneurship Monitor 2007.* Babson Park, MA: Babson College.
Bosma, N. and V. Schutjens (2008) 'Mapping Entrepreneurial Activity and Entrepreneurial Attitudes in European Regions', *International Journal of Entrepreneurship and Small Business* (forthcoming).
Bosma, N., E. Stam, and V. Schutjens (2006), *Creative Destruction and Regional Competitiveness*, Zoetermeer: EIM Business and Policy Research.
Braaksma, R. and J. Meijaard (2007) *Entrepreneurship and innovation. Trends and patterns in the Netherlands*, Zoetermeer: EIM.
Buenstorf, G. (2001) 'Designing Clunkers: Demand-Side Innovation and the Early History of the Mountain Bike', Max Planck Institute of Economics, Papers on Economics and Evolution #0105.
Callejón, M. and A. Segarra (1999), 'Business Dynamics and Efficiency in Industries and Regions: The Case of Spain', *Small Business Economics,* 13: 253-271.
Carree, M. and R. Thurik (2007) 'The Lag Structure of the Impact of Business Ownership on Economic Performance in OECD Countries', *Small Business Economics* (forthcoming).
Casson, M. (2003) *The Entrepreneur. An Economic Theory*,2nd edition, Cheltenham: Edward Elgar.
Coalitieakkoord (2007) *Samen Werken, Samen Leven. Coalitieakkoord tussen de Tweede Kamerfracties van CDA, PvdA en ChristenUnie.*
CBS (2006) *Kennis en economie.* Voorburg: CBS.
CBS (2007) *Het Nederlandse ondernemingsklimaat in cijfers 2007.* Voorburg: CBS.
Chesbrough, H. (1999) 'The Differing Organizational Impact of Technological Change: A Comparative Theory of National Institutional Factors', *Industrial and Corporate Change,* 8(3): 447-485.

Chesbrough, H. (2003) 'The Governance and Performance of Xerox's Technology Spinoff Companies', *Research Policy*, 32(3): 403-421.

Chesbrough, H. and R. S. Rosenbloom (2002) 'The Role of the Business Model in Capturing Value from Innovation: Evidence from Xerox Corporation's Technology,' *Industrial and Corporate Change*, 11(3): 529-555.

Choi, Y.R. and P. H. Phan (2006) 'The Influences of Economic and Technology Policy on the Dynamics of New Firm Formation', *Small Business Economics*, 26: 493-503.

Christensen, C. (1997) *The Innovator's Dilemma*, Boston: Harvard Business School Press.

Christopherson, S. and J. Clark (2007) 'The Politics of Firm Networks: How Large Firm Power Limits Small Firm Innovation', *Geoforum* 38: 1-3.

Connell, D. (2006) 'Secrets' of the World's Largest Seed Capital Fund: How the United States Government Uses its Small Business Innovation Research (SBIR) Programme and Procurement Budgets to Support Small Technology Firms', Centre for Business Research, University of Cambridge: Cambridge, UK.

Davidsson, P. (2004) *Researching Entrepreneurship*. New York: Springer.

De Jong, J. and B.H.G. Jansen(2007) *Innovatie in het MKB. Ontwikkelingen 1999-2007.* Zoetermeer: EIM.

Dirks, F., M. Rosenbrand and N. Bosma (2003) *De overstap naar het ondernemerschap: levensloop, beweegredenen en obstakels.* Tilburg: Organisatie voor Strategisch Arbeidsmarktonderzoek.

Dosi, G. and D. Lovallo (1997) 'Rational Entrepreneurs or Optimistic Martyrs? Some Considerations on Technological Regimes, Corporate Entries and the Evolutionary Role of Decision Biases' in: R. Garud, P.R. Nayyar and Z.B. Shapira (eds.) *Technological Innovation Oversights and Foresights*, Cambridge/New York: Cambridge University Press.

Edquist, C. (2001)' Innovation Policy – A Systemic Approach' in: Archibugi, D. and B.A. Lundvall (eds.) *The Globalizing Learning Economy*, Oxford: Oxford University Press, 219-238.

EIM (2006) *Internationale Benchmark Ondernemerschap 2005*. Zoetermeer: EIM Business and Policy Research.

Engelen, E. (2002) 'How Innovative are Dutch Immigrant Entrepreneurs: Constructing a Framework of Assessment', *International Journal of Entrepreneurial Behaviour & Research*, 8(1/2): 69-92.

EOS Gallup Europe (2004) 'Flash 160 "Entrepreneurship"', survey for Directorate General Enterprise – European Commission. Brussels: EOS Gallup.

Ernst, H., P. Witt and G. Brachtendorf (2005) 'Corporate Venture Capital as a Strategy for External Innovation: An Exploratory Empirical Study', *R&D Management*, 35(3): 233-242.

European Commission (2006a) *European Trend Chart on Innovation. Annual Innovation Policy Trends and Appraisal Report - The Netherlands, 2006*, Brussels: European Commission Enterprise Directorate-General.

European Commission (2006b) *Report on the implementation of the Entrepreneurship Action Plan*, Brussels: European Commission – DG Enterprise and Industry.

Fallick, B., C.A. Fleischman, and J.B. Rebitzer (2006) 'Job-Hopping in Silicon Valley: Some Evidence Concerning the Microfoundations of a High-Technology Cluster', *Review of Economics and Statistics*, 88: 472-481.

Fischer, E. and A.R. Reuber (2003) 'Support for Rapid-Growth Firms: A Comparison of the Views of Founders, Government Policymakers, and Private Sector Resource Providers', *Journal of Small Business Management,* 41: 346-365.

Foray, D. (2004) *The Economics of Knowledge*, Cambridge, MA: MIT Press.

Frey, B.S. and M. Benz (2003) 'Being Independent is a Great Thing: Subjective Evaluations of Self-Employment and Hierarchy', CESifo Working Paper #959, Munich, Germany.

Funke, M. and R. Ruhwedel (2001) 'Product Variety and Economic Growth: Empirical Evidence for the OECD Countries', *IMF Staff Papers*, 48(2): 225-242.

Garnsey, E. (2001) *Entrepreneurship in the Netherlands. New economy: new entrepreneurs!* EIM: Zoetermeer.

Garnsey E. and P. Heffernan (2005) 'High Tech Clustering through Spin Out and Attraction: The Cambridge Case', *Regional Studies,* 39(8): 1127-1144.

Gilson, R.J. (1999) 'The Legal Infrastructure of High Technology Industrial Districts: Silicon Valley, Route 128, and Covenants Not to Compete', *New York University Law Review,* 74(3): 575-629.

Hamilton, B.H. (2000) 'Does Entrepreneurship Pay? An Empirical Analysis of the Returns to Self-employment', *Journal of Political Economy,* 108: 604-631.

Harhoff, D., K. Stahl, and M. Woywode (1998) 'Legal Form, Growth and Exit of West German Firms – Empirical Results for Manufacturing, Construction, Trade and Service Industries', *Journal of Industrial Economics,* 46(4): 453-488.

Hebert, R.F. and A. N. Link (1989) 'In Search of the Meaning of Entrepreneurship', *Small Business Economics,* 1(1): 39-49

Heger, D. (2004) 'The Link Between Firms' Innovation Decision and the Business Cycle: An Empirical Analysis', ZEW discussion paper No. 04-85, ZEW: Mannheim.

Hekkert, M.P., R.A.A. Suurs, S.O. Negro, S. Kuhlmann, and R.E.H.M. Smits (2007) 'Functions of Innovation Systems: A New Approach for Analysing Technological Change', *Technological Forecasting & Social Change,* 74: 413-432.

Helpman, E. (2004) *The Mystery of Economic Growth*, Cambridge, MA: The Belknap Press of Harvard University Press.

Henderson, R.M. and K.B. Clark (1990) 'The Reconfiguration of Existing Product Technologies and the Failure of Established Firms', *Administrative Science Quarterly,* 35(1): 9-30.

Hoffmann, A.N. and M. Junge (2006) 'Documenting Data on High-growth Firms and Entrepreneurs across 17 Countries', mimeo, Fora Copenhagen.

Hoffmann, A. (2007) 'A Rough Guide to Entrepreneurship Policy' in: Audretsch, D.B., I. Grilo, and A.R. Thurik (eds.) *Handbook or Research on Entrepreneurship Policy*, Cheltenham: Edward Elgar, 140-171.

Holbrook, D., W.M. Cohen, D.A. Hounshell, and S. Klepper (2000) 'The Nature, Sources, and Consequences of Firm Differences in the Early History of the Semiconductor Industry', *Strategic Management Journal*, 21: 1017-1041.

Kenney, M. (1986) 'Schumpeterian Innovation and Entrepreneurs in Capitalism: A Case Study of the US Biotechnology Industry', *Research Policy,* 15: 21-31.

Intermediair (2007) 'Ik ben maar een uithangbord' interview with Raymond Spanjar, *Intermediair,* August 10, 26-29.

Jensen, M.C. (1993) 'Presidential Address: The Modern Industrial Revolution, Exit, and

the Failure of Internal Control Systems', *Journal of Finance*, 48, 831-880.
Kirzner, I.M. (1973) *Competition and Entrepreneurship*, University of Chicago Press: Chigago.
Kirzner, I.M. (1997) 'Entrepreneurial Discovery and the Competitive Market Process: An Austrian Approach', *Journal of Economic Literature*, 35(1): 60-85.
Klepper, S. (2001) 'Employee Startups in High-Tech Industries', *Industrial and Corporate Change*, 10(3): 639-674.
Knight, F. H. (1921) *Risk, Uncertainty and Profit*. Boston, MA: Hart, Schaffner & Marx.
Kortum, S. and J. Lerner (2000) 'Assessing The Contribution Of Venture Capital To Innovation', *Rand Journal of Economics*, 2000, 31(4): 674-692.
Landes, D. S. (1969) *The Unbound Prometheus: Technological Change and Industrial Development in Western Europe from 1750 to the Present*. Cambridge: Cambridge University Press.
Langlois, R.N. and P.L. Robertson (1995) *Firms, Markets, and Economic Change*, London: Routledge.
Layard, R. (2005) *Happiness: Lessons from a New Science*, London: Penguin Books.
Lerner, J. (1999) 'The Government as Venture Capitalist: The Long-run Effects of the SBIR Program', *Journal of Business*, 72(3): 285-297.
Lester, R.K. and M.J. Piore (2004) *Innovation – The Missing Dimension*, Cambridge, MA: Harvard University Press.

Library House (2007) *Spinning Out Quality: University Spin-out Companies in the UK*, Cambridge: Library House.
March, J.G. (1991) 'Exploration and Exploitation in Organizational Learning', *Organization Science*, 2(1): 71-87.
Marshall, A. (1890) *Principles of Economics*, London: Macmillan.
Moore, G. and K. Davis (2004) 'Learning the Silicon Valley Way' in: T. Bresnahan and A. Gambardella (eds.) *Building High-tech Clusters: Silicon Valley and Beyond*, Cambridge: Cambridge University Press, 7-39.
Mowery, D.C. and N. Rosenberg (1993) 'The US National Innovation System' in: R.R. Nelson (ed.) *National Innovation Systems: A Comparative Analysis*, New York: Oxford University Press, 29-75.
Mueller, P., A. Van Stel, and D.J. Storey (2008) 'The Effects of New Firm Formation on Regional Development Over Time: The Case of Great Britain', *Small Business Economics* (forthcoming).
Murphy, K.M., A. Schleifer, and R.W. Vishny (1991) 'The Allocation of Talent: Implications for Growth', *Quarterly Journal of Economics*, 106 (2): 503-30.
Nooteboom, B. (1992) 'Towards a Dynamic Theory of Transactions', *Journal of Evolutionary Economics*, 2:281-299.
Nooteboom, B. (1994) 'Innovation and Diffusion in Small Firms: Theory and Evidence', *Small Business Economics*, 6: 327-347.
Nooteboom, B. (2000) *Learning and Innovation in Organizations and Economies*, Oxford, Oxford University Press.
Parker, R. (2001) 'The Myth of the Entrepreneurial Economy: Employment and Innovation in Small Firms', *Work, Employment & Society*, 15(2): 373-384.
Parker, R. (2006) 'Small Business and Entrepreneurship in the Knowledge Economy: A Comparison of Australia and Sweden', *New Political Economy* 11(2): 201-226.

Parker, S.C. (2004) *The Economics of Self-employment and Entrepreneurship*, Cambridge: Cambridge University Press.
Pasinetti, L.L. (1993) *Structural Economic Dynamics*, Cambridge: Cambridge University Press.
Perren, L. and P.J. Jennings (2005) 'Government Discourses on Entrepreneurship: Issues of Legitimization, Subjugation, and Power', *Entrepreneurship Theory and Practice*, 29(2): 173-184.
Phillips, B.D. (1985) 'The Effect of Industry Deregulation on the Small Business Sector', *Business Economics*, 20: 28-37.
Porter, M. (1980) *Competitive Strategy*, New York: The Free Press.
Porter, M.E., K. Schwab, and X. Sala-i-Martin (2007) *The Global Competitiveness Report 2007-2008*, London: Palgrave MacMillan.
Praag, C.M. van and P.H. Versloot (2007) 'What is the Value of Entrepreneurship? A review of Recent Research', *Small Business Economics* 29(4): 351-383.
Rekenkamer (2002) *Stimulering van midden- en kleinbedrijf en Ondernemerschap*. The Hague: SDU Uitgevers.
Rogers, E.M. and J.K. Larsen (1984) *Silicon Valley Fever*, New York: Basic Books.
Romer, P. (1990) 'Endogenous Technological Change', *Journal of Political Economy*, 98: 71-102.
Rosenberg, N. and L. Birdzell (1986) *How the West Grew Rich*, New York: Basic Books.
Rosenbloom, R.S., and W.J. Spencer (1996) *Engines of Innovation: U.S. Industrial Research at the End of an Era*, Boston: Harvard Business School Press.
Sarasvathy, S. D. (2000) 'Report on the Seminar on Research Perspectives in Entrepreneurship', *Journal of Business Venturing*, 15(1): 1-58.
Saxenian, A. (1994) *Regional Advantage: Culture and Competition in Silicon Valley and Route 128*, Cambridge, MA: Harvard University Press.
Saxenian, A. (2006) *The New Argonauts: Regional Advantage in a Global Economy*, Cambridge, MA: Harvard University Press.
Schultz, T. (1975) 'The Value of the Ability to Deal with Disequilibira', *Journal of Economic Literature*, 13(3): 827-846.
Schumpeter, J.A. (1934) *The Theory of Economic Development*, Cambridge MA: Harvard University Press.
Schumpeter, J. A. (1942) *Capitalism, Socialism, and Democracy*, New York: Harper and Brothers.
Sen, A. (1999) *Development as Freedom*. Oxford: Oxford University Press.
Shahid Y., and N. Kaora (2007) *How Universities Promote Economic Growth*, Washington, DC: World Bank.
Shane, S.A. (2000) 'Prior Knowledge and the Discovery of Entrepreneurial Opportunities', *Organization Science*, 11(4), 448-472.
Shane, S. (2004) *A General Theory of Entrepreneurship: The Individual-Opportunity Nexus*, Cheltenham: Edward Elgar.
Shane, S. and J. Eckhardt (2003) 'The Individual-Opportunity Nexus' In: Z. Acs and D. Audretsch (eds.) *Handbook of Entrepreneurship Research: An Interdisciplinary Survey and Introduction*, Boston: Kluwer Academic Publishers.
Sharma, P. and J.J. Chrisman (1999) 'Toward a Reconciliation of the Definitional Issues in

the Field of Corporate Entrepreneurship',. *Entrepreneurship Theory and Practice,* 23(3): 11-27.
Smallbone, D., R. Baldock, and S. Burgess (2002) 'Targeted Support for High-growth Start-ups: Some Policy Issues', *Environment and Planning C* 20(2): 195-209.
Smith, D.K. and R.C. Alexander (1988) *Fumbling the Future: How Xerox Invented, Then Ignored, the First Personal Computer,* New York: William Morrow.
Stam, E., K. Suddle, J. Hessels, and A. Van Stel (2007) '*High Growth Entrepreneurs, Public Policies and Economic Growth',* Jena Economic Research Papers #2007-019, Max Planck Institute of Economics: Jena.
Stel, A. van and D.J. Storey (2004) 'The Link Between Firm Births and Job Creation: Is There a Upas Tree Effect?', *Regional Studies,* 38: 893-909.
Stel, A. van and K. Suddle (2008) 'The Impact of New Firm Formation on Regional Development in the Netherlands', *Small Business Economics,* 30(1): 31-47.
Storey, D.J. (1997) *Understanding the Small Business Sector.* London: International Thomson Business Press.
Storey, D. (2002) 'Innovative Entrepreneurship in the Netherlands: Observations from a UK Perspective' in: EIM (eds.) *Entrepreneurship in the Netherlands. Innovative Entrepreneurship. New Policy Challenges!* Zoetermeer: EIM Business & Policy Research.

Te Peele, G.W. and A.M.J. Brummelkamp (2007) *Succes in Nederland: Casestudies van zeven succesvolle bedrijven.* EIM: Zoetermeer.
Thurik, A.R. and B. Van Dijk (1998) 'Entrepreneurship: visies en benaderingen' in: *Handboek ondernemers en adviseurs in het MKB,* D.P. Scherjon and A.R. Thurik (eds.), Kluwer Bedrijfsinformatie,127-147.
Thurik, A.R., M.A. Carree, A. Van Stel, and D.B. Audretsch (2008) 'Does Self-employment Reduce Unemployment?', *Journal of Business Venturing,* (forthcoming).
Ucbasaran, D., G. Alsos, P. Westhead, and M. Wright (2008) 'Habitual Entrepreneurs', *Foundations and Trends in Entrepreneurship* (forthcoming).
Waasdorp, P. and M. Bakkenes (2006) 'High-growth Companies: A Policy Perspective' in: M. Bakkenes and J. Snijders (eds.) *Entrepreneurship in the Netherlands. High growth enterprises; Running fast but still keeping control,* The Hague/Zoetermeer: Ministry of Economic Affairs/EIM,33-48.
Wennekers, S. and R. Thurik (1999) 'Linking Entrepreneurship and Economic Growth', *Small Business Economics,* 13: 27-55.
Wennekers, S., R. Thurik, A. Van Stel, and N. Noorderhaven (2007) 'Uncertainty Avoidance and the Rate of Business Ownership Across 21 OECD Countries, 1976-2004', *Journal of Evolutionary Economics,* 17: 133-160.
Winter, S.G. (1984) 'Schumpeterian Competition in Alternative Technological Regimes', *Journal of Economic Behavior and Organization,* 5: 137-158.
Wong, P., Y. Ho and E. Autio (2005) 'Entrepreneurship, Innovation and Economic Growth: Evidence from GEM Data', *Small Business Economics,* 24(3), 335-350.

6 BARRIERS TO INNOVATION

Leo van der Geest and Lars Heuts

6.1 INTRODUCTION

In all industrial countries governments embrace innovation as a source of future wealth. The European Union sees innovation as a key factor in its aim to become the most dynamic and competitive region in the world by 2010, as stated in its 'Lisbon agenda'. In the Netherlands an 'Innovation Platform' has been established by the government, under the direct guidance of Prime Minister Balkenende, in 2003 to promote innovation in Dutch society.

This is not surprising. Countless studies have led to the conclusion that innovation is the main source of productivity rise and wealth creation, not only nowadays, but throughout the centuries (Baumol 2002). However, history also shows that innovative developments often provoke strong economic and political resistance. Acemoglu and Robinson describe how the political establishment in the Russian and Austro-Hungarian Empires tried to hold back the industrial revolution, because they perceived it as a threat to their political power (Acemoglu and Robinson 2000). In Islamic lands, religious leaders forbade the printing press as a source of blasphemy and heresy; according to Landes this is the main reason why the Islamic world started to lag behind the West in economic development (Landes 1998). And even today, access to the Internet is restricted in China and other countries to secure political stability and protect those in power.

In modern Western economies too, technological change does not emerge without provoking resistance. Innovation changes established economic and political relations, so there are winners and losers. The mainstream view among economists is that the 'losers' oppose technological change in order to protect their economic interests. For instance, a monopolist may try to stop the introduction of a new, superior technology by a competitor, in order to protect his market share. Societies that want technological and economic progress, must arm themselves against this, for instance with effective regulation of competition.

According to Acemoglu and Robinson, however, the 'political-loser' hypothesis is not completely satisfactory and certainly not complete (Acemoglu and Robinson 2000). It fails to explain why the economic losers do not use their power to capture the gains of innovation for themselves. This means that the power distribution must be accounted for as well. Groups or individuals without political power, disadvantaged by economic and technological change, cannot stop it. Those with economic and political power can. Hence, we have to consider economic and political institutions[1] if we want to understand how innovation is obstructed by important actors in society.

6.1.1 ECONOMIC AND POLITICAL LOSERS

Economic analysis usually assumes that the market determines the level of technological development. Spurred by competitors, companies are developing new technologies all the time, creating new attractive products or more efficient production methods. This is the simple Darwinistic model of competition.

However, there are nearly always other factors besides market forces, namely the institutions constraining them. In the public sector, where the market fails, decisions on technological developments are obviously made within the political process. On the other hand, political interference in the private sector is not unusual. When public health and safety are at stake, there are good reasons to curtail the market for goods and services. Examples are new types of medicinal drugs and cars. Inevitably a 'political market' emerges, where rent seeking and lobbying by interest groups can determine the outcome of the innovation process or – even more effectively – where interest groups try to control the decision-making process – the economic and political institutions (Buchanan et al. 1982) There can be different reasons for individuals or groups in society to resist innovation (Mokyr 2000).

Loss of jobs

People may fear to lose their jobs. The best known examples were the Luddites in Britain, who fought tooth and nail against the introduction of mechanised looms during the Industrial Revolution. They did not have the power to stop progress and some of them ended up hanging from a gallows. There is little historical evidence that fear of job loss can halt technological progress, but it can slow it down. To what extent this fear can slow down the innovation process, depends on the institutionalised decision-making process. Think of the current discussions about the influence of 'private equity' on corporate decisions.

Direct economic damage

New efficient production methods and technologiesa replace existing ones. Established companies lose market share or go out of business. Valuable knowledge and skills are lost. This is 'creative destruction' as defined by Schumpeter. The owners of those companies, knowledge, and skills will try to slow down or block the introduction of new technologies, especially if it is difficult or expensive to switch to new technologies. Baldwin and Robert-Nicoud (2007) noted the asymmetrical nature of lobby activities for technological change: high sunk costs and exit barriers ensure that losers lobby harder than winners. As a consequence, government policy tends towards protection of vested interests.

Negative external effects

New technologies can also provoke resistance from groups without direct financial interests, because unfavourable external effects are feared. Nuclear energy, genetically modified food, the bio-industry, prenatal screening, megastores, and numerous other innovations cause fear in society at large, because they could potentially somehow damage the quality of the society. Again, the extent to which resistance of external groups can block innovative developments depends on economic and political institutions.

Information problems and uncertainty

Every innovation, especially a fundamentally new one, creates uncertainty about the social consequences and who is to gain and who is to lose. That is why it will always be a matter of dispute if the market is the appropriate mechanism to decide whether an innovation is to be adopted. Precisely because the uncertainty is high, political debate, political pressure, and propaganda are bound to influence decisions on technology. A remarkable example is nuclear energy, which is fully accepted in France, while there is widespread resistance against it in the Netherlands. Invoking technical experts is only a partial solution, since they usually have opposing views and expectations on the social consequences of important technological changes.

Distrust

Resistance against technological changes can also be the consequence of politically or culturally determined distrust. The introduction of quinine in England was banned at the end of the seventeenth century because the Jesuits had brought it to Europe; it was called 'Jesuit powder' (Duran-Reynals 1946). There are many examples of major, established companies ignoring the major inventions of small companies, assuming 'if it was really any good, we would have invented it ourselves – the "Not Invented Here" syndrome'.

6.1.2 MARKET FAILURES AND COORDINATION PROBLEMS

Obstacles to innovation do not always come from individuals or groups. They may also be the result of the nature of the market or the 'system'.

Lacking markets

For some inventions there is little if any market. This applies, for instance, to sustainable energy. As long as the costs to the natural environment and other external effects are not accounted for in the energy prices, the development of sources of sustainable energy is bound to remain slow and difficult. Nobody is willing to invest in a product if there is no prospect of commercial application. In theory it is the task of the public sector to correct this failure of the market. However, this has to be done cautiously, since government intervention can disturb the situation, creating new obstacles.

High costs of adaptation

Some innovations that are technologically superior, fail to break through because a large number of potential users has already chosen inferior technology, making the costs for adaptation too high. Economists call this 'path dependence' or hysteresis. The classical example is the DVORAK keyboard, which is considered superior to the standard QWERTY system, but has not broken through because all keyboards in the world would have to be replaced and everybody would have to learn to type again (David 1986)

System repercussions
More generally, the problems of the introduction and implementation of an innovation grow larger as the number of different actors, activities, and technologies grows larger, in other words: if the system repercussions are greater. A switch from petrol to hydrogen or bio-ethanol car fuel does not only require different engines, but also far-reaching and therefore costly changes in the production and distribution system (refineries, filling stations), quality and safety standards, technical knowledge in garages, and fiscal arrangements (taxation of cars and fuel). Many forms of innovation, especially in the public sphere, can have far-reaching system repercussions.

Coordination problems
In some cases different actors are unable to agree on a common approach, bringing the innovation process to a halt. Uncertainty about the outcome, distrust of other parties or disagreement on the sharing of the costs and of the rewards can stand in the way of cooperation. In that case a moderator can play a useful role. This could be the government, but does not necessarily have to be.

6.1.3 COUNTERPRODUCTIVE POLICY

All these factors can be reasons to reject the market mechanism for decision making on the development and introduction of new products and technologies. A 'policy' must be devised for the correction of failures of the market mechanism and to promote and guide innovation. This policy can involve many things: subsidies as an incentive to innovation; permits, procedures, and quality control to assess the social impact of innovations; organisations to widen bottlenecks and remove obstacles; and mediators and moderators to bring parties together and promote cooperation. However, all this does not guarantee optimal results. Not only can there be failure of the market, but also failure of policy. The government is susceptible to lobbying and rent seeking. The government too has to cope with uncertainty and incomplete information, and tends to avoid risks. A government can be manipulated by parties whose cooperation is necessary but who are pursuing their own interests. The need for transparency and accountability results in detailed regulation and bureaucracy which discourage the innovators. The policy can even create new obstacles to innovation.

6.1.4 CASE STUDIES

The following sections describe four cases of innovative developments meeting a variety of obstacles. These four cases are:
- Wind energy in the Netherlands;
- The Whisper, low-noise bus;

- The Electronic Patient Dossier;
- Energy houses.

6.2 BLOWING IN THE WIND

In the 1970s interest in the Netherlands for alternative sources of energy – one of them being wind energy – grew, partly spurred by the 1973 oil crisis. Since then, the Dutch government has tried to promote the development of wind turbines, with considerable subsidies for research institutes and enterprises. While the Netherlands initially had a leading position in wind energy, it has clearly fallen back in the successive decades. Why has wind energy not been as successful as in other countries?

6.2.1 NETHERLANDS MUST BECOME FRONT RUNNER (KAMP 2002)

The Dutch government's 1974 'White Paper on Energy' contained an outline of a plan for the development of alternative sources of energy; with the government subsidising and coordinating projects. Two new organisations were created: the LSEO, the Dutch acronym for a national coordinating committee for energy research, and the NEOM, the Dutch energy development company. The LSEO was responsible for, among other things, the selection of projects qualifying for a subsidy. The task of the NEOM was to promote projects and knowledge transfer between research institutions and commercial enterprises.

Although the country had no experience with the construction of wind turbines for energy production, the LSEO did not foresee problems in this area, because of a lengthy experience in the construction of classical windmills. This committee even expected a leading role for the country in the development of wind turbines. To this end, the government invested a large amount of money in R&D.

However, the construction of wind turbines appeared more difficult than expected in the 1980s. Due to technical problems and high maintenance costs, the necessary investments turned out to be much higher than expected. The energy efficiency was disappointing. As a consequence, Dutch energy companies and major turbine manufacturers, such as Fokker and Stork, withdrew from the project. The government then changed its policy and concentrated on smaller manufacturers. In the late 1970s they had started with the production of smaller wind turbines and could not expect much in terms of government subsidies in those days. However, this change in government policy has not led to success either. Before 2000 most Dutch wind turbine manufacturers had gone out of business or had been taken over by foreign competitors.

In contrast to the Netherlands, wind energy in Denmark has proven to be very successful. Currently several Danish wind turbine companies have a strong position on the world market (Kamp and Mulder 2005). Countries like Spain and Germany also have relatively successful wind energy policies. Table 6.1 indicates

Table 6.1 Wind energy in nine countries in 2006

	Capacity (MW)	% Wind energy / Total electricity produced
Denmark	3,136	20%
Germany	21,283	9%
Ireland	866	5%
Italy	2,285	1,5%
Netherlands	1,655	3,2%
Austria	965	3%
Portugal	1,874	9%
Spain	12,500	8%
USA	31,000	0,8%

Source: Wind Service Holland, 2007

that the total wind energy capacity in these countries is much higher than in the Netherlands.

6.2.2 OBSTACLES TO WIND ENERGY

Lack of market demand

Although the price gap is narrowing, wind energy is still more expensive than fossil fuel, partly because the costs to the natural environment and other external effects are not included in the price of 'grey' electricity. Because the (artificially) high price didn't make wind energy very interesting for private investors, the government had to create a market. It tried to do so by subsidising R&D, investments, and energy 'resupply' to the power grid.

At present, in 2007, wind energy is still not profitable in the Netherlands and the market still depends on government subsidies. According to the Dutch Wind Energy Association the installation of land-based wind turbines without subsidies will not be possible before 2014; at sea it will not be possible before 2021. Wind park subsidies until 2020 will amount to €2,400 million (€500 million on land and €1,900 million at sea)[2]. Creating a wind energy market does not only take much time (several decades), but also a large amount of public means (several billion euros).

Lack of interest of energy companies

As a consequence of the lack of a real market for wind energy in the Netherlands, private investors were initially hardly interested in small scale, sustainable energy production from wind. It just did not fit into the business models of any commercial enterprise. Energy companies were focused on large-scale energy production from natural gas, coal, oil, or nuclear fuel. They were not involved in the development of wind turbines. This attitude of Dutch energy companies changed a little in the 1990s, when subsidies for owners of wind turbines became available.[3]

No level playing field

Wind turbine owners produce wind energy for their own use, but supply their surplus production to the power grid (i.e., the energy companies), for which they receive financial compensation. The 'resupply' tariffs for this compensation have been kept low by the energy companies for a long time, leaving the wind turbines unprofitable in many cases. This unfair competition only came to an end in 1996 when the government started to subsidise these tariffs[4]. This sort of subsidy had already existed in other countries for quite some time, for example in Denmark since 1984. Table 6.2 indicates that the 'buy-back' tariffs in the Netherlands in 2003 were higher than in Denmark, Germany, and Spain, but the country had lagged behind in the production of turbines so much, that it could not catch up with these countries anymore.

Table 6.2 Buy-back tariffs in four countries in 2003

	State subsidy since	Buy-back tariff (€ cent/kWh)
Denmark	1984	4.4
Germany	1991	6.2-9.1 *
Netherlands	1996	7.8-9.7 *
Spain	1994	6.3

* Depending on location
Source: Van Giesel and Van der Veen, 2004

Coordination problems

Many parties (Kamp 2002) were involved in the development of turbines in the 1970s and 1980s: the government, research institutes, and major companies such as Fokker and Stork. Although they did cooperate, they did not reach a successful common approach to any sufficient degree. An important problem was the minor role of the energy companies, who were the obvious potential market for wind turbines; the 'market' was not involved in the development.

The minor manufacturers of wind turbines were also ignored when it came to handing out subsidies. This only changed in the late 1980s, when the government adjusted its policy, after the major companies (Stork, Fokker) had withdrawn. Although small companies had sought cooperation with the research institute ECN, this cooperation had its limitations, since the ECN operates on a far more academic level than the more application-oriented manufacturers. Mutual cooperation between minor manufacturers was practically non-existent.

Because of these coordination problems, little if anything was learned, preventing technological innovation and the practical application of wind energy in the Netherlands. In Denmark, in contrast, there has been close cooperation between manufacturers of wind turbines, owners, and research institutes, as well as small companies. Moreover, the energy companies played a major role in the development of wind turbines; the Danish government created a production company for wind turbines in cooperation with the energy companies. The good cooperation between technological research and application resulted in important learning effects. The success of the Danish wind energy policy is, according to Kamp (2002), due to this learning process in particular.

6.2.3 POLITICAL WINDINESS

Lacking a market for wind energy, the Dutch government had an important task to do in creating one. In practice, the government has not been very successful in doing this. This section focuses on the reasons for this lack of success.

R&D bias

As in many other countries, the development of wind turbines in the Netherlands strongly focused on the stimulation of demand, partly through technological research (technology-push). This stress on the supply side resulted in generous subsidies for research institutes (ECN, Universities of Technology in Delft and Eindhoven) and (major) companies (Stork, Fokker, Holec), who were active in the development of turbines. During the first year of the National Development Program for Wind Energy (NOW), 15 million guilders (1 Euro = 2.20 guilders) was made available from March 1976 until March 1977. In the following years, projects received many millions of guilders more in subsidies (Kamp 2002).

The government's focus on technological research reduced the attention for the demand side of the market. There were no subsidies for wind turbine manufacturers, making it unattractive to invest in their development. It was not until 1986 that the first investment subsidies for wind turbine production were introduced, enabling a certain growth in the turbine market. However, this was not enough to save the Dutch manufacturers; most of them went out of business in the 1990s or were taken over by foreign enterprises (Kamp and Mulder 2005).

Denmark chose right from the start for a policy that included both a *technology-push* and attention to the demand side of the market. Investment subsidies for buyers of wind turbines were established in the 1970s, enabling the creation of many small wind turbine companies (Kamp 2002). Initially Germany was strongly focused on research, like the Netherlands, but because of lack of success, attention shifted to a more market-driven policy. In 2000 a bill was passed (*Einspeisegesetz*) to stimulate the use of sustainable energy in general. The law compelled energy companies to pay an acceptable price for 'resupplied' wind energy.

Erratic policy

A consistent long-term government policy is very important for the creation of a successful wind energy market. However, in the Netherlands this is what has been lacking only too often. In particular subsidies for wind energy was subject to varying political criteria.[5] Since the introduction of investment subsidies for wind energy producers in the late 1980s, many arrangements have been abolished and/or modified. For instance, subsidies were initially related to the installed generator power in kilowatt hours (kWh), encouraging the installation of relatively heavy generators. This had a negative effect on the quality of the turbines, making them unsuitable for export. Later the subsidy was related to the energy actually produced in kWh.

Until 2004 there was also a subsidy for the import of 'green' electricity from abroad (Regulatory Energy Taxation, REB), that frustrated rather than stimulated the domestic production of wind energy. Since Dutch enterprises were allowed to invest in foreign wind parks with Dutch subsidies, they invested less in their own country. The creation of the MEP-subsidy (concerning the ecological quality of the electricity production) had to remedy this (Mast et al. 2007). Although the MEP was very popular, it was ended in August 2006; the subsidy for off-shore wind parks had already been scrapped the year before. This continual changing and/or abolition of subsidy regulations made it very difficult for manufacturers and energy producers to develop a long-term strategy.

Learning process for the government

The Dutch government has learned from some of its bad policy choices. When it became apparent that focusing on technology had not brought about the desired results, more attention was paid to the demand side of the wind energy market, resulting in subsidies for 'resupplying' to the power grid and for investments. Moreover, in the mid-1980s the government began to work with small manufacturers of wind turbines, who had previously been ignored.

6.2.4 CONCLUSION

The ambitious plans in the 1970s for a leading Dutch role in wind energy have had very little success. Thirty years later the Netherlands still lags behind. One of the causes of the lack of success of wind energy is the poor cooperation between different actors, who learned little from each other as a consequence. Investment subsidies and subsidies on buy-back tariffs were introduced only at a late stage, making investment in wind energy unattractive for commercial enterprises. The government also has (unwittingly) created obstacles itself, by unduly stressing technological research, while neglecting the demand side. Its policy on subsidies was erratic. Certain regulations were actually counterproductive.

An important lesson is that the 'market' must be involved from the beginning. Interactive learning is greatly improved by good cooperation between researchers, manufacturers, and users. If the government wants to support innovative developments, it is necessary to have a consistent long-term policy that partners in the innovation process can rely upon. For the development of an effective policy thorough market knowledge is necessary. The government must invest in this policy, otherwise the government is pushed around instead of being a coordinating force.

6.3 WAITING FOR THE BUS

A couple of years ago the company e-Traction in Apeldoorn developed a bus called the Whisper; it is far more silent, environmentally friendly and cheaper than normal buses. A first prototype was commissioned in Apeldoorn in 2005. In spite of the advantages, plans to use the buses on a regular basis throughout the Netherlands have not yet materialised. The question is why the Whisper is still not being used all over the country.

6.3.1 ADVANTAGES OF THE WHISPER

The Whisper has a conventional diesel engine and an electromotor. When it brakes or moves downhill, the released energy is stored in batteries. This energy can be used again for the electromotor driving the back wheels. The batteries can be supported, if need be, by the diesel engine when moving uphill.

A study by TNO concluded that the Whisper has a number of advantages. First, its engine produces less noise than that of ordinary buses. Its typical noise level is 58 dB instead of 78 dB (*Nederlands Dagblad* 2005). Another great advantage of the electromotor is that it saves a great deal of energy. A normal bus with a diesel engine gets 2.2 kilometres to the litre, the Whisper gets 6.3 kilometres to the litre. Thirdly, the Whisper is very environmentally friendly: its CO_2-emissions are 85 percent less than a regular bus (*De Stentor* 2004). Finally, the energy advantages can help to keep the ticket price low.

Although the purchase price of the Whisper is higher than that of a normal bus, the difference can be quickly earned back. The first five Whispers will each cost about €280,000, while the price of a normal bus is about €200,000. If the production of the Whisper goes up, scale advantage will reduce the price to €220,000 (*Nederlands Dagblad* 2005). Because the Whisper saves about 20,000-25,000 litre diesel oil per year with its electromotor, the price difference would be earned back after a short period of time.

6.3.2 BUS STOP

In 2005 the prototype of the Whisper was used for the first time in the municipality of Apeldoorn. Only test drives have been made ever since. E-Traction currently depends on subsidies in order to continue the production of the Whisper. The costs of a test phase with five prototypes amount to €1.8 million. The Netherlands Ministry of Housing, Spatial Planning and the Environment (VROM) is willing to pay a subsidy of €350,000 but the remaining costs are still too high for e-Traction. The company is seeking subsidies from various provinces, municipalities, and/or public transport companies to enable them to develop the additional prototypes and start testing.

Three Dutch municipalities have shown interest in the Whisper so far: Apeldoorn, Amersfoort, and Enschede. The province of Utrecht intended to grant a subsidy for five Whispers in October 2005, to be used on the bus line from Central Amersfoort to the Amersfoort district of Vathorst. In this way, the province wanted to stimulate new development in public transport and reduce inconvenience for the population. When the Whisper buses had lived up to expectations, all buses in Amersfoort would be replaced by Whispers. Anno 2007 none of the three municipalities has made a start yet with such a Whisper project, but in the meantime plans have been made in cooperation with the urban region of Rotterdam to convert five normal buses to Whispers and use them in regular service.

6.3.3 OBSTACLES FOR THE WHISPER

Catch 22: Everybody is waiting for everybody else

E-Traction is dependant upon several parties for the introduction of the Whisper: municipalities, provinces, transport companies, and coach builders. In practice none of them is willing to bear the initial costs. Every party has its reasons – often understandable – to look to other parties to jump first. As a consequence, nothing happens. The innovator (e-Traction) is caught between these different parties and so far has not been able to convince any of them to make the first move.

Although a bus like the Whisper has widely acknowledged value for society, there is no real market for it yet. None of the parties involved is willing to invest in a product without proven technological quality and commercial value. Because of uncertainty about the market potential of the Whisper, manufacturers hesitate to join forces with e-Traction. It is too risky for them to invest a few million euros in the development of five Whispers without certainty that the market is actually willing to pay for it. Moreover, adaptation of their existing production lines is costly.

Uncertainty is also an obstacle in getting transport companies to participate in the project. Since they have a responsibility to their customers, they are unwilling to experiment; they do not want to risk that the new Whisper technology might break down.

The government: A silent partner

In the past the province of Gelderland has given subsidies to e-Traction for the development of the prototype of the Whisper, but it is not willing to invest more; the provincial authorities think that market partners must take responsibility for introducing the Whisper to the market. The municipality of Apeldoorn, the home town of e-Traction, is also not willing to invest more, because it does not see this as the task of a local authority. The arguments of both of these authorities are understandable. They must be cautious with their market interventions, because unfair competition can create new obstacles.

Moreover, a local authority can only intervene in the market in a limited way. Municipalities hand out concessions to transport companies, for a temporary transport monopoly. Once a concession has been given, a municipality cannot easily change the conditions. That is why a municipality cannot compel local transport companies to buy the Whisper, since the conditions of the concession do not require that a specific type of vehicle be used.

The so-called innovation concession will probably allow municipalities to adapt their concession policy from the end of 2007. This enables local authorities to introduce innovative products because of their public interest. Transport companies that have a concession already can be compelled from then on to purchase some buses based on ecological arguments. That would brighten the prospects for the Whisper.

Quality control

An additional obstacle for e-Traction is that the Whisper has to be inspected by several quality control institutions on safety aspects. An important example of a certificate to be obtained is the CE benchmark, indicating that EU-safety guidelines have been observed. In order to obtain a certificate, a new product must undergo several tests; the certification process can sometimes be lengthy and costly. Although this certification is essential, it can be an obstacle for innovative companies, especially small ones, without much time or money.

6.3.4 CONCLUSION

The main problem of e-Traction with the introduction of the Whisper is that none of the actors in the innovation process can break out of this 'Catch 22' situation; everybody is waiting for somebody else. A real market for the Whisper is effectively non-existent, although its importance in saving fossil fuel is widely acknowledged. It is understandable that potential subsidising authorities are hesitant, since it is not their task to buy or use such vehicles. Existing bus safety requirements require that vehicles be tested by several organisations, which is another obstacle for the company in Apeldoorn.

The role of the authorities as 'launching customer' can be particularly important to promote innovative developments. The innovation concession enables the authorities to hand out concessions based on criteria other than price. To be sure, we must be on our guard that this kind of concession does not evolve into a form of subsidy from taxpayers.

6.4 VIRTUAL PATIENT

In 1997 the former Dutch health minister Borst announced that a start would be made with the introduction of the Electronic Patient Dossier (EPD), to facilitate the exchange of medical records between health care workers. Ten years later there is still no working countrywide information system for medical records. Why is it taking so much longer than expected?

6.4.1 ADVANTAGES OF THE EPD

The EPD is intended to facilitate the exchange of medical records between health care workers. Authorised personnel must be able to query and view up-to-date information on a patient anytime and anywhere in the country. Such a system would be of great use for example to a doctor substituting for a general practitioner (GP) or in a first aid centres. With the EPD a doctor would be able to see a patient's medical history, including past illnesses, chronic diseases, allergies, current medication they are taking and any counterindications. This would be a helpful tool in preventing medical errors and enabling doctors to provide better quality medical care.

Another advantage of the EPD is that it would facilitate more effective medication monitoring. Incorrect medication is the cause of approximately 90,000 hospitalisations per year in the Netherlands. The main reason is that different pharmacists, doctors, and hospitals do not know about each others prescriptions. The direct costs of this amounts to about €300 million per year (*Pharmaceutisch Weekblad* 2005). The total nationwide costs of incorrect medical information transfer are even higher: over €1 billion annually, with an additional €1.6 billion in incidental costs (TNS NIPO 2004).[6] Although an EPD would not rule out all such errors, it could reduce them drastically.

The EPD could also result in higher efficiency, because with such a system medical records would not have to be gathered again and again by different health care workers and institutions, and lab work would not have to be repeated. Finally, the EPD can contribute to demand-oriented care. When patients have their own medical data at their disposal, they are better able to discuss decisions on care and treatment with their doctors and other health care workers.[7]

6.4.2 DELAYED PROGRESS

After receiving the go-ahead of Health Minister Borst in 1997, the associations of hospitals, GPs, and specialists signed a declaration of intent for the realisation of a nation-wide electronic infrastructure for all health care services. In 2001 Minister Borst estimated that the EPD would be operational by 2004; a few years later this appeared not to be feasible. In 2003 the National ICT Institute for Health Services (NICTIZ), which is coordinating the development of the EPD under the guidance of the Ministry of Health, announced a delay of three years in making the EPD fully operational (KPN 2006) Later, even this prediction had to be adjusted.

The initial plan was that the health services would take responsibility for the implementation of the EPD. When it appeared that this was slowing down the whole process, late in 2005 Parliament forced the Ministry of Health to take over this responsibility. The current target date to make two components of the EPD , the Electronic Medication Dossier (EMD) and the Dossier for stand-in GPs (WDH), operational nationwide is 2009.
The introduction of the EMD and WDH has several phases. The basic infrastructure (national switch point, LSP) is now ready. After a 'proof-of-concept' phase and a 'front-runner' phase the nation-wide implementation will take place. The front-runner phase is now in progress; the aim is to test the functionality in a limited working environment with GPs, group practices, hospitals, and pharmacists in terms of safety, reliability, and user-friendliness (Ministerie van Volksgezondheid, Welzijn en Sport 2007). In the eastern region of Twente a local WDH will be made accessible via the LSP and in Rotterdam and Amsterdam the already-operational EMD will be linked to the LSP.[8] When all these field tests have been completed and evaluated, the regional and the nation-wide implementation of both systems can start. After that, other components of the EPD can be added.

6.4.3 OBSTACLES FOR THE EPD

Complexity of the operational environment

Many different health care services and people are involved in the implementation of the EPD: GPs, dentists, hospitals, home care organisations, pharmacists, psychotherapists, physiotherapists, obstetricians, and nurses. The Ministry of Health has to coordinate its policy with about 25 branch organisations. Since they all have their own ideas and interests, it is not always easy to reach consensus. This is one of the reasons why a rapid and simple implementation of the EPD has been difficult to achieve.

For example, the implementation of the EPD in hospitals appears to be a laborious process. All hospitals have their own board of governors whose consent is required. Moreover, hospitals are complex organisations, since they often host no less than twenty specialisms, which have their specific ways and methods of record keeping. This adds greatly to the complexity and reduces the efficiency of the EPD. Negative experiences with automatisation in hospitals has bred a lot of scepticism (Health Sciences Digest 2005). In general, Dutch hospitals have no intention of connecting to the LSP in the near future; in 2007 less than 30 percent had allocated funds for this purpose, and 75 percent indicated it was not attractive at the moment (*Digitaal Bestuur* 2007).

GPs, pharmacists, hospitals, municipal health services, nursing homes, and home care institutions all have their own strategies and automatisation policies; their ICT infrastructures are also different. This means that strategy, organisation, coding of information, and the technical infrastructure have to be made mutually compatible. Figure 6.1 gives an overview.

Figure 6.1 The EPD has to be standarddised at different levels*

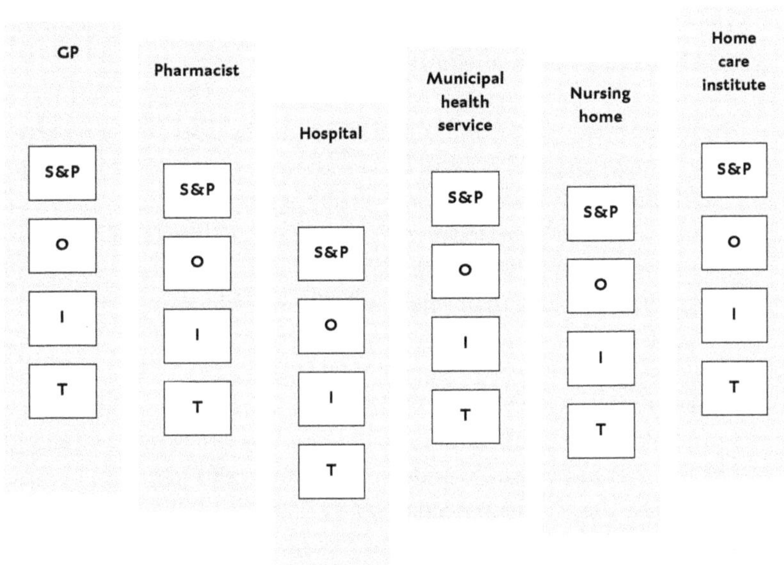

S&P = Strategies and Policy; O = Organisation; I = Information; T = Technical infrastructure

Source: NICTIZ

Resistance of the actors

Most health care services already have a system to store medical records. Most GPs have already an electronic patient dossier containing consultation reports, inquiry results, and medication. Communication between some of these existing systems is possible already, for instance between the systems of GPs and pharmacists. Communication is usually limited to some health care services in the same region. Health care services will have to hook up their existing systems to the new EPD. Health care workers and hospitals still without systems of their own will have to introduce a completely new system. Many of them are not keen to do that, especially if they are content with their existing system. GPs find the EPD time consuming, since they have to 'translate' their activities into diagnostic codes (*Automatisering Gids* 2004). There is also substantial lack of trust in the functionality of the new system; the perception is that it will not meet the same requirements as the existing ones.

There are also financial objections to the introduction of the EPD. Connecting the existing information systems to the WDH will cost a GP about €4,500; this could be earned back in five years, according to calculations from Squarewise (an innovation management bureau).[9] Other health care workers also face similar initial costs, that could be earned back in a few years. Although the EPD has little if any net costs over a period of a few years, the initial costs can create an extra hurdle for new users.

The EPD can also result in loss of income for some health care workers. Since it contains information about earlier inquiries from a patient, those inquiries, like taking X-rays and blood samples, do not need to be done again when a patient is transferred to another medical practice.

The introduction of the EPD implies that medical records stored by a doctor or other health care workers will be accessible to others. This means loss of an information monopoly and autonomy, enabling a doctor's colleagues and patients to get more detailed knowledge of their professional practice. The (compulsory) sharing of patient information means loss of territory and power, for instance within the organisation of a hospital. The introduction of market forces only adds to the anxiety.

A survey of the Royal Dutch Medical Association (KNMG) in 2004 shows that 80 percent of the respondent doctors are unwilling to give direct access to electronic data to other health care workers or institutions for clinical purposes. About 76 percent say they have no need to query data from colleagues (Ministerie van Volksgezondheid, Welzijn en Sport 2005).

Uncertainty

A major issue related to the EPD is the authorisation to access medical records. Especially health care workers value the confidentiality of the medical records they maintain; patients themselves are often less concerned about confidentiality of their medical records. The law on the protection of personal information (WBP) and the law on medical treatment agreement (WGBO) stipulate that a health care worker can get access to medical records only if it is necessary for a patient's treatment and if the patient does not object. Health care workers who are registered in the existing BIG system have a right to view medical records. These are doctors, dentists, pharmacists, psychologists, psychotherapists, physiotherapists, obstetricians, and nurses. The WDH-component of EPD is accessible only for GPs, their other doctors who temporarily stand in for them, and their medical assistants. The EMD component is, for the time being, accessible only to GPs, medical specialists, hospital and general pharmacists, and their assistants.[10]

There are, to be sure, good reasons for all this attention to privacy. For example, when a company doctor can view the medical record of a job applicant, he might stop the hiring procedure if he finds a psychiatric record in the applicant's file. However, privacy rules that are too strict also have a downside: the obvious advantages of the EPD would be lost. The fear of privacy violation can also be used as a pretext to thwart or delay the introduction of the EPD.

Box 6.1 Experiences abroad

In spite of all the delays in the EPD project, the Netherlands is still one of the three front runners when it comes to exchange of medical data, the others being Britain and Canada. Most other countries run into similar difficulties, such as privacy protection and safeguarding and interoperability with existing systems (Ministerie van Volksgezondheid, Welzijn en Sport 2005).

The introduction of an EPD in Britain was also delayed for several years. Phased introduction was scheduled to begin in 2007. In contrast to the Netherlands, it will be introduced in a top-down manner: health care workers will not be involved in the implementation of the new system. The National Health Service has signed a contract with one ICT supplier for each region in Britain for the delivery of one integrated system for each hospital. Health care workers and patients have strongly criticised this model, since they, the future users, have not been consulted. Privacy is also a matter of concern, since the police and the secret services must also have access to combat crime. Britain also has to cope with costs that have gone way over budget; the initial budget was about €3 billion, but it has already more than tripled (*de Ziekenhuiskrant* 2007). The costs of the Dutch EPD are 'only' a few dozen million euros.

The implementation of a nationwide EPD in the US is still not forthcoming. A National Health Coordinator appointed in 2004 has yet to realise such a system. It is expected to take about ten years.[11]

Another concern about the EPD is protecting the system against hackers. An incident in 2005, when hackers tested the Dutch system by breaking into the data of 1.2 million medical records of a Dutch hospital (*de Volkskrant* 2005), sparked a heated debate on the protection of medical records. In order to prevent unauthorised access to medical records in the EPD, national guidelines for protection have been defined. Health care workers have to identify themselves with a special 'UZI pass', and the computers of health care workers, hospitals, etc. and the LSP must meet high standards of protection. A health care worker who does not meet the requirements cannot connect to the LSP.[12]

The emphasis on protection of the EPD illustrates a phenomenon that can be observed in other innovation processes as well: the safety standards for new systems are much higher than for the existing ones. However, absolute safety does not exist, and striving for watertight guarantees can be an effective obstacle to the introduction of innovations.

Lack of market demand

A major problem of the EPD is that there is no effective demand for it. Patients, who would ultimately benefit most from it, are not actors in the Dutch market and cannot exert pressure to introduce the new system. This means that the social benefits are much larger than the private benefits of the parties involved directly. Moreover, the lack of competition in the medical sector ensures that health care workers have little financial incentive to realise optimal safety for their patients. Because of lack of transparency, medical errors often go unnoticed and market forces have little effect. As a result, the government had to take up the role of initiator and play a guiding role in the development and implementation of the EPD. Because the government pays and controls the development of the central facilities of the system almost completely, health care workers and ICT suppliers are not stimulated to think pro-actively about an EPD that is as effective and efficient. Since the government represents the supply side and controls the demand side, it is susceptible to lobbying and rent seeking by market parties who want to profit from the introduction of the EPD or who want to prevent its introduction.

Since many health care workers do not have expertise with ICT, they are dependant on the ICT suppliers for the implementation of the EPD. This gives these suppliers strong leverage over their customers. Moreover, cooperation with a chosen ICT supplier has 'lock-in' effects: it is very difficult to switch to another supplier. Since the data in a nationwide EPD must be available anywhere anytime, the system cannot be switched off during the search for another supplier.

6.4.4 CONCLUSION

In spite of the obvious advantages of the EPD, the innovation process is a laborious one. There are several causes. The numerous parties involved in the development and implementation of the EPD are very diverse, and have different interests. There are also considerable initial costs for the customers in the introduction phase, putting up barriers to many would-be customers. The EPD also results in a loss of autonomy for some, because medical records must be among health care workers and can be viewed by the patients themselves. The lack of an 'articulated demand' from patients has also slowed down the introduction of the EPD. Finally, protection of privacy against unauthorised users and hackers creates more delays.

A complex system innovation like the introduction of the EPD can necessitate a mediating role for the government. An absolute requirement for successful implementation is that the objectives are clear and that a realistic and feasible time schedule is projected, enabling market parties to adapt their strategies and investments to it. Besides this, a carefully administered combination of 'carrots' and 'sticks' is necessary to ensure sufficient progress is made. The government as a moderator must monitor the whole process and not get trapped in a quagmire of contradictory demands and interests.

6.5 ENERGY NEUTRAL HOUSES

Sustainable energy plays an increasingly important role in our society. Fossil fuel is bound to be exhausted some day, so there is growing attention for alternative sources of energy, like wind and solar heat. The increasing dependency on major energy companies and governments and the decreasing number of countries exporting fossil fuels, are also reasons for this growing attention.

As early as 1974 the Stichting De Kleine Aarde ('Little Earth Foundation') built an energy neutral house, meeting its energy needs from alternative sources of energy, including wind and solar heat. Several initiatives of this kind have been developed in the Netherlands by companies and individuals since 1974. In spite of public interest in the reduction of energy consumption, such initiatives run into all sorts of obstacles.

6.5.1 FROM RECYCLING HOUSE TO ENERGY NEUTRAL HOUSE

Energy neutral houses function partly or (in rare cases) completely separate from the power grid and use alternative sources of energy, such as wind and solar heat. In 1974 the Little Earth Foundation built an energy neutral house in the town Boxtel, in the southern part of the country. This house had a windmill generating the necessary electricity, a solar collector supplying heat, and a methane gas installation for cooking. Drinking water was obtained from a well.

In the Netherlands and elsewhere, numerous initiatives of this kind have been realised. A special type of energy neutral house is currently under development by the OTB Group in Eindhoven. This 'Energy House is, in principle, completely independent from the power grid. The generated energy can be stored for later use. Energy is stored in batteries or by producing hydrogen through electrolysis, which can be used in a fuel cell for heat production. Currently, the development is in the 'proof-of-principle' phase, in close cooperation with the Eindhoven University of Technology and the Fontys Hogeschool (OTB Group 2007)

6.5.2 ADVANTAGES OF AN ENERGY NEUTRAL HOUSE

Energy neutral houses use sustainable energy only, so a major advantage is the absence of CO_2-emission. Moreover, solar heat and wind energy are inexhaustible sources of energy, a good alternative for fossil fuels that are bound to be exhausted some day. The independence from dubious suppliers is another advantage. Because energy is generated right where it is consumed, there are fewer losses in transport and conversion; energy production in large power stations makes transport of fuel and of generated electricity over long distances necessary; more than 60 percent of the generated power (European Commission 2006) is dissipated before it reaches the consumer. Of course there are no energy bills to be paid, since sun and wind are free.

6.5.3 OBSTACLES TO ENERGY NEUTRAL HOUSES

Unfair competition
There is no fair competition between sustainable energy and fossil fuel, because the costs to the natural environment and other external effects are not accounted for. The generation of 'green' electricity is more expensive in the Netherlands than that of 'grey' electricity. Taxation on fossil fuel with high CO_2-emission is relatively low in the Netherlands. That is why there is currently no such thing as a

'level playing field' between 'grey' and 'green' electricity. In order to realise fair competition, all market parties should carry their own financial burden, which must include the damage done to the natural environment and by contributing to global warming. Nowadays this is only partly the case.

Resistance of vested interests

Energy neutral houses are a potential threat to vested interests in the energy market. Since some houses are not connected at all to the power grids, this means loss of income for oil companies and energy suppliers. It is understandable that they are trying to protect their interests. Although they do not overtly resist alternative energy projects, their cooperation, if any, is half-hearted, especially if their own market position is in peril. Political lobbying enables them to influence political decisions in their favour. Suppliers of fossil energy do cooperate when natural gas retains its importance and when energy is 'resupplied' to the grid. In this way oil companies and energy suppliers remain in control.

Dutch construction companies are equally hesitant about integrating renewable elements into new houses, as was experienced by the Little Earth Foundation. One reason is the uncertainty among contractors and project developers, since renewable materials and new building concepts remain to be proven in the construction sector. Some municipalities in the province of Brabant have compelled local project developers to integrate solar panels or heat pumps into newly built houses.

6.5.4 ENERGETIC AUTHORITIES?

Strict regulation

Construction companies have to cope with a plethora of regulations and certificates. These are an obstacle to innovation. Some components used in energy-neutral houses require special regulation. An example is the use of hydrogen in the OTB Group's Energy House, which is used for the storage of surplus energy. Since hydrogen is a fuel with unique properties, its use necessitates special laws, regulations, and permits. The often laborious administrative processes lead to delays in the innovation process. Since most people are ignorant about the technicalities of the application of hydrogen, there is much uncertainty about the safety aspects.

Erratic policies

The social and environmental advantages of sustainable energy are acknowledged by the Dutch government. In practice, however, this results in erratic policies, since different political parties have different ideas about sustainable energy. Although the government wants to stimulate the use of sustainable energy, it has abolished several 'sustainable' subsidies in recent years. In 2003 the Energy Premium Regulation (EPR)[13] a subsidy for energy-saving appliances such as solar boilers and solar panels – was scrapped; the subsidy for off-shore wind parks was scrapped two years later, and in 2006 the subsidy promoting the ecological quality of the electricity production (MEP) – came to an end as well. The lack of consistent government policy makes it very difficult for enterprises to devise a long-term business plan. The abolition of the EPR has led to a collapse of the Dutch market for solar boilers and solar panels. Some companies had invested for many years in these products, but saw their home market shrink and had to turn to foreign countries. Moreover, the buy-back tariffs of energy are low, discouraging Dutch consumers from using solar panels, for instance. The Dutch tariff for resupply is 6 cents per kWh; in Germany it is 46 cents.[14]

Research done by Simona Negro at the University of Utrecht confirms that Dutch policy on sustainable energy is erratic and inconsistent, making it very difficult to develop a reliable market. In 1998 the Netherlands started with gasification of biomass in the Amer power station near Geertruidenberg. Initially, everybody was enthusiastic because its efficiency was much higher than that of the burning of biomass. But when some difficulties arose because the technology appeared to need some further development, and obtaining biomass for an affordable price became difficult, enthusiasm declined and the experiment was ended. That led to the collapse of the entire innovation system for gasification technology (Negro 2007).

Devising subsidy conditions

When companies request a government subsidy, they often run into practical difficulties. The regulations are often very complex, or it is hard to meet all requirements. Especially small companies – without specialised staff that can concentrate on lobbying for subsidies – are the first to give up, leading to the end of their innovative activities.

For the development of the Energy House the OTB Group had requested a subsidy in the past; eventually this request was rejected. One of the reasons given for the rejection was that no other relevant market parties – such as architects, constructors, and project developers – were involved. Another reason was that the chance of success was slim, according to the subsidising agency Senternovem, since apparently no professional party with the required expertise was involved in the project. Because the OTB Group Energy House is still in the development phase, active involvement of such professionals is not yet possible; they are only willing to commit themselves when the 'proof-of-principle' phase is completed. After successful completion of the first phase of the Energy House, the follow-up will be pursued by a separate enterprise, with a number of parties as shareholders. These partners could bring in financial, building/developing, and possibly technical/scientific expertise (OTB Group 2007). One more reason for rejection was that the applicability and the environmental advantages of the Energy House were perceived as meagre, because the technology is designed first and foremost for newly built, free-standing, up-market privately owned houses. However, the OTB Group had good reasons to test the technology first in a limited environment; after proven success, the technology can be applied to other segments of the housing market.

6.5.5 CONCLUSION

The importance of a sustainable economy is acknowledged by an ever- increasing number of people in the Netherlands. Although the building of energy-saving houses in the Netherlands could make an important contribution to such an economy, developers of the necessary technology have run into a number of obstacles. There is no fair competition between suppliers of sustainable energy and 'traditional' energy companies, since environmental pollution is not included in the energy prices. Established energy companies and oil companies have no financial interest in energy- saving houses and for that reason are not eager to cooperate in their development. The government also creates obstacles, by failing to devise reliable and consistent policy. Finally, the development of energy-saving houses is delayed because of uncertainty about the application of sustainable components.

When a social issue acquires more prominence in political agendas, as is the case with sustainable energy, political parties want to promote themselves with 'new policy' to achieve various objectives. However, for market parties it is far more important to have clear, consistent long-term objectives than having to repeatedly respond to policy changes and initiatives.

6.6 GENERAL CONCLUSION

In spite of all the high-minded rhetoric, innovation is often a laborious process, as evidenced by the experiences documented in this chapter.Supporting innovations can be an uphill struggle against uncertainty, disbelief, and risk avoidance, as well as a difficult fight against vested interests, arcane conventions and procedures, and rigid institutions.

Innovation is no 'manna from heaven'; it causes some to lose, some to gain, economically or politically. Fear of losing market share or political influence are major obstacles. When the development of sustainable sources of energy appeared on the political agenda in the 1970s, the established energy companies stayed focused (with few exceptions) on fossil fuel and nuclear energy. The use of wind energy was considered unprofitable and was discouraged with very low resupply tariffs. This may be rational from a corporate point of view, but it shows once more that innovations often have to be initiated by economical and political outsiders, who should not count on support from those with the most powers.

Uncertainty and distrust also frustrate innovation. The Whisper is, in spite of the obvious advantages of reducing fuel consumption, CO_2-emission, and noise pollution, still not in use because manufacturers of buses and transport companies are not sure about its technical reliability and commercial feasibility, and do not want to take the economic risk. Authorities are – justifiably – not keen to intervene directly in the market, which would upset the 'level playing field' for transport and manufacturing companies. A 'Catch 22' situation prevents innovation from materialising. It illustrates how the innovation process falters if the corporate world and public authorities do not create space for experiments and do not demonstrate the courage needed to force a breakthrough.

Moreover, market failures can hinder innovation. The failure of the Whisper and the Energy House to succeed is partly due to the absence of positive external effects in the calculations of market prices. The implementation of the Electronic Patient Dossier, meanwhile, has been frustrated because the group who would profit most from its implementation (patients) is absent from the market. The authorities can adopt the role of regulator, market supervisor, 'launching customer' or moderator, trying to create openings, but all the time it runs the risk of stepping on economic and political land mines, 'blowing up' the project or at best causing delays.

The more comprehensive the consequences of an innovation are for existing methods and organisations, the more difficult and slower the introduction will be. The EPD is a case in point. It affects the policy, the organisation, and the financial interests of a large number of individuals and organisations. Without clear guidance, based on clear objectives, and a consistent strategy for implementation, a project like this can easily get bogged down in a lengthy and costly operation.

Last but not least, the government itself can become a part of the problem of innovation. Conflicting policy objectives, lack of knowledge of the market, limited policy horizon, risk avoiding behaviour in government bureaucracies, and fear of loss of political prestige can lead to interventions that frustrate rather than promote innovation.

As pointed out in this chapter, obstacles for innovation result partly from the nature of our institutions. The organisation of the energy market, the transport market, or health care services are not neutral entities, but reflect economic and political interests and powers. The same applies to taxation on pollution of the environment, investment subsidies, or rules for awarding contracts intended to regulate the market. Innovation policy thus requires willingness to adapt the structure of institutions if they obstruct innovation. Willingness to brave vested interests, to give outsiders a chance, to make experiments possible, to reward risk taking, to accept failure, to simplify procedures, to put reputations on the line, and to pursue a consistent policy. To swim against the current, so to speak.

NOTES

1 We use the word 'institutions' here as defined by North (1991: 97-112): 'Institutions are the humanly devised constraints that structure political, economic and social interaction. They consist of both informal constraints (sanctions, taboos, customs, traditions, and codes of conduct) and formal rules (constitutions, laws, property rights).'

2 See website Nederlandse Wind Energie Associatie: www.nwea.nl.

3 Interview with Linda Kamp, August 23, 2007.

4 In the province of Zeeland a subsidy on resupply to the main grid existed in the late eighties already.

5 Interview with Gijs van Kuik, August 13, 2007.

6 These include societal costs of sick absence and disability.

7 See website Ministerie van VWS: www.minvws.nl.

8 Until now this is restricted to personal data, medication history and applications for electronic expenses claims. Treatment history and potential allergies are not included yet.

9 Ministerie van VWS, 2007.

10 See www.minvws.nl.

11 Ministerie van VWS, 2005.

12 See www.minvws.nl.

13 See website Senternovem: www.senternovem.nl.

14 Interview with Cees Collart, Augustus 6th 2007.

REFERENCES

Acemoglu, D. and J. Robinson (2000) 'Political Lsers as a Barrier to Economic Development', p. 126-130 in *American Economic Review*, 90(2).

Baldwin, R. and F. Robert-Nicoud (2007) 'Entry and Asymmetric Lobbying: Why Governments Pick Losers' p. 1064-1093 in *Journal of the European Economic Association*, 5(5).

Baumol, W.J. (2002) *The free-market Innovation Machine: Analyzing the Growth Miracle of Capitalism*, Princeton, NJ: Princeton University Press.

Buchanan, J., R. Tollison and G. Tullock (eds.) (1982) *Toward a Theory of the Rent-seeking Society*, College Station, TX: Texas A&M Press.

Brenninkmeijer, R. (2005) 'Naar groepsgebonden, rationeel voorschrijven' p. 164-168 in *Pharmaceutisch Weekblad*, February 4.

David, P.A. (1986) 'Understanding the Economics of QWERTY: The Necessity of History', p. 30-49 in W.N. Parker (ed.) *Economic History and the Modern Economist*, New York: Basil Blackwell.

Duran-Reynals, M.L. (1946) *The Fever Bark Tree: The Pageant of Quinine*, quoted by Mokyr (2000).

European Commission (2006) *Action plan for energy efficiency: Realising the potential*, Brussels: EC.

Health Sciences Digest (2005) 'Elektronisch patiëntendossier (EPD)' in *Medisch specialist: 'What's in it for me?'*

Kamp, L. (2002) *Learning in Wind Turbine Development: A Comparison Between the Netherlands and Denmark*, Utrecht: University of Utrecht.

Kamp, L. and K. Mulder (2005) *Technological Promises Based on Old Traditions: The Introduction of Wind Turbines in the Netherlands*, Delft: TU Delft.

KNMG (2004) *Eindrapportage Onderzoek naar wensen en ervaringen van artsen op het gebied van autorisatie voor toegang tot en gebruik van patiëntgegevens*, Utrecht: KNMG.

KPN (2006) *Zorg voor innovation! Snelle Beter – Innovation en ict in de curatieve zorg*, The Hague: KPN.

Landes, D.S. (1998) *The Wealth and Poverty of Nations: Why Some are So Rich and Others So Poor*, New York: W.W. Norton.

Mast, E., G. van Kuik, and M. Zaaijer (2007) 'Offshore Wind Energy Policies and Their Effects: Experiences in Denmark and the UK and Prospects for the Netherlands' in *Proceedings of the European Wind Energy Conference 2007*, Milan, April.

Mokyr, J. (2000) 'Innovation and Its Enemies: The Economic and Political Roots of Technological Inertia', p. 61-91 in M. Olson and S. Kähkönen (eds.) *A Not So Dismal Science*, Oxford: Oxford University Press.

Ministerie van VWS (2005) *ICT in de zorg. Van Elektronisch Medicatie Dossier naar Elektronisch Patiënten Dossier*, The Hague: Ministerie van VWS.

Ministerie van VWS (2007) *Voortgangsrapportage ICT inzake de invoering van het Elektronisch Patiënten Dossier (EPD)*, The Hague: Ministerie van VWS.

Negro, S. (2007) *Dynamics of technological innovation systems. The case of biomass energy*, Utrecht: University of Utrecht.

North, D.C. (1991) 'Institutions', p. 97-112 in *Journal of Economic Perspectives*, 5(1).
OTB Group (2007) *Energyhouse. Independent zero emission solutions*, Eindhoven: OTB.
OTB Group (2007) *Memo energyhouse*, Eindhoven: OTB.
Provincie Utrecht (2005) *Aanvraag stimuleringsfonds luchtkwaliteit – schone bussen*, Utrecht: Provincie Utrecht.
Schop, E. (2004) 'Huisarts vindt EPD tijdrovend' in *Automatiseringsgids*, no. 40.
Spaink, K. (2005) 'Het medisch geheim is in gevaar' in *de Volkskrant*, September 3.
De Stentor (2004) 'TNO meet laag gebruik fluisterbus', October 27.
TNS NIPO (2004) *Fouten worden duur betaald. Een onderzoek naar medische overdrachtsfouten (deel 2)*, Amsterdam: TNS NIPO.
Ubels, R. (2005) 'Fluisterstil door de binnenstad' in *Nederlands Dagblad*, April 23.
Verweij, A. (2007) 'De patiënt is de baas' in *Digitaal Bestuur*, January.
de Ziekenhuiskrant (2007) 'IT-ontwikkelingen in zorg lopen vast', June 26.

7 COLLABORATION, TRUST, AND THE STRUCTURE OF RELATIONSHIPS

Bart Nooteboom

7.1 INTRODUCTION

The literature on 'open innovation' and the analysis in chapter 3 recommend collaboration between organisations as a source of innovation. Dutch and EU innovation policy, on all spatial levels (regional, national, supra-national), have caught on to the trend of 'open innovation', in the stimulation and even organisation of collaboration, networks, and local 'clusters'. However, collaboration is often risky and difficult, and frequently fails. So, we need to consider the governance of collaborative relationships.

This, I propose, entails a shift in the notion of governance. Traditionally, concepts of governance have been oriented towards control. The World Bank defined governance as the way in which power is exercised, the UNDP defined it as the exercise of economic, political, and administrative authority, and the OECD defined it as the use of political authority and exercise of control (Weiss 2000). In a world of relationships and networks for collaboration, approaches of implicitly or explicitly unilateral control, authority and power are no longer adequate, because there is no central authority or controller. The economic perspective of 'principal-agent' relationships and the business perspective of 'marketing warfare' have become counterproductive. In relationships of collaboration, players are each others' agents as well as principals. Operation in markets is not warfare but alliance management. Governance in networks must be multilateral, in equilibration of power or dependence, somehow. If we can still talk of control, it must be mutual control.

It is routinely recognised that collaboration requires trust. Especially in innovation, uncertainty is too large to allow for complete contractual control. The uncertainty of innovation makes it difficult to foresee what needs to be contracted for (tasks, rights, duties, penalties, responsibilities, goals, competencies), the novel opportunities that will require a change of direction, the pressures that may tempt people to renege on commitments, and the avenues available to conduct such escape. Beyond technical uncertainty for contracting, there is the more fundamental uncertainty in exploration, or in 'interpretation' as Lester & Piore (2004) called it, that goals, means, and their causal connections and resource requirements are not yet known, so that the actors involved must take the time to deal with ambiguity, and to iterate between goals and means, without knowing where they will wind up. Here, one must resist inclinations or pressures to go for the quick fix of doing what is known, arguable, and measurable. Acceptance and utilisation of such uncertainty in the form of ambiguity of goals and means requires trust.

However, while there is widespread recognition of the importance of trust for innovation, it remains unclear what the meaning, conditions, and limits of trust are. Trust is a complex notion that has caused much misunderstanding and confusion. What do we mean by trust, what is the basis for it, and what are its limits? What is the relationship between trust and incentives? The unravelling of trust requires some space, in this book, since an understanding of trust is needed for insight into collaboration within organisations, which forms the micro-level structure of innovation systems (analysed in chapters 8 and 9), and in collaboration between organisations, in networks and clusters, which forms the meso-level structure of innovation systems (analysed in chapters 10 and 11). Conditions for building trust depend on network structure.

The analysis in this chapter has implications for policy, in particular the issues of whether and how government should pick economic activities and design or facilitate their configuration. There are several different aspects to these issues, which are not all taken up in this chapter. Here, we focus on the role of brokerage in building trust between actors: is this a role for government?

7.2 TRUST IN WHOM AND WHAT?

First, what is the object of trust: to what sorts of things does trust apply? We can trust objects, but trust becomes more complex when we trust things that have intentionality, such as people, organisations, institutions, and socio-economic systems, in 'behavioural trust'. In collaboration between organisations people need to trust the people they are dealing with, as well as their organisations, and relevant institutions, such as legal systems, that regulate ownership, contracts, etc. If we trust our personal contacts, but they do not enjoy the support of their superiors or associates, such trust is not very reliable. Vice versa, we can trust an organisation, perhaps because of its reputation, but if its trusted policies are not reliably executed by its personnel, such trust is also not very reliable. In the case of small firms one may go a long way with trust in the owner-entrepreneur, but in larger organisations we should be able to trust the people, the organisation, and the relations between them. Those relations depend on the positions, roles, and functions of people, organisational procedures, and organisational culture. Commercial organisations may come under commercial pressures to renege on commitments. Public organisations can be benevolent and have integrity, but their conduct depends on politics, which can be volatile. Civil servants with high personal integrity can be institutionally unreliable. In government there are many horizontal, vertical, and lateral connections relevant for trust.

Within the chain of justice, distrust in one link can spill over to another. Distrust in the justice department can affect trust in the police, and vice versa. Distrust in the police can spill over into distrust in politics and in the rejection of the state's monopoly on violence.

The next question is in what, i.e., what aspect of conduct, we can trust. We can have trust in *competence*, i.e., the *ability* to act according to agreements and expectations. We can also have trust in *intentions*, i.e., the *will* to act 'properly', with attention, commitment, and benevolence (no opportunism, no cheating, no free riding). The distinction between competence and intentional trust is important. When something goes wrong because of lack in competence, we react differently from when something goes wrong because of opportunism, or because of neglect or lack of care. In the first case we can invest in an improvement of competence |(e.g., by training, advice, and help) while in the second case we may tighten a contract. We can trust competence but not intentions, and vice versa. If it concerns a partner in collaboration, we should be able to trust both competence and intentions. If it is an adversary, and we do not trust his intentions, we hope he is *not* very competent. Trust suffers from 'causal ambiguity': if something goes wrong we often do not know the cause, or even what exactly went wrong. It may be due to a mishap or accident, a shortfall of competence, lack of attention or commitment, or opportunism. This causal ambiguity has serious consequences for dealing with trust, and in particular for the openness needed for trust.

When we trust the integrity of the police we would like to also trust their competence. Surveys in the Netherlands indicate that the Dutch trust the integrity of the police, but not so much their competence. That is better than the other way around.

7.3 TRUST AND CONTROL

What is the difference between trust and control? Can we speak of trust when reliable behaviour is enforced by contract or hierarchy? And if it is based on material incentives such as profit? If we take trust broadly as the expectation that, for whatever reason, 'things will be all right', despite risks of dependence, then control is part of trust. Is this what we mean by trust, or can we speak of trust only when control no longer applies? Are people trustworthy not because they are forced or rewarded, but because they choose on the basis of the intrinsic motivation of ethics or solidarity? Then trust is to be defined as the expectation that 'things will be all right' even if the partner has both the opportunity and an incentive to cheat. Trust then entails that we do not assume that partner will defect as soon as he sees a more profitable opportunity ('exit'), but that he will give us a chance to jointly improve the relationship ('voice').

To avoid confusion, I (Nooteboom 2002) proposed to speak of 'trust' and 'trustworthiness' only beyond control, and to employ the terms 'reliance' and 'reliability' when there is also or is only control. In other words, reliance can be based on trust, control, or a combination of the two.

Trust may be based on psychological mechanisms that move people to trust or distrust others (causes of trust), and on a rational assessment of trustworthiness (reasons for trust). Even while trust is often based on information (one has good reasons to trust), there is never certainty concerning future conduct. This uncertainty needs to be bridged by a 'leap in the dark' and willingness to extend the benefit of the doubt. It is very fruitful to employ insights from social psychology into the non- or partly rational motives and decision heuristics that lie behind trust and trustworthiness (Nooteboom 2002), but in the following we focus on the rational reasons why people may be reliable. A survey is given in table 7.1 In the two columns of the table reasons for reliability are categorised on two levels: within a relationship (micro) and in the institutional environment of the relationship (macro). A distinction is made between control and trust.

Table 7.1 Sources of reliability

	macro; institutional	**micro; relational**
control		
opportunity control	contracts, law	hierarchy, 'fiat' of management
incentive control	reputation	dependence, incentives, hostages
trust	values, social norms, moral imperative	empathy, identification, routines, affect, friendship

Source: adapted from Nooteboom (2002)

One form of control is *opportunity control*. Here action space is constrained, either by legal contracts (macro) or by hierarchy (micro). Another form of control is *incentive control*, where actions are elicited by rewards. Within a relationship this can be due to dependence: when a partner is more dependent he is more inclined to take my interest to heart. This can be due to the unique value that I offer to him, or due to his relation-specific investment in the relationship. The latter notion, derived from transaction cost economics, suggests that the investment has (full) value only in the specific relationship, and when the relationship breaks an investment has to be made anew in another relationship, which entails switching costs. Also taken from transaction cost economics, the partner may also be dependent because I own a hostage from him: something that he values but that I do not, which I can destroy without hesitation when the partner defects. This can take the form of a minority shareholding that may be sold in the hostile takeover of the partner. Mostly, it concerns commercially sensitive information that can be surrendered to a competitor of the partner. Another form of incentive control is reputation. That also is a matter of self-interest: the partner behaves well, because bad behaviour not only jeopardises the focal relationship, but also the development of fruitful relationships with others.

Trust and trustworthiness need to begin where control ends, or control begins where trust ends. Trustworthiness can be the result of established codes of conduct, based on widely shared norms and values or habits. Within a relationship trust can be based on values and routines or empathy developed in the course of a relationship. Empathy is the ability to put oneself in the shoes of a partner to gain insight into his needs, views, expectations, fears, and his strong and weak points. This can lead on to identification, i.e., the experience of a shared destiny and a merging of needs and goals. For fruitful collaboration, identification is often not needed, and empathy may suffice. Trust and trustworthiness can also emerge from sheer routinisation, where things are taken for granted because nothing ever went seriously wrong. In identification and routinisation trust can go too far, where flexibility disappears and opportunities for innovation are missed.

Trust and control are both complements and substitutes. With more trust less control is needed. But since both trust and control have their limits the one may be needed where the other ends. Contracts should not be so strict, extensive, or adversarial that the basis for trust is destroyed. Trust can both precede and follow contracts (Klein Woolthuis et al. 2005). Excessive and strict contracting tends to obstruct the development of trust, but limited contracting can set a relationship going, and when trust grows contractual slack can increase. Without some prior trust one may not want to take the trouble and risk of crafting a contract.

A certain balance of dependence is generally desirable. Unbalanced dependence is not hopeless but it is more difficult to sustain. There is no need for balance in every single determinant of reliability, and imbalance in one may be offset by counterbalance elsewhere. If one partner is more dependent due to a greater share in specific investments, balance may be achieved by redistributing their ownership, or by offsetting the imbalance with another instrument, such as a hostage.

As discussed above, third parties or go-betweens can play important roles in the management of collaboration. A policy question is who plays such roles: (local) governments or private go-betweens in commercial business services? They can offer arbitration or intermediation in conflicts, but there are more potential roles to play. There is one aspect that merits special attention. As noted earlier, trust suffers from causal ambiguity: when something goes wrong it is often not clear what caused it. One can draw the wrong conclusion that the cause was opportunism, because that is what one fears most, while in fact there was merely an accident. But how can one know? When detected, opportunism may be masked as an error. A go-between may disambiguate the situation, explaining what really happened, eliminating misunderstandings, and side-stepping emotions due to fear and suspicion – which yield a vicious circle of mistrust that may escalate beyond repair – and propose remedial actions. Go-betweens can also contribute to an efficiënt and reliable reputation system by filtering just from unjust accusations of opportunism or incompetence and broadcasting the results. Go-betweens may be found at banks, trade or professional organisations, knowledge transfer agencies, or might be lawyers or private consultants.

There are limits to trust. While reliance can go beyond control, to include trust, the room for trust depends on pressures of survival. When a manager is pressured to catch any profit he can or else the firm will go down, his loyalty will likely lie more with his firm than with external partners, and he may be forced to renege on agreements with them.

In 1983, when due to a recession there was a strong pressure on price, in the highly competitive car industry, car producers reneged on commitments to suppliers, which had arisen earlier in the then fashionable rhetoric of co-makership and long-term supplier relationships. At that time I gave a lecture to an organization of buyers and suppliers in that industry, proposing that there is both trust and opportunism, limiting each other. I was told to be naive, talking about trust, since for them, opportunism ruled all. A few days later I gave a similar lecture to the organization that exploits natural gas. There, I was told I was cynical, talking about opportunism, since in their relationships with suppliers there was ample loyalty and give and take. Driving home I realized that was easy to say, for a largely state-owned monopolist.

In chapter 3 we quoted the proposition from Lester and Piore (2004) that competition can eliminate the conditions and resources needed for the 'interpretation' that is part of exploration. Here we see a specific aspect of that: pressures of competition can eliminate the trustworthiness needed for innovation.

7.4 CONDITIONS AND THE ROLE OF GOVERNMENT

As indicated, there are various ways to manage relational risk for the sake of innovation. None of them is universally the best way, and a proper mix should be found to fit the circumstances. Table 7.1 offers a basis to find such a mix, where one must take care that different instruments complement and do not operate against each other. If appropriate external institutions (legal system, reputation mechanisms, shared norms and values of conduct) are lacking, the basis must be sought within relationships. If there is no basis for trust, one can only fall back on control.

A classic case is Italy, where a paucity of reliable institutions and government necessitated, as a substitute, the development of skills and traditions for networking of private relationships, for which Italy became a proverbial success, which others then wanted to imitate even if they do not suffer from comparable institutional weaknesses.

Contracts make no sense in countries where there is no adequate legal basis, or where the police or judiciary are incompetent or corrupt. Contracts also make no sense if one cannot reliably monitor contract compliance, as may be the case in highly specialised professional work that does not yield a deliverable that can be judged. Reputation mechanisms are not automatic, and reliable go-betweens may not be available. If economic or political volatility is so high as to discourage a longer-term perspective in the development of trust, or pressures of survival are so harsh as to preclude solidarity, the only option may be to fall back on family, friends, or the clan, limiting the perspective for prosperity by division of labour, long-term investment and innovation from cognitive distance. There lies some of the tragedy of underdeveloped countries.

Thus, a primary task for government is to maintain an adequate institutional basis for trust and control (Nooteboom 2000). That is an important location factor, particularly for innovation. With a weak institutional basis for ethics and empathy, one will have to expend more on legal control and control by incentives, with corresponding monitoring. That is costly and bad for innovation.

The Netherlands has a tradition of deliberation, in what has been called the 'polder model', as a basis for building trust and consensus, notably between employers, employees and government, in negotiating wages and conditions for labour, but also more widely, as part of organizational culture. Here, the Netherlands may even be at an advantage over the US, with its more exit- and litigation-oriented mode of operation. Deliberation for consensus can be carried to excess, in the inability to make any move without consensus, but it does provide an advantage in the exercise of 'voice' for the building of trust. Decision making may take a little longer, but execution may be faster and more productive, also, and perhaps even especially, in innovation. There is all the more reason for surprise that labour has not been represented in the 'Innovation Platform' to participate in the deliberation on innovation policy.

While a certain amount of flexibility of relationships, in labour, ownership, and inter-organisational relationships, is needed for innovation, present market rhetoric that pleads for maximum flexibility goes too far. Relationships require a certain amount of investment in mutual understanding, intellectually and morally, to cross the cognitive distance that is beneficial for innovation (see chapter 3), and to invest in the building of trust. Such investments tend to be relation-specific, and by the logic of transaction cost economics this entails that relationships must be expected to last sufficiently long to recoup such investment. Hence we should aim for optimal, not maximal flexibility: long enough to warrant specific relational investment, but not so long as to produce rigidities.

Do governmental agencies have a task in the building or support of relationships for innovation? In present Dutch policy concerning 'key areas' (*sleutelgebieden*) the assumption is that such a role of government is indeed needed. Concerning the policy of key areas, there is much more to be discussed, and we will turn to these issues later, in chapter 12. Here we focus on the potential role for government as a go-between in building trust. As indicated in chapter 1, in the discussion of system failures in innovation, we can see reasons for that, and government may then use table 7.1 for an analysis of the situation and a choice of instruments. However, we also have reservations. In particular, governments should ask themselves if they are the appropriate actors to play the role of go-between. They should be aware of the dangers, which can also be seen from the table. Governmental agencies are more vulnerable to loss of reputation than most firms. Earlier we indicated that there are many connections between layers and chains of government, where distrust can spill over from one part to another (we took the chain of justice for an example). Also, actions, pronouncements and mistakes can be used as hostages, for blackmail, with the threat to publicise them unless demands are met. Government itself is more constrained than private actors in such naming and shaming. The government must be wary in building trust on the basis of personalised empathy and identification, in view of the risk of real or apparent corruption. In view of democratic accountability and transparency, governments must exercise more openness, which they cannot reserve only for preferred partners.

The possibilities are larger and the risks smaller for local/regional authorities than for central ones. At the central government level, distrust spills over more widely. Local government has better access to knowledge of local, historically grown specificities and peculiarities, and is better embedded in local reputation mechanisms. It can more easily offer a wider, integrated whole of subsidies, permits, location, housing, and facilities of schooling and training. The downside, of course, is that precisely because of these conditions there is a larger risk of clientism or even corruption. All the more reason for central government to stay aloof and maintain supervision of decent local government.

Let us reconsider the case of Italy. The institutional shortcomings that left the need to fall back on network relationships opened up the opportunity for the mafia to move in, and for public servants and politicians to move in, in a mix of public and private relationships that easily evoke corruption. Anthony Pagden (1988) gave an analysis of the breakdown of trust in the kingdom of Naples in the 17th and 18th century, when it belonged to the Habsburg empire of the Spanish king Carlos and later Philips II. The Spaniards opted for a cheap way to suppress the country from a distance, by breaking down the social structure and the culture of trust. In the Netherlands they had learned that sheer military suppression is costly and does not suffice, and that for control from a distance the institutions of a society must be broken down, in a strategy of *divide and rule*. The nobility was divided by creating a new, upstart nobility that owed its position to the Spanish masters. These were given the task of collecting exorbitant taxes, of which they could keep a large share for themselves provided that it be spent on idle, economically useless and politically harmless activities such as duelling and the defense of personal honour. This destroyed trust in the nobility as defenders and personification of order and reliability. The critical role of the intelligentsia was destroyed by a relaxation of academic standards. Universities were obliged to continue the teaching of Aristotelian logic "because it never accounted for anything". Excessive attention to religious ceremony was required. Academic requirements for the legal degree were lowered, and the degree was also awarded as a token of honour to the upper class, which contributed to the undermining of the legal system. Arbitrary and unpredictable exceptions to legal rules were granted. Through the oversupply of incompetent lawyers their price was lowered, yielding an excess of worthless and inconclusive litigation. Social ties were replaced by mutual suspicion, and people were thrown back on themselves or close family. Trade became a game of mutual cheating. Exchange was reduced to immediate quid pro quo, without credit or investment. Gambetta (1988) showed how this breakdown of institutions as a basis for trust in the kingdom of Naples, which included Sicily, allowed the mafia and the Gamorra to insinuate themselves, stealing into the holes created by institutional breakdown, and to fill the vacuum with their perverse order. Of course we are not suggesting that Dutch policy for 'key areas' will inevitably lead to such dire conditions, but the Italian case does yield a lesson.

In view of the difficult capability of competent support as a go-between, and the risks for government, both local and central, it is recommended to stimulate the emergence of commercial services of go-betweens, who may compete in the art of the go-betweenit. Some existing 'knowledge intensive business services', such as consultants in engineering, marketing, and finance are orienting themselves to this emerging market, but there are opportunities for new services that are more dedicated to the issues involved. The availability of reliable go-betweens contributes to the attractiveness of a country or region.

7.5 TRANSPARENCY AND TRUST

As discussed in chapter 3, in innovation one should not try to do everything oneself, and one should utilise collaborative relationships with others, in 'open innovation'. That requires openness of communication, and many firms struggle with that, from fear that others will hi-jack knowledge in order to compete. As a result, in negotiation people are wary of openness. However, good negotiation is looking for the solution of problems for the partner that mean much to him and cost you little. If the partner acts in the same way, everyone benefits.

In teaching, health care, and other professional work there are increasing complaints of an excess of control that eliminates motivation and the room for action that are needed especially for innovation. Hence there is a call for less control and more trust. At the same time, people demand transparency, to enable trust, but what is the difference between transparency and control? If one demands or imposes transparency that is proof of distrust. But can one trust without transparency? Trust and transparency are both needed, and they both require and enable each other, but neither can be imposed: they must be earned. This requires 'voice' rather than 'exit' (Hirschman 1970). In voice one expresses one's weaknesses and fears, expecting others to react constructively, with the commitment to jointly solve problems. Only when that fails persistently one falls back on exit: one walks out, fires personnel, sells shares or a business, or forces a cabinet crisis. There are several reasons for voluntary transparency, but a condition is that the partner responds constructively.

One must give room to a partner to contribute his creativity and competence. One needs the partner because he knows or masters something that you do not. Then is is odd to pretend that you can tell him what precisely he must do, and that you can reliably monitor and assess his actions. Monitoring and control are limited by cognitive distance, which, as argued in chapter 3, is needed for innovation. That applies to firms as well as professional work in education and health care. Partners should be open in terms of what they want and expect, and in terms of what concerns or worries them: what dangers or risks they see, if only because otherwise the partners cannot know what they can do to help.

Within teams, in firms, or in inter-firm collaboration, openness requires 'psychological safety' (Edmonson 1999), i.e., the confidence that in reporting an error one will not be made a fool of in public. Further conditions for innovation in teams within organisations are discussed by Anderson and Gasteiger, in chapter 9. For instance, they report that "leader-member-exchange (LMX)" in terms of supervisor support, in relationships characterised by mutual trust and respect, is positively related to creative and innovative behaviour (Harrison et al. 2006).

In collaboration between firms, a danger that is often seen is that in openness you may surrender commercially sensitive information with which the partner can become a competitor, or which via the partner can reach a competitor, so that one cannot 'appropriate' the rewards for innovation. That danger may be real, but in innovation is often smaller than people are inclined to think. The issue there is not whether sensitive information reaches a competitor, but whether he can absorb it, i.e., understand and implement it in his organisation, given his knowledge, assets, and competencies, and whether he can achieve that so fast that competition is still effective. If by that time the knowledge involved is already superseded, there is no risk.

Nevertheless, in collaboration one is vulnerable to mishaps, mistakes, lack of dedication, and opportunism. As indicated earlier, that yields causal ambiguity: one does not know which is at work. And when something goes wrong people may all too readily jump to the conclusion that opportunism is at play, especially when the possible loss is large or when one has little self-confidence. In the 'Calimero syndrome' the weak or dependent are overly sensitive to risks and threats, and see bad will wherever something goes wrong. Given all this, openness is the best strategy. It is in one's own interest, when something is seen to go wrong, someone has not paid attention, or falls short in competence, to report this to the partner and to help to preempt the problem or solve it in an early stage. It is difficult to admit error, but in trying to hide one's failures one jeopardises one's own interests. When you don't report it, you are suspected of bad will when later the problem appears anyway, as it usually does. If it was an error and not bad will, why didn't you report it when something could still be done about it? Mistrust, once rooted, is difficult to eradicate. Trustworthiness must be shown time and again; untrustworthiness can appear from a single action. Hence the saying that 'trust comes on foot and departs on horseback'. Also, by remaining silent one robs oneself of the opportunity to learn, with the help of the partner, and that is what counts in innovation. If one doesn't like that, then one should not start a partnership for innovation.

Openness cannot be imposed, and must be earned by a constructive response to reports of error. Such reports are valuable, in order to repair the error and to prevent a repeat, in a shared effort, with pooled resources. If, on the contrary, an error report is used to impose guilt and punishment, then transparency will disappear, and we fall back on straightjackets of control that smother the movement and motivation of innovation.

Trust and transparency require empathy: the ability to imagine oneself in the position and perspective of the other, and to look at one's own actions from there: how would one interpret them, and how would one feel and respond? Empathy increases the ability to collaborate, in crossing cognitive distance. It aids insight into how to help the partner. It also reduces causal ambiguity: one can better see what is at play when something goes wrong. One can better foresee forced failures: how would one act if one were under the same pressures? Earlier we noted that trust has its limits due to pressures of survival. One may be forced to renege on commitments. However, one has the option to be open about that as well, warning the partner in time, and explaining one's predicament.

Earlier, we noted that the Netherlands has a tradition of deliberation, as part of political and organisational culture. Hence, one might expect that the Dutch are better at this game of openness, transparency, and trust. To some extent that may be the case, but like everybody else the Dutch are caught in the momentum of 'accountability' on the basis of strict and measurable performance on 'deliverables', in order to ensure control, which comes at the expense of risk taking, of room for professional discretion and experimentation, and of openness, which are all needed for innovation. This drive towards accountability by strict performance measurement is accentuated by considerations of efficiency, in that limited resources and time preclude the more time-consuming judgement of less-easily measured forms of quality.

As scientific director of a 'Ph.D. school/research institute' of the faculties of economics, business and spatial sciences at Groningen University, the present author was confronted, in the 1990s, with the need to evaluate research performance, and pressure was high to do this simply by counting journal publications ranked according to their impact score (average number of times a paper in that journal is cited). One objection to this practice is that for any journal the variation of citations around the mean is enormous. Why not count citations of specific publications, then? That also, of course, has its problems. A more fundamental problem was that books and new journals were not counted, while in some disciplines cases books are more important than journal articles. A monograph can yield an opportunity to develop a new synthesis. The problem with books and some journal articles was that to judge their quality one would have to have them actually read by some competent readers, and for that there were no resources. We did it anyway, of course, but how many institutes do?

In its subsidy programmes for innovation, in particular the 'Framework Programme' (FP) for collaborative research between research organisations and firms, the EU appears to be caught in an upward cycle of control that is becoming tighter, in the sequence of the fifth to the sixth and now the seventh FP. Earlier on, in FP5, EU programme managers were personally involved in the progress of projects. This allowed them to informally assess progress and problems, employing tacit knowledge, and to judge proposals for changes of direction as they emerged. Increasingly, they are being pulled out, and control has become more distant, and as a result more codified, explicit, and detailed, in the form of criteria, rules, and regulations. This has the adverse effect of increasing transaction costs, which is problematic especially for smaller firms, and reducing the flexibility needed to allow for the openness to surprises and unpredictable twists and turns that are characteristic for innovation.[1]

7.6 TRUST AND NETWORK STRUCTURE

The structure of networks has effects on innovation in two ways: on the competence side (learning) and on the governance side (managing relational risk) (Nooteboom 2004). On the competence side, the number, diversity, density, centrality, and strength of ties have effects on the diversity and accessibility of network nodes as sources of information, and on the flexibility of making and breaking relationships between them. On the governance side, they affect conditions for trust and control, such as monitoring, reputation mechanisms, bonding, and coalition formation to constrain conduct.

On the competence side, Granovetter (1973) and Burt (1992, 2000) argued that dense and strong ties are bad for innovation, and more distant, weak, and less-dense ties generate the higher level of variety that is needed for innovation. This is related to the notion of structural holes, discussed in chapter 3. Bridging a structural hole taps into variety for innovation, but in dense networks there are no promising holes left. Dense and strong ties keep players from moving out and new players from moving in. Dense ties also yield redundancy: If you are connected to A and to B who are also mutually connected, one of the ties is redundant. If you only had a tie to A you could reach B through A. Since ties are costly to set up and maintain, this is more efficient. From the analysis of cognitive distance it follows that ties that are strong in the sense of durable and exclusive will in due course reduce cognitive distance and the dynamic potential of the ties. However, note that less exclusiveness yields greater density of ties. Gilsing and Nooteboom (2005), argued that in exploration, where knowledge is embryonic, density is needed for agents to complement their absorptive capacity (in the collaboration with A you also use B to help you understand A), and in order to 'hedge bets' concerning the availability of partners. If in exploration there is much volatility of entry and exit, you cannot count on partners yielding access to other partners, since intermediary partners may soon drop out, so that one must also have one's own, direct access to other partners. This also increases density. Furthermore, in exploration the cost of maintaining a relationship is often less, and in any case in exploration cost matters less (than in exploitation), so that the cost of redundant relations matters less. An environment of multifarious interactions, in business and recreation, in 'local buzz', contributes to the utilisation of diversity on the basis of planned and unplanned encounters (Bathelt et al. 2004).

On the 'governance' side of relationships, Coleman (1988) argued that strong and dense ties enhance control and bonding. Strong ties, in the sense of high frequency and intensity, and long duration, yield shared experience, which reduce cognitive distance, and enable the development of empathy and identification.[i] Multifarious personal encounters, in sufficiently dense networks, can be indispensable for building trust, particularly in the more personal foundations of trust, such as empathy and identification. Business trustworthiness may be assessed in part by how people treat a waiter during a business lunch. Second, local 'buzz' may be needed for reputation mechanisms. Third, reliable, locally embedded go-betweens may help to forge and manage collaboration – in the bridging of cognitive distance and in the management of relational risks – and this is supported in a milieu that is rich in diverse and frequent interaction. To generate and utilise local buzz, opportunities for collaboration, and reputation mechanisms, networks in local clusters tend to be fairly dense, with many ties between actors. Ties also tend to be fairly strong, with frequent interaction, and investments in mutual understanding and trust. As indicated earlier, the building of mutual understanding and affinity, to cross cognitive distance and to build and maintain trust, requires investments that are to some extent specific to the relationship. In local networks an investment in a relationship will sooner have value in another local relationship, i.e., will be less specific, which encourages such investments. In so far as the investment is still specific its utilisation requires some continuity of the relationship or frequent interaction, which are both more easily achieved at a small distance.

These arguments confirm the value of agglomerations with fairly dense and strong ties between actors that offer related variety, at sufficient but not too large cognitive distance. However, earlier (chapter 3), from the cycle of discovery, we indicated the need for connections between local clusters and clusters elsewhere in the world: an outside avenue is needed for the processes of differentiation and reciprocation that yield exploration. Earlier, we also indicated that a lengthy and exclusive relationship can yield too large a decrease of cognitive distance, whereby its innovative potential dissipates. Then one needs either 'channels' out of the agglomeration, a lively entry and exit of new players, or a combination of the two, in order to replenish the variety of inspiration and insight, based on a variety of experience, and thereby maintain cognitive distance.

In other words, agglomerations should be open, with outside connections, domestically and internationally, to other agglomerations, and should promote entry and exit of firms. Concerning entry and exit, Knoben (2006) showed that for automation services the relocation of firms has a positive effect, especially in the longer term, on the innovativeness of firms.

Concerning external linkages, similar arguments were given earlier by others (Asheim and Isaksen 2002, Boschma and Lambooy 2002, Oinas and Malecki 2002). Here, we wonder whether perhaps universities can provide the connecting nodes of small worlds, connecting regional innovation systems to similar systems elsewhere in the world (Benneworth et al. 2006 Kitagawa 2005). Universities are geared to access such internationally dispersed communities.

Combining the arguments for and against agglomerations with dense and strong ties, we arrive at a theoretical argument for what is called 'small worlds', illustrated in figure 7.1. There, different local networks with high density and strength of ties are mutually connected by less dense and weaker ties. The strong and dense local ties enable the utilisation of opportunities for novel combinations in related variety, flexible reconfigurations of ties, high joint absorptive capacity, and support governance by reputation, go-betweens, and trust building. On the other hand potential weakness due to lock-in, too much stability of ties, shrinking cognitive distance, and reduced variety, are compensated by much entry and exit of players and the weaker and less dense ties to other communities that extend the scope of variety and maintain cognitive distance.

Figure 7.1 Small worlds

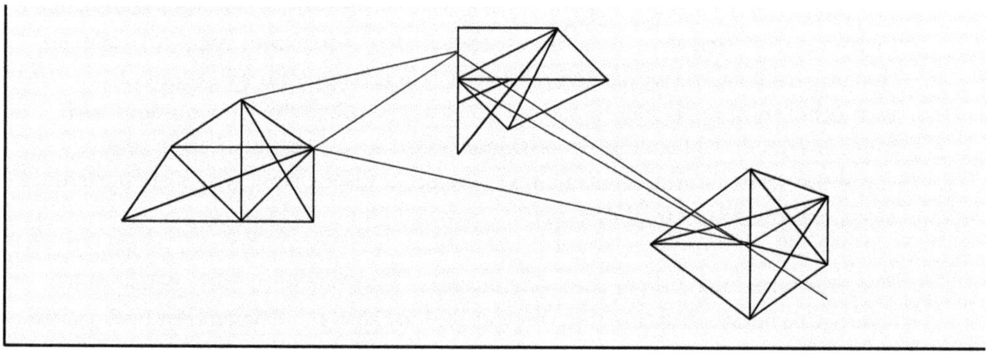

Gilsing and Nooteboom (2006) found this structure for Dutch biotechnology, with dense local clusters with strong ties, around universities, supplemented with more sporadic and variable ties with similar clusters elsewhere in the world. The importance of these external linkages is illustrated also by the finding that in biotechnology 70 percent of co-publications are with scholars abroad (Ponds et al. 2006).[2]

In the development of small world structures we see a reflection of the phenomenon that the economic significance of the national level declines and that of the global and regional levels increases. That does not say that there is no role left for the national level. Perhaps its crucial role is to see to it that the connections between the regional and the global are indeed made.

Small world structure is also emerging in global networks of political governance (Soh 2007). There, the locally dense clusters are national governments, but nodes should be added that intermediate in connections between governments, or independently connect with several or all of them. In the present context of business or regional networks, such nodes arise in the form of various intermediaries such as universities, professional service firms, and professional or trade associations. Such intermediaries can help to create mutual understanding (crossing cognitive distance), and can play the other roles of third parties in building and maintaining trust that were discussed before.

As an agglomeration increases to a larger size and diversity of activities, it can remain innovative longer without connections to outside agglomerations, although one may expect that it may then break up into a set of smaller agglomerations. In a small country, like the Netherlands, agglomerations will be smaller than for example in the US or China, and therefore especially in a small country the need for outside connections is great.

The analysis has implications for the dynamics of clusters or networks (Nooteboom 2006). As discussed earlier, in the early stage of exploration one would expect a relatively high need for local embedding, in strong and dense ties. Later, one would expect a certain amount of disembedding, to utilise the potential of emerging innovations in distant markets, and to achieve access to novel sources of novelty to replenish local variety and restore cognitive distance. This raises considerable complications for a policy for local clusters and regional innovation systems. Are policymakers able to correctly identify the stage of development that a local cluster is in, and are they able to implement policy on time, before development has reached the next stage, where the policy may be counterproductive? One may wind up furthering local embedding by the time that disembedding is needed.

In sum, we see a trade-off between central and local/regional government in the promotion of innovative clusters. Local/regional government is superior in its local knowledge and governance of embeddedness and related issues of infrastructure of various kinds – traffic, zoning, housing, education, and training – that have an impact on knowledge ecologies. They can intervene faster to ensure that intervention is in tune with the stages of development. However, local/regional government may get too entangled in local embeddedness, which may yield clientism or even corruption. There is a danger that an agglomeration might become disconnected from other agglomerations, cutting off access to sources of new knowledge, to complementary resources, and to distant markets. Local or regional authorities may try to keep firms locked into their regions or municipalities, thus obstructing healthy relocations. Central government has a role to monitor developments and to ensure outside linkages and disembedding where needed.

7.7 CONCLUSIONS

For collaboration, trust is needed, especially in exploration, where high uncertainty limits the scope for contracts and monitoring of contract compliance. To eliminate misunderstandings concerning the notion of trust it is useful to distinguish between reliance and trust. Reliance includes control, by contractual enforcement or hierarchy, and by incentives of dependence, reputation, and hostages. Trust goes beyond control, in norms and values of ethical conduct, and in routinised conduct and personalised empathy and identification. However, trust has its limits, which depend on pressures of survival. Intense price competition reduces the scope for trust, and hence is part of the overall obstacle that competition may present to exploratory activities.

Trust, openness, and voluntary transparency based on trust that openness will be met constructively, in a culture of 'voice', are needed to give professional work the autonomy and room that it needs to exercise the discretion of professional judgement, and the space needed for the experimentation of innovation. This requires a transformation of the current drive towards accountability on the basis of the strict measurement of 'deliverables' (even where the latter cannot be clearly specified), which can be detrimental to more difficult to measure dimensions of quality. This connects with earlier propositions, in chapter 3, concerning the inherent unpredictability, variety, and risk of innovation (particularly of exploration) that need to be accepted for innovation to take place. The connection between innovation and organisation, including the issue of accountability and performance measurement, particularly in the public sector, is further discussed in chapter 8.

Concerning the structure of networks for innovation, there are arguments in favour of dense and fairly strong ties, based upon considerations of both competence (learning) and governance (trust and control). However, such networks carry the danger of getting locked into insufficient variety, cognitive distance that is too short, and insufficient flexibility. To repair for that, and to complement local dense and strong ties, weaker and sparser ties to other, outside networks are needed, to yield a 'small world' structure. This yields a call for 'open agglomeration', i.e., an agglomeration that is open to outside ties, and to the entry and exit of players. This connects with our general call for an open innovation policy in chapter 3. Perhaps universities can play a role in providing such connections to outside sources of knowledge and expertise.

In current Dutch innovation policy, central government, having recognised the importance of collaboration for innovation, in its policy for 'key areas' goes quite far toward getting involved in crafting collaborative relationships between firms and knowledge institutes. Arguments for local embedding of such collaborative structures suggests that local authorities, with knowledge of local specificities and attuned to local reputation mechanisms, seem to be in a better position to do that. Also, local authorities are likely to be faster in identifying the stage of development that a local cluster is in, and faster in implementing a policy of embedding or disembedding that is appropriate to that stage of development. In view of the complexities of collaboration and of the relational risks and complexities of managing reliance and trust, the question is whether govermental agencies are equipped for this task, and whether they might become too involved, with risks of being taken hostage by private interests. Preferably, an increase in private sector go-betweens will occur to fulfill the emerging demand for go-betweens in the emerging network economy. If government does have to play the role, at least temporarily, local government appears to be better equipped.

On the other hand, the close involvement of local/regional government in local dense and strong ties carries a risk of clientism or even corruption. All the more reason for central goverenment to step back and concentrate on preventing that. Local/regional government may neglect or be unable to craft the outside linkages needed for open agglomeration, and may be tempted to try and keep firms from moving out when that is best for innovation. Central government has a role to ensure openness and outside connections. The role of networks is further analysed in chapter 10, and regional innovation systems are further analysed in chapter 11.

On the national level, we might consider opportunities for the Netherlands to develop into a 'knowledge ecology' for connecting exploration and exploitation on a global scale. We refer to the notion of a knowledge ecology from chapter 3, as a system where exploration and exploitation build upon each other, in a 'cycle of discovery'. The Netherlands has traditionally functioned as a place for trading, combining, and distributing goods, as a 'portal to Europe', with the Rotterdam harbor and the commercial and financial hub of Amsterdam. The question now is whether in the future this may be extended to include more 'trade' in knowledge. Other countries might find, in the Neterlands, not only logistical access but also knowledge of knowledge and technology, and of supply and demand, for access to Europe. The Netherlands might function as a place for meetings, of shorter or longer duration, between explorers and exploiters of knowledge of many kinds: scientists (keeing in mind the Netherlands' international reputation in the fields of agriculture, food, flowers, astronomy, and some fields of engineering, e.g., water management), producers and users of technology, designers and artists, traders and businesspeople, politicians, diplomats, and lawyers (e.g., the International Court of Justice, located in the Hague), security (with Interpol, also in the Hague), and certain areas of publishing. To support such a system, we would need a variety of supporting services, in law, finance, transport, distribution, conferencing, communications, translation, publishing, accommodation and housing, with attractive spatial, recreational, and cultural environments, and, hopefully, a renaissance of traditional openness to other cultures, as a 'hub of buzz'. Hopefully, the Netherlands might be a place where trust is built; a place where identity matters little and processes of identification take place (WRR 2007). Perhaps the Dutch can again, even more extensively than in the past, assume the many roles available for 'go-betweens' to help other people cross their cognitive distances.

NOTES

1 McAllister 1995, Lewicki & Bunker 1996, Hansen 1999.

2 In figure 7.1, the thickness of a connecting line represents the strength of the tie, with a number of dimensions: the scope or 'multiplexity' of the content of the tie, the volume of business or communication involved, duration, frequency of interaction, specific investment in understanding and trust, and personal bonding.

REFERENCES

Ashheim, B.T. and A. Isaksen (2002) 'Regional Innovation Systems: The Integration of Local 'Sticky' and Global 'Ubiquitous' Knowledge', forthcoming in *Journal of Technology Transfer.*

Benneworth, P., G.J. Hospers, and B. Jongbloed (2006) 'New Economic Impulses in Old Industrial Regions: The Role of the University of Twente in Regional Renewal'. In: A. Prinz, A. E. Steenge & J. Schmidt (eds).*Innovation: Technological, Economic and Institutional Aspects,* Münster: Lit Verlag, 1-24.

Bathelt, H., A. Malmberg and P. Maskell (2004) 'Clusters and Knowledge: Local Buzz, Global Pipelines and the Process of Knowledge Creation', *Progress in Human Geography* 28(1): 31-56.

Boschma, R.A. and J. G. Lambooy (2002) *Knowledge, Market Structure and Economic Co-ordination: The Dynamics of Italian Industrial Districts*, Utrecht: University of Utrecht.

Burt, R. (1992) 'Structural Holes: The Social Structure of Competition', in: N. Nohria and R. Eccles (eds.) *Networks and Organizations: Structure, Form and Action*, Boston: Harvard Business School Press, 57-91.

Burt, R.S. (2000) 'The Network Structure of Social Capital', in: R.I. Sutton and B.M. Staw (eds.) *Research in Organizational Behavior*, 345-423.

Coleman, J.S. (1988) 'Social Capital in the Creation of Human Capital' *American Journal of Sociology* 94: 95-120.

Edmonson, A. (1999) 'Psychological Safety and Learning Behaviour in Work Teams', *Administrative Science Quarterly* 44: 350-383.

Ministry of Economic Affairs (2004) *Peaks in the Delta (Pieken in de delta)*, The Hague.

Gambetta, D. (1988) 'Can we trust trust?', in D. Gambetta (ed.) *Trust: Making and Breaking of Cooperative Relations*, Oxford: Blackwell, 213-237.

Gilsing, V.A. and B. Nooteboom (2005) 'Density and Strength of Ties in Innovation Networks: An Analysis of Multi-media and Biotechnology', *European Management Review* 2: 179-197.

Granovetter, M.S. (1973) 'The Strength of Weak Ties', *American Journal of Sociology* 78(6): 1360-1381.

Harrison, M.M., N.L. Neff, A.R. Schwall and X. Zhao (2006) 'A Meta-analytic Investigation of Individual Creativity and Innovation', paper presented at 21st Annual Conference for the Society for Industrial and Organizational Psychology, Dallas.

Hirschman, A.O. (1970) *Exit, Voice and Loyalty: Responses to Decline in Firms, Organisations and States*, Cambridge MA.: Harvard University Press.

Kitagawa, F. (2005) 'The Fukuoka Silicon Sea-belt Project – An East Asian Experiment in Developing Transnational Networks', *European Planning Studies* 13(5): 793-800.

Klein Woolthuis, R., B. Hillebrand and B. Nooteboom (2005) 'Trust, Contract and Relationship Development', *Organization Studies* 26(6): 813-840.

Knoben, J. (2007) 'A Spider and the Stickiness of its Web: The Causes and Consequences of Spatial Firm Mobility', Ph.D. dissertation, University of Tilburg.

Lester, R.K. and M.J. Piore (2004) *Innovation – The missing dimension*, London: Harvard University Press.

Nooteboom, B. (2000) 'Institutions and Forms of Co-ordination in Innovation Systems', *Organization Studies* 21(5): 915-939.

Nooteboom, B. (2002) *Trust: Forms, Foundations, Functions, Failures and Figures*, Cheltenham: Edward Elgar.

Nooteboom, B. (2004) 'Competence and Governance: How Can They Be Combined?', *Cambridge Journal of Economics* 28(4): 505-526.

Nooteboom, B. (2006) 'Innovation, Learning and Cluster Dynamics', in B. Asheim, P. Cooke and R. Martin (eds.) *Clusters and regional development*, London: Routledge, 137-163.

Oinas, P. and E.J. Malecki (2002) 'The Evolution of Technologies in Time and Space: From National to Regional to Spatial Innovation Systems', *International Regional Science Review* 25(1): 102-131.

Pagden, A. (1988) 'The Destruction of Trust and its Economic Consequences in the Case of Eighteenth-century Naples', in D. Gambetta (ed.) *Trust, the Making and Breaking of Cooperative Relations*, Oxford: Blackwell, 127-141.

Ponds et al. (2006) *Knowledge Hubs in the Netherlands: Spatial Patterns of Collaboration in Research (Kennishubs in Nederland: Ruimtelijke patronen van onderzoekssamenwerking).*

Soh, C. (2007) 'Governance in Global and Regional Context: The Korean Case as a Study of Metagovernance', paper presented at the Korean-Dutch governance forum, Leiden, August 30.

Weiss, T.G. (2000) 'Governance, Good Governance and Global Governance', *Third World Quarterly* 21(5): 795-814.

WRR (2007) *Identification with the Netherlands* (in Dutch), Amsterdam: Amsterdam University Press.

8 INNOVATION AND ORGANISATION

Bart Nooteboom and Robert Went

8.1 INTRODUCTION

In this chapter we look at how to organise for innovation and how to innovate organisation.
In a review of the existing literature on organisational innovation, Lam (2005) notes that "(t)here is no single coherent conceptual framework for understanding the phenomenon of 'organisational innovation'. This is partly due to the great conceptual ambiguity and confusion surrounding the term organisational innovation." To begin with, there is no consensus definition of the term organisational innovation. Lam holds that this conceptual indeterminacy may reflect the fact that organisational innovation embraces a very wide range of phenomena. She notes that "at present, research on organizational change and adaptation is fragmented: the different levels of analysis are disconnected and often rooted in different theoretical paradigms that use different research methods." She finds that "many innovation studies continue to be dominated by an economic approach that allows little room for the analysis of creative change and innovation within the organization itself", and argues that "treating the organization as an interpretation and learning system directs our attention to the important role of internal organizational dynamics, actor cognition, and behaviour in shaping the external environment and outcomes of organizational change."

As discussed in chapter 3, a fundamental problem of innovation processes is the combination and relation between exploration – i.e., the development of new ways of doing things (also referred to as second order learning) – and exploitation, i.e., improving on existing ways of doing things (also referred to as first order learning). The first is necessary for companies in order to survive in the short term, the second for their survival on a longer term. The challenge is to do both. That is not simple, because they raise different and sometimes contradictory demands. For exploitation, meanings have to be clear and stable norms, divisions of labour, and know-how are all necessary. For exploration, it is necessary to cut across existing denotations, norms, and divisions of tasks. This leads to tensions between control and flexibility, with consequences for the motivation and coordination of labour and the measurement of performance. The task to nevertheless combine the two may well be the biggest challenge for companies seeking to be innovative

In chapter 9, Anderson and Gasteiger report evidence that conscientiousness, which is known to be the most consistent predictor of job performance was shown to have a zero or even slightly negative relationship with creativity and innovation (Harrison et al. 2006). The former is associated with exploitation and the latter with exploration. Conversely, Miron et al. (2004) found that creativity had a significant negative effect on performance quality when the task required accuracy and adherence to rules.

The key differences between exploitation and exploration have neatly been summarised by Lester and Piore (2004) in table 8.1, using their own terms to differentiate between 'analysis' (exploitation) and 'interpretation' (exploration).

Table 8.1

ANALYSIS	INTERPRETATION
The focus is a project, with a well-defined beginning and end	The focus is a process, which is ongoing and open-ended
The thrust is to solve problems	The thrust is to discover new meanings
Managers set goals	Managers set directions
Managers convene meetings and negotiate to resolve different viewpoints and eliminate ambiguity	Managers invite conversations and translate to encourage different viewpoints and explore ambiguity
Communication is the precise exchange of chunks of information (bits and bytes)	Communication is fluid, context-dependent, undetermined
Designers listen to the voice of customers	Designers develop an instinct for what customers want
Means and ends are clearly distinguished, and linked by a causal model	Means and ends cannot be clearly distinguished

Source: Lester and Piore 2004, pp. 97-98

More companies nowadays try to develop links downstream and upstream, with suppliers and customers, to improve on existing products and help to develop new applications or products (Von Hippel 2005). As discussed in chapter 7, this requires organisational innovation in governance and in relations between producers, users, and suppliers. A large literature has developed on collaboration for innovation between organisations (Nooteboom 2004). Here, we focus on organisation within firms, in relationships with users, and in collaboration between users.

In this chapter we look at three aspects of the non-technical side of innovation. First, we consider the deployment of people and the organisation of work, or what is sometimes called 'social innovation' or 'smarter work'. The following section deals with the more fundamental problem of finding organisational forms that can manage the relationship between exploitation and exploration, in what are sometimes, rather modishly, called 'ambidextrous organisations'. Next, we turn to new forms of organisation that are facilitated by the use of ICT and Internet, such as user platforms, innovation by lead users, and open source communities. Finally, given these new forms of organisation, we focus on opportunities for innovation in the organisation of government services.

8.2 DEPLOYMENT OF PEOPLE AND THE ORGANISATION OF WORK

Adam Smith argued already 230 years ago that employees of companies can be a major source of innovation.

"A great part of the machines made use of in those manufactures in which labour is most subdivided, were originally the inventions of common workmen, who, being each of them employed in some very simple operation, naturally turned their thoughts towards finding out easier and readier methods of performing it. Whoever has been much accustomed to visit such manufactures must frequently have been shown very pretty machines, which were the inventions of such workmen in order to facilitate and quicken their particular part of the work" (Smith 1776).

For some years, Dutch ministries have widened attention from technology to non-technological aspects of innovation. However, Volberda and van Den Bosch (2005) argue that the Dutch innovation debate is still too concentrated on technology-related macro-variables such as private investments in R&D and the number of available scientists and engineers. Also, the Dutch Review Committee ('Visitatie Commissie') on Emancipation noted that the way in which 'social innovation' was discussed in the Ministry of Economic Affairs was rather technological (Visitatiecommissie Emancipatie 2005). Until recently, not much attention has been given to characteristics of management and organisational features of successful organisations. This is reason for concern, because "it is widely proven that the adaptation of concrete organizational concepts has a paramount impact on the ability of a company to improve its performance" (Armbruster et al. 2007). Volberda and Van den Bosch (2005) even claim that the drop of the Netherlands in the World Economic Forum rankings of most innovative and competitive countries , "can be mainly attributed to the present lack" of managerial capabilities and novel organising principles in Dutch organisations.

While the measurement of technical product innovations is based on commonly agreed definitions and has methodologically been harmonised and standardised, this cannot be said of the definition and measurement of social innovation. Armbruster et al (2007) compare how organisational innovation has been measured in a number of surveys in Europe, and report that "different approaches lead to significantly different results regarding the organizational innovativeness of companies within one and the same sample." They conclude that more research is needed to bring "light into the black box of measuring organizational innovation in large scale surveys", and to develop a sound methodology for monitoring systems of organisational innovation. Their research suggests that the following four elements have to be taken into account when measuring organisational innovation:

- *Complexity of social innovation*: different organisational innovations have different effects on performance indicators;
- *Life-cycle of organisational innovation*: because organisational innovations do not age as quickly as product innovations, it is important to determine the proportion of firms that has actually implemented an organisational innovation (when you ask if such innovations have been implemented during the last three years you only classify latecomers as innovative);
- *Extent of use of organisational innovations*: it is necessary to identify the extent to which organisational innovations have been implemented into business processes;
- *Quality of organisational innovation*: merely using labels when inquiring about organisational innovations biases the diffusion of organisational innovations across companies, thus it is crucial to know how terms like 'team work' or 'total quality management' are used in the respective company.

Notwithstanding methodological qualifications about the measurability of organisational innovation, there are indications of a positive correlation between organisational (or 'social') innovation and the success of innovation within Dutch companies. Volberda et al. (2006) have surveyed directors and members of management teams of 9,000 Dutch companies with more than 25 full-time employees for their *Erasmus Competition & Innovation Monitor 2005*. They claim that R&D and investment in ICT accounts for 25 percent of the ultimate innovation success of companies, while social innovation is three times as important (75 percent). 'Social innovation' is defined here as "the development of management skills (dynamic management), the implementation of innovative principles for organisation (flexible organisation forms) and the realisation of high-grade forms of work ("smart work" and the development of talent) to improve capacities for competition and productivity" (Volberda et al. 2006). These authors argue that a number of aspects and characteristics of flexible organisation forms, dynamic management, and 'smart work' are positively correlated with the innovation successes of Dutch organisations:
- Flexible organisation forms
  - High internal speed of change
  - Self-organising by decentralised, well-built social networks
  - Balancing innovation and efficiency in separate organisational units
- Dynamic management
  - High absorptive capacity of management
  - Entrepreneurship of visionary leadership
  - Cross-functional internal cooperation and integration
- 'Smart work'
  - Development of talent through a deep knowledge foundation
  - Variety in management expertise
  - Remuneration on the basis of team performance

According to Volberda (2007), companies that have invested in these forms of social innovation are 27 percent more profitable than other companies, with 19 percent higher growth of sales. It is not clear whether this correlation reflects causality, and if so in which direction this causality goes.

In light of such evidence and other literature (Volberda and van Den Bosch 2005) it is hard to understand why until recently the Dutch Innovation Platform rather one-sidedly concentrated on technological innovation, while ignoring the non-technological dimensions of innovation. Recently the Innovation Platform has supported the institution of the Netherlands Centre for Social Innovation (NCSI).[1] But the lack of attention to organisational aspects of innovation is still reflected and reproduced in the composition of the Innovation Platform, which has no trade union representatives s among its ranks. A possible motive for excluding trade unions from the Innovation Platform is that they represent specific vested interests, or may even be "natural enemies of process innovation" (de Nooij and Poort 2005). But this argument apparently does not hold for other members of the Platform, currently including representatives of the Dutch corporations Philips, Corus, and DSM. More importantly, there are even indications that trade unions at the central and local level may be a positive factor for the development of new forms of work organisation and skills development in firms, and that "European traditions of concertation between government, business, and labour may prove to be a comparative advantage if there is a willingness on all sides to take up these challenges" (Archibugi and Lundvall 2002). There are no indications that this does not hold for Dutch unions, which participate in the NCSI, have collaborated in the Social and Economic Council of the Netherlands to produce recommendations on social innovation (SER 2006), and are involved in projects to stimulate and facilitate what they call 'smarter work' (FNV Bondgenoten 2006).

"*Smarter work is a long term investment.* With that we deliberately take a risk. If smart work becomes a success and in such a way labour productivity increases, there may be job losses unless more products are sold. Still, we take this course ... The trade union does however ask something in return. In our view, a company is only healthy when it not only performs economically but also socially" (FNV Bondgenoten 2006).

'Social innovation' has gradually become more of an issue – or even a 'hype', according to some observers (Biermans and Poort 2006) – in the Netherlands, as evidenced by the attention given to it by employers organisations, trade unions, and an increasing number of companies, as well as from the launch of the NCSI in June 2006. However, without playing down the importance of this initiative, it has to be noted that so far attention is predominantly given to the stimulation of what is sometimes called 'operational excellence', i.e., improving on the way exploitation is organised.

"FNV Bondgenoten, the largest trade union in the Netherlands, has begun to stimulate the concept 'draw up your own roster' with among other things conferences, software packages and pilot projects. The basic idea is that employees draw up their own roster, within the precondition that demands are being met for the number of employees needed to do the work. Employees thus can get a better match between their working hours and private life and get more responsibilities, while experience in Sweden shows that labour productivity will be increased" (FNV Bondgenoten 2006).

Representatives from the biggest Dutch trade union in the market sector (FNV Bondgenoten) and from one of the biggest Dutch employer's organisations (AWVN) estimate that so far in just approximately 50 out of 800 collective agreements in the market sector have taken up some element(s) of social innovation. Employees and employers will have to cooperate and interact within the existing structures to further social innovation and to foster the necessary good labour relations, space for experiment, possibilities to learn and develop talents, and diversity of teams. Resistance and obstacles to such changes may be vast, and can come from layers of management, as a consequence of prevailing short-termism, a fixation on measurable targets ('spreadsheet management'), or a fear of loosening control mechanisms. Although they are of course not solely responsible for organisational innovation, managers do have the main responsibility to initiate the desired innovative processes and changes, to clear the way for experiments, and to create a climate in which employees do not have to fear making mistakes. But resistance to social innovation may also come from employees, out of (understandable) fears for insecurity and job losses when routines and traditions are challenged. From an analysis of aggregate data the OECD (2002) concludes that the cross-market interactions between innovative activity and labour market policies and institutions are "complex", and that "there seems to be a U-shaped relationship between innovative activity and labour market regimes, defined by different combinations of Employment Protection Legislation (EPL) and coordination in industrial relations."

The organisation of exploration for new applications and products is even more challenging and demanding than aiming for 'operational excellence', and at least as important in the longer term. As Volberda et al. (2006) have argued, both *over*-exploitation of existing opportunities via reorganisations, cost cutting, and short-term accomplishments ('exploitation herd behaviour') and *over*-exploration, by exaggerating the importance of local changes and oversensitivity for hypes and fashions, are dysfunctional for companies, leading to either competence or modernisation traps. How exploitation and exploration can be combined in a productive and sustainable way is therefore a challenge for every company.

"Lely Industries in Maassluis (250 employees) is a global market leader in milking robots and also produces machines for working the soil for dairy farmers. The Lely Group has a separate company – Lely Technologies – where ten to fifteen engineers permanently work to develop new innovative products. Lely Technologies is strictly separated from the rest of the company, so that the creative and innovative processes do not interfere with the production phase, which demands much more discipline.

Bronckhorst High-Tech in Ruurlo (250 employees) is a European market leader in flow meters. No less than 45 of its employees work in the company's R&D department, and 15 percent of annual sales is invested in R&D and the development of new products. R&D employees at Bronckhorst High-Tech are allowed to work one day per week on their own projects, the so-called 'Friday afternoon projects'" (Van der Geest and Heuts, 2006).

Although exploration is indispensable for the survival and continuous renewal of companies and economies, it generally attracts much less attention than exploitation. A great number of comparative and case studies have been made to try and find the 'Holy Grail' for the correct balance between exploitation and exploration, but as we will see in the next section there is no general answer or magic solution to this problem.

8.3 AMBIDEXTROUS ORGANISATION

In the past, whether companies were oriented towards exploitation or exploration was dependent on the phase in the life cycle in which the company or sector of industry was situated. That no longer holds true. Everywhere now companies have to carry out exploitation to survive in the short term, plus explorative activities for their survival in the longer term. Firms may use external networks to externalise exploratory or exploitative innovations, thus outsourcing potential problems associated with the tension between exploitation and exploration (Nooteboom 2000; Jansen 2005). Combining exploitation and exploration may well be the biggest challenge for companies (and organisations in general), and is far from being a trivial exercise. The fashionable term for an organisation combining the two is 'ambidextrous organisation'. Note that this is not a solution or formula, but a mere label of the problem.

Exploitation and exploration build on each other, and are dependent on highly educated knowledge workers and the shop floor. Li et al. (2007) conclude from a review of organisational learning and innovation studies from the last 15 years, that exploitation and exploration are complementary rather than substitutes for each other: extreme explorations (exploitations) score high (low) in all dimensions of knowledge search and recombination, and both kinds of extreme situations are rare. Exploitation and exploration require different competences, organisation, and mentalities. This has major implications for the organisation of companies in general, and for human resource management, motivation, and the coordination of work in particular. Research by Birkinshaw and Gibson (2007) shows that "ambidextrous individuals" can be very useful for companies, because they tend to "take the initiative and are aware of opportunities beyond their own jobs; ambidextrous individuals co-operate, seeking opportunities to combine their efforts with those of others; ambidextrous individuals are brokers who build internal networks; ambidextrous individuals are natural multi-taskers." Table 8.2 schematically summarises a number of key differences between exploratory and exploitative innovations.

Table 8.2 Exploratory and Exploitative innovation

	EXPLORATORY INNOVATION	EXPLOITATIVE INNOVATION
Definition	are radical innovations and are designed to meet the needs of emerging customers or markets	are incremental innovations and are designed to meet the needs of existing customers or markets
Outcomes	new designs, new markets, and new distribution channels	existing designs, current markets, and existing distribution channels
Knowledge base	require new knowledge and departure from existing knowledge	build and broaden existing knowledge and skills
Result from	search, variation, flexibility, experimentation, and risk-taking	refinement, production, efficiency, and execution
Performance indications	distant in time	short-term benefits

Source: Jansen 2005, p. 19

Small firms have fewer resources for innovation, in the form of specialised personnel, absorptive capacity, slack time, and resources for R&D and experimentation. However, due to a smaller size, fewer hierarchical levels, less separation between management, work, and customers, and greater informality they do seem to have an advantage in becoming 'ambidextrous' by solving the problem of combining exploitation and exploration. The following case gives an example.

"OTB, short for 'Only the Best', run by owner-manager Ron Kok in Eindhoven, the Netherlands, develops and produces a variety of new, mostly high-tech products. One entails the use of polymer sheets as the substrate for solar cells, which increases their energy efficiency and flexibility, and reduces their price. Second is a fully automated production 'street' to produce the cells, yielding a low production price. At OTB, developers and producers are sitting in the same large space, at tables adjacent to each other. There are frequent hassles, heckles, and quibbles between them. When those accumulate, Ron Kok takes both sides out into the country, in an informal setting, where under his leadership they thrash out disagreements until solutions are found. Ron is a bit concerned about what will happen when his firm grows in scale, with increased specialisation of labour, and when separation of the two groups may be unavoidable" (Kok, author interview 2007).

In larger organisations, the problem of combining exploration and exploitation is expected to be particularly difficult when exploitation forms the overriding strategic task of the organisation, and the structure of exploitation is highly systemic, i.e., there is a complex of many components that are densely and tightly connected. This means that most component activities are constrained by narrow tolerances on interfaces with other components. By definition, this yields little room for the experimentation or trial and error that is needed for exploration. When you are caught in many strong ties there is little room to move. Small deviations from the established order reverberate in the system, yielding widespread upheaval and inefficiency. This systemic feature is generally accompanied by a corresponding culture of conformance to established roles, tasks, and corresponding standards, and precision in the execution of work. Then, there is pressure to separate exploration and exploitation, and allow for more looseness, variability, and ambiguity in a separate unit for exploration. Then the problem arises of how to connect the two, so that exploration is inspired by experience in exploitation, yields results that are exploitable, and exploitation is willing and able to absorb the results. Here we present an illustration of an attempt to solve the problem.

"The 'Central Book House', in the Netherlands, buys and supplies books for the book trade. It has three core divisions, which need to be tightly integrated:

1. the assortment of books, which must offer a high degree of differentiation to customers, in terms of composition and size of packages;
2. logistics for efficient and fast delivery;
3. information technology to enable 2 and to collect and provide strategically vital information as to what books are read by whom where at what price, for 1 and for customers (both publishers and shop keepers).

Clearly, efficient exploitation is crucial, and tends to predominate. But exploration also is important, to keep up with developments in reading habits, technologies of information processing and dissemination, changes in publishing. Due to the systemic coherence of the three divisions, innovation must occur in step. To achieve this while maintaining efficient exploitation, the following solution was found. A permanent R&D team was instituted, with ten people, recruited at different moments (two new people every half year), from the three different departments, and moving out after two years, but not necessarily to their own, previous division, not knowing to which division they would go. Moving out from exploration, back into exploitation, they carry responsibility for implementing innovations they helped to develop.

One can see the advantages. One is that exploration is based on experience in exploitation, and is conducted with a view to the expected responsibility for implementation to be taken on later. The different divisions each contribute both to exploration and its implementation, to protect systemic coherence. There is turnover in the team to maintain variety and ongoing influx of experience from exploitation. Since people may not return to their original division. Not knowing to which division they will go, they have no incentive to protect the interests of their 'own' division, keeping things easy for 'their' people. At any one moment there are recent and seasoned members of the development team.

Are there problems? It seems that the assumption in the system is that the cycle of innovation is two years. If it turns out to be longer, a person moving out of the R&D team has nothing new to implement in the division he joins, at least not an innovation to which he contributed. Is this an incentive or a disincentive to see to it that something new comes out within the two years? What if people are valuable but do not want to move out of exploration and back into exploitation, and if their competencies support that preference? What guarantees that one will always find people who have the skills and interest in both exploitation and exploration? What guarantees are there that rivalries between the divisions will not be carried into the development process? Is this prevented by the prospect of having to move into another division after the two years in exploration?" (Nooteboom 2000).

By contrast, when exploitation is already exploratory, with custom made products for a diversity of customers, with needs that are diverse and changing, and the product is produced in highly autonomous teams, with limited or weak connections to a back office or other production teams, then the problem may disappear. This yields a strong proof of the problem: if even here there is a problem of combining exploration and exploitation, the problem is indeed pervasive. A corresponding case is the following.

"The consultancy firm Arthur Andersen, when it still existed as an independent firm, was known as a paradigm example of a flexible firm. Its consultants were highly autonomous, employing their individual knowledge, skill and creativity to provide custom-made advice. Exploration was automatic, and built into day to day practice. However, a problem appeared with regard to exploitation. Consultants did not share their experience and new ideas, because there were no incentives to do so, and consultants also saw each other to some extent as competitors within the firm for careers and good customers. To remedy this, a computer-supported pool of ideas was instituted, to which consultants were encouraged to contribute their ideas on the basis of an incentive system. At first, part of their remuneration was related to the number of ideas contributed to the pool. Later, this was shifted to a reward proportional to the number of times their ideas were picked up from the pool by colleagues, not unlike the shift in the evaluation of scholars from their number of publications to their number of citations. This, however, caused a problem for exploration. The common pool required a certain minimal amount of standardization of concepts and procedures in a thesaurus. That fixity hampers exploration. The more radically new ideas would not fit established meanings in the thesaurus, and would hence be neither used nor rewarded. To remedy this, a committee was instituted to collect ideas that did not fit, and to periodically revise the thesaurus to assimilate them. This still caused a delay in the reward for radical novelty that also had to be compensated" (Nooteboom 2000).

While the solution at Arthur Andersen was formal and top-down, an alternative, more informal and bottom-up way of dealing with the same problem was developed by another consultancy firm, Cap Gemini, as follows.

Here, it was recognised that it was in the self-interest of consultants, without the need for any extra rewards, to share good ideas and experiences with colleagues, since the improved quality of advice would itself yield better performance and more and better clients, and the rewards attached to them. This triggered a process of self-organisation, in groups where consultants sought each other out to share ideas. Problems of free-ridership were avoided by social control, in the form of exclusion of free riders from the group in case of lack of competence or commitment. However, a problem now arose for exploitation because groups saw each other as competing, which blocked exchange of ideas and experience, but now not between individuals but between groups. To avoid this, management had to intervene by periodically disbanding the groups to effect diffusion between them.[2]

Another approach is to engage in 'stage specialisation'. Here one focuses on one stage of the innovation cycle, and seeks partners who engage in the other stages. Thus we can have firms that specialise in generating novel innovations; in testing them; in bringing them up to large scale and systemic, rationalised production, and global distribution; or in differentiating them. This stage specialisation may be more important and fundamental than specialisation in specific products or markets. Differentiation and to a greater extent reciprocation, accommodation, and early consolidation require disintegrated structures, and here industrial districts have a comparative advantage over tightly integrated large corporations. Different firms that are specialised in different stages can enter a relation of symbiosis.

An example is the relation between small, diverse, specialised biotechnology firms and large pharmaceutical firms. The first yield the novel combinations and the second yield the lengthy process of regulatory approval of new drugs, the systemic, large scale production and marketing of novel products, and the spread of risk across a portfolio of products.

However, it can be very difficult to separate exploration and exploitation between firms, or even within an organisation.

Brusoni (2006) gave an example in the chemical engineering industry. He distinguishes between 'division of labour' and 'division of knowledge'. While chemical plants are highly modular, and hence decomposable, and hence partially outsourceable, there are different ways of doing decomposition, with important differences in outcomes. The complexity lies in the coupling between systems of reaction processes, separation processes, energy, and control and safety. To optimise design requires a thorough understanding of systemic effects of alternative decompositions. Given a choice of decomposition, the elements can be outsourced, but the capability to decompose and synthesize elements in different ways cannot. As a result, some re-integration of capability and activity took place after chemical producers had fully outsourced the production of chemical plants.

Pisano (2000) gave an example in biotechnological process technology compared to chemical synthesis technology, in (1) ... theoretical understanding of the basic processes (2) ability to precisely and fully characterize intermediates and final products, and (3) knowledge of the second order effects of scaling up from experimental to operational production. As a result, "... biotechnology process development relies ... on trial and error and iteration of the process design *after* the process is transferred out of the laboratory" (op. cit: 135). Given these conditions, "At some organizations, the plant was viewed as an integral part of the development process and a critical venue for experimentation. Others kept the plant relatively isolated from the development process; preferring instead to do most development in the laboratory and pilot plant" (op cit: 137).

One company had previously focused on product development, and had mostly outsourced process development, and now started to integrate the entire development process. Then they ran into the familiar problem, in connecting exploration and exploitation, of "... the huge gulf between the interests of research scientists focused on finding novel products and the capabilities required to get a process up and running in an actual plant" (op cit: 138). The company next instituted a process development group, where a pilot plant is integrated with the commercial manufacturing plant. This required the build-up and integration of a broad base of scientific capabilities in process technology disciplines. For that, they could build on the knowledge accumulated by the initial process development group. A second company resisted disruptions of ongoing manufacturing for the sake of development, not to disturb the continuity of clinical testing to satisfy demands from regulatory agencies, and face d the consequences of lack of integration between exploration and exploitation. A third company went further in separating production and development, by outsourcing, but ran into the problem of sticky knowledge, with the contractor lacking the subtleties of process knowledge.

A similar problem was identified by Appleyard et al. (2000), in the semiconductor industry. There also, product innovation is highly intertwined with process innovation, "... requiring that the receiving fab (fabrication unit) have an equipment set that is identical to that on which a new process is developed in the development facility. Even stringent requirements for equipment duplication... cannot eliminate all significant differences" (op cit: 189). A solution was attempted, here also, in integrating development in a hybrid development/production facility where the pilot is eventually implemented at a commercial scale, and was transferred only then to a operational high-volume production facility, with substantial personnel rotation and equipment duplication (op cit: 198), and the process remains in parallel operation in the hybrid facility for at least 12 months after the transfer, as a back-up for any unforeseen problems of transfer. However, a remaining problem of this arrangement lies in disincentives for experimentation in the manufacturing environment, which may eventually impair its performance (op cit: 201).

These examples serve to illustrate how difficult it may be to separate exploration and exploitation, within and between firms, and the need, in that case, to establish processes that connect them. This is typically supported by rotation of personnel across the two to support mutual understanding and ability to collaborate. Here we are back at the type of solution implemented by the Central Book House.

A further, more radical way to deal with the problem is not to be entangled in production at all, but to act as an orchestrator of the productive activities of others. ICT increasingly yields the opportunities for this. The orchestrator conducts exploration by flexibly exploiting the productive competencies of different companies, in shifting configurations, and thereby tries to escape the need and the dangers of inertia. The risk of inertia due to standardised, more or less fixed systems for efficient production is separated and hived off to other players. But even here, the focal, orchestrating firm must be careful to both maintain and develop his core competence of orchestration.

Quinn (1992) gives the example of a company in custom-made ASICS (Application Specific Integrated Circuits). They interface with clients directly by means of ICT, to determine functional specifications. They then employ their own specialized software to convert this into photo masks, which are sent by ICT to a company in Japan for etching, next to a company in Korea for dicing and mounting, next to Malaysia for assembly, from where the chip is flown directly to the customer. A similar example in sports shoes is Nike. Another example in fashion clothing is Benetton: it also performs the task of orchestration, by means of ICT, of a decentralized network of individual producers and retailers.

Concerning this case, the question is: If orchestration becomes easy, with accessible design software and coordination of producers by internet, couldn't customers take over? This is precisely what has been proposed by Von Hippel (1999, 2005), and has meanwhile happened in the custom semiconductor industry (Von Hippel 2005: 16). Here we see the mergence of platform organisations, where producers offer modules for users to configure or develop products to their individual preferences. This is further discussed in the next section.

8.4 PLATFORMS

Now we turn to novel forms of organisation, enabled, in particular, by ICT and Internet. In platform organisations, producers offer a platform of modules from which users can configure their own goods or services or can route their questions or submit information or a platform for outsourcing, where potential suppliers can scan and analyse requirements and submit their tenders. Platforms are not necessarily restricted to the products of a single producer.

A company like e-Bay, for instance, hosts many different types of producers, some fully-fledged businesses, some family firms, and others amateurs or enthusiasts.

In the cycle of discovery, the platform principle offers product differentiation by users in the form of customised configuration. It does not include the reciprocation stage of the cycle of discovery. As shown by Von Hippel (2005), some producers go further and offer users 'toolkits' for exploration, which offer means for product-design, prototyping, and design-testing. For example: software for users to design their own chips. Here, users can reciprocate existing elements offered by producers with elements added from their idiosyncratic context of use.

We can analyse organisation *for* innovation and innovation *of* organisation, also for goals other than product innovation, for example increased efficiency. Here, we will first focus on the latter, and later we will return to organisation for innovation. Recall the analysis, in chapter 2, of total factor productivity. We reported attempts to explain differences between the US and the EU as a result of the use of ICT. While the analysis was not entirely conclusive, there was an indication that the US has achieved higher productivity growth on the basis of more advanced use of ICT. Perhaps this lies in the utilisation of ICT for new forms of organisation that increase productivity. And indeed, many organisations, not least in public services, can be innovated to make large improvements, especially in the efficiency, speed, responsiveness, and customisation of service delivery. A number of examples are found in an investigation, sponsored by the UK National Accountant's Office (NAO 2004), of innovations in the public sector.

The fundamental logic of platforms lies in the asymmetry of knowledge between users and producers (Von Hippel 2005). Users have idiosyncratic and highly tacit knowledge of their experiences and their needs, while producers have knowledge and experience in providing solutions to needs. Transfer of need information from users to producers, in market analysis, is costly, partial, reductive, slow, and faulty. It is costly in transaction costs of selecting potential customers and communicating with them. It is partial in that much of the knowledge is tacit and for that reason difficult to identify and capture. It is reductive in that it tends to yield averages that filter out the outliers in experience and ideas that may be the most interesting and innovative. It is faulty because in interpretation information is added and knowledge is transformed to fit the absorptive capacity of the producer. The totality of these problems has been labeled as 'stickiness' of information (Von Hippel 1999). As a result, in terms of communication and interpretation it is much more efficient to leave information on needs where it originates, and to supply modules from which users can configure their own custom-made product. This presumes, of course, that the product modules yield the required potential, and can indeed be consistently combined in unforeseen ways, and that users know how to configure them. For the latter, one will need to supply instructions, and they, in turn, may need to be customised for different users. For this, one needs continual feedback on user experience, in the development of modules and instructions. Some products can be more easily broken down into modules than others.

The tipping point between traditional forms and the platform lies in the relative complexity of user- and producer innovation. User production is more efficient when user information is more complex, with greater variety of user needs, and more sticky, than producer information on how to configure a product. The logic applies not only for the initial configuration of a product, but also for its subsequent maintenance and repair. For that, producers would have to keep records of what exactly it supplied to a given customer, including any special parts, while users have direct access to that information (Von Hippel 2005: 49).

In service firms there is a familiar distinction between the *front-office* where the contact with customers takes place, in researching, marketing, and delivering services, and the *back-office* that is concerned with production and making things happen (for example, providing IT and accounting systems, managing human resources, and investing in production). Traditionally, for many services the front-office is large and closely involves customers in interaction with the firm, who in many cases have to be physically present, at a desk or treatment facility. Now let us consider the switch from the traditional form to the platform principle, where firms no longer deliver fully specified products, but instead provide access to internet-based facilities such as databases or admission to particular networks. This is illustrated in figure 8.1. Customers can then use these databases, websites, or other facilities in a wide range of different ways, which the provider firm does not have to fully anticipate in the initial design of the service. This kind of change builds on customer *self-service* and *co-production* of services, allowing the firms involved to shrink down their front-office – because they are no longer seeking to deliver fully-specified traditional services and interact less with customers. In these kinds of firms the back office also *grows* relatively in size, because the extended high-tech facilities involved, and the development and improvement of modules for customers to configure, and requisite toolkits, need close support and continuous technical innovation and development.

Figure 8.1 The systems switch in services

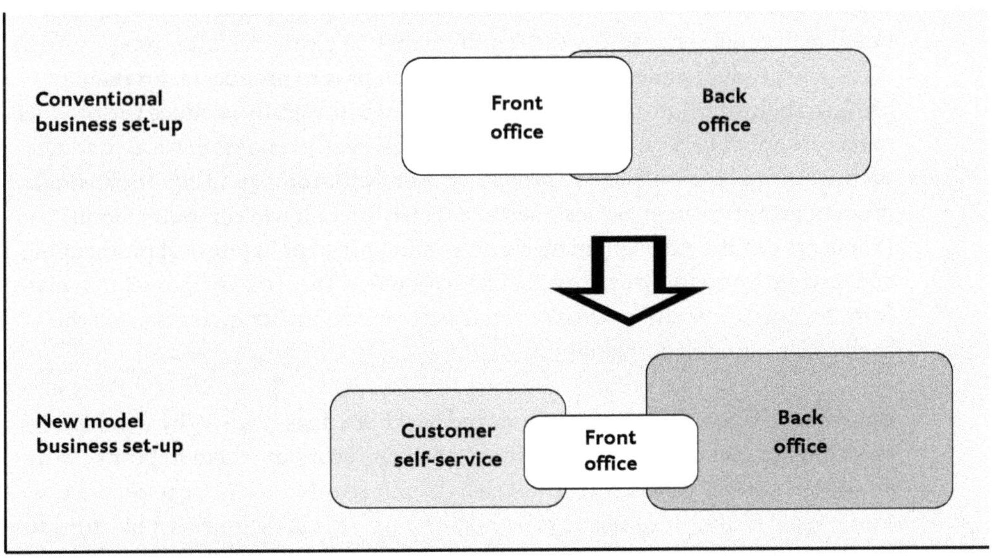

8.5 INNOVATION BY LEAD USERS

If it is true that exploration requires exploitation as a source of insight from experience with successes, failures, and new opportunities, as argued in chapter 3, and that exploitation is largely conducted by users, including citizens that use government services, then it follows that users and citizens are a source of innovation, as has been argued by Von Hippel (2005). This is the case, in particular, if a 'lead user' connects a familiar field of application with an unfamiliar but still related one. Thus, for more radical innovation one has to look for lead users beyond one's present customers, in different but relevantly related areas or 'analog fields' as Von Hippel (2005: 134) calls them.

An example in the car industry is that of braking system for airplanes, which in their landing make much more extreme demands on the brakes than cars. From this connection the car industry adopted the abs system (antilock braking system).

In addition to sources of knowledge and experience, there are other arguments for user-led innovation. Producers may be wary about making custom-made products with specialties or add-ons that they cannot easily guarantee, in terms of reliability and safety, for fear of incurring legal risks of liability and affecting their product's image for reliability, while for users it is up to them to decide what risks they will take (Von Hippel 2005: 50). A crucial condition for user-led innovation is that it is freely offered to other users, either directly or through a producer. If users had to separately develop each novelty, there would be a lot of duplication and waste, with different users repeatedly developing the same innovation.

While innovation of services is easiest, innovation of physical products is also possible, though it has additional requirements. In particular, after innovation by users, the product under development needs to pass to producers, for them to bring in their solution-related knowledge, and to physically produce the product or its components (Von Hippel 2005: 126). However, even for physical products producers can go a long way in providing users with tools and help. Increasingly, product prototypes can be designed and tested virtually, in computer simulation. Producers can provide help in machine-tooling parts, or in physical prototyping and testing, by opening up their facilities for users. In terms of figure 8.1, in platform organisations the volume of activities of users further increases, and the back-office of the firm shrinks.

But where does the profit come from and to whom does it go? Why would users be willing to make development efforts without recouping them in profit? Why would they freely yield the fruits of their labour to other users, or to producers? Von Hippel (2005) specifies a variety of reasons. One is the intrinsic pleasure and satisfaction of developing something new and seeing it work. A second is the prestige that diffusion of their achievement yields. A third is that the end result for them is still faster, cheaper, or better fitting to their needs than alternatives. Users may still need producers for sticky solution knowledge not available in the toolkits provided to users, and, as indicated, for physical production they still need producers. A fourth is that by offering their contribution to a wider field of users they benefit from economies of scale or scope (or network externalities), in production and distribution. A fifth is that by offering their contribution to others, those proceed to improve upon it, which, when others reciprocate in openness, benefits them further.

Why would producers play this game? As indicated, they may benefit from sticky and user information on needs, and on unexpected possibilities, which arise, in particular, from connections between the product and idiosyncratic needs or sources of knowledge and insight into possibilities from novel combinations.

An example given by Von Hippel (2005) is a cross-country biker who also was an orthopedic surgeon, and from that expertise developed a new shock-absorbing saddle to facilitate his particular preference for bike-jumping.

Von Hippel (2005: 126) summarises options for producer strategies as follows:
- produce user-developed innovations for general commercial sale;
- offer a custom manufacturing service to specific users;
- sell kits of product-design and/or 'product platforms' to ease innovation-related tasks;
- sell products or services that are complementary to user-developed innovations (and are more difficult or less efficient for users to develop or produce).

8.6 OPEN SOURCE COMMUNITIES

In open source communities, a further step is taken: here there are no longer separate producers, and users together take care of production. In terms of figure 8.1, both front and back offices of firms virtually disappear, apart from marginal coordination activities, and only the space of user action remains. This arises especially in the production of non-physical products, such as software (e.g., Linux, Apache, Sendmail) or information (e.g., Wikipedia), for two reasons. First, one no longer needs producers with specialised hardware, and users together have the solution knowledge needed in the system. Second, without physical goods one can use the full potential of the Internet and does not need anything beyond that. Users produce ongoing additions or improvements to an existing system, enabled, when needed, by free access to the underlying source code. The condition usually is that they attach their name to the contribution, their contribution can be used and modified freely by other participants, and that they may not incorporate the contributions from others in products they sell. This is guarded by a 'General Public License' (GPL), supported by internet mailing lists (IML) whereby infringements are tracked and communicated. Why would users produce without payment? The reasons are mostly as indicated earlier: intrinsic satisfaction, reputation and professional recognition, and collective improvement of the system they use.

Lakhani & Wolf (2005; Von Hippel 2005: 60) studied individuals (n = 684, response rate 34 percent) who had contributed to open source software projects, asking for their main reasons, and found that for 45 percent one of their top three reasons was intellectual stimulation, and that for

41 percent one of their top three reasons was to improve their programming skills. 61 percent said that participation in the open source project was their most creative experience or as creative as their most creative experience.

Why would people *use* products from such communities? Are they reliable?

"[Sixty-seven] percent of all active internet websites are supported by the open source programme Apache, and 80 percent of all e-mail traffic is enabled by the open source product Sendmail. Amazon, the New York Stock Exchange, IBM, the municipality of Munich and the Brazilian central government use open source software" (Wendel de Joode 2005: 221).

The basic source of added value that cannot be obtained otherwise is this: A system of user-producers that attracts contributors by self-selection opens up to expertise and ideas that no central planner or controller would ever think about and would not know where to find. In the Linux lore: "Many eyeballs make all bugs shallow". How can independent contributions come together in a consistent whole? This requires modularity, with independent 'patches' attached to each other, on the basis of standardised interfaces or connections developed by specialised developers and distributors.

Why isn't the system affected by misuse and bad quality from opportunistic or incapable contributors? From his study of the Linux and Apache communities, Wendel de Joode (2005) proposed the following logic. Particularly competent and useful contributions get nominated to a 'hall of fame'. Incompetent or low-quality contributions do not survive for long, and are quickly identified, reported, and replaced. This yields a penalty, in loss of face or reputation. Blatantly bad behaviour may even be relegated to a 'hall of blame', and may get punished by getting 'spammed to death' by insulting e-mails. The user community forms an efficient reputation system. Consequently, contributors make an assessment of their capabilities, in a trade-off between aiming at a contribution with high impact, with fundamental improvements, wide repercussions, and hence visibility, but higher chance of failure, and low-impact, low-risk contributions. They select the most difficult task they can still handle. Thus, the system yields an efficient allocation of talent. This is not unlike scholars trading off high-prestige journals with low chance of acceptance of a paper, and lower-level journals with a higher chance of acceptance. Thus, with a variety of capability the system yields an automatic, unplanned, emergent division of labour, with both incremental and radical improvements, and limited failure.

In case of disagreement on content or quality, no time is spent on meetings or committees to deliberate and make a judgment. People can freely develop parallel approaches to existing ones, thus creating redundancy. The resulting complexity and problem of search and selection among alternatives is tackled on the basis of 'tags' attached to alternative patches, which serve as a basis for selection. These tags include the name and reputation of the contributor, and the frequency with which a patch is used. Also, distributors make selections among alternatives in packages they offer. To get a contribution accepted and used in the community, there is not only an incentive for substantive quality, but also for elegance and ease of understanding. In sum, the issue is solved by competition: bad contributions fall into disuse by an increasing majority of the community. In other words, this is an evolutionary system, running on wide variety generation, with room for redundancy, plus survival in competition. Here we are back at chapter 3, where we argued that an evolutionary system is a paradigmatic system for combining exploitation and exploration.

There is also a phenomenon of 'swarming': in their contributions, participants tend to latch on to high reputation areas and the activities of high-prestige members, to ride on the tails of their success. Thus, prestige is self-reinforcing: contributors with better reputations attract more followers and thus further improve their reputations. This is called the Matthew effect, and appears also in citations of scientific publications.[3] This process will often cause the convergence to an emergent dominant style and ethic that subsequent new entrants will have to adopt. Could this convergence yield conservatism, in waves of fashion and idolatry, yielding a reduction of cognitive distance, with an excess of exploitation at the expense of exploration? The dynamics of the structure of open source communities has been simulated in agent-based computational modeling by Muller (2006). He confirmed such processes of convergence and cumulative reputation. However, disagreements and radically new ideas that are incompatible with the established scheme of things can lead to 'forking', where a community splits up into different 'schools' that proceed to compete. That may be seen as a waste, but also as an opening up to exploration. Such forking generally requires a leader within the community who deviates from other leaders but has enough of a reputation to take along a following in the forking process. Newcomers that have not yet accumulated enough of a reputation will have to either latch on and conform to existing leaders or face a long and lonely struggle out in the deserts of dissent.

One difference between the Apache and the Linux communities is that the former had several leaders while the latter had one central leader. The former is more likely to yield forking.

Here we can link back to the cycle of discovery from chapter 3. Deviations may be triggered by shifts to new contexts of application, with new types of users/contributors, generating new challenges, opportunities, and hints for novel elements and directions. In evolutionary terms, new schools of thought may then generate their own selection environments, with new communities to generate and select contributions from a different perspective. In science, new schools of thought may set up their own journals as a dedicated selection environment for papers.

Do these communities completely self-organise, or are there organisational measures concerning entry and exit, acceptance or rejection of contributions, division and coordination of labour, and conflict resolution? Wendel de Joode noted that sometimes (e.g., in the case of Linux) officially there are formal instruments of governance, such as entry selection, voting systems, and project leadership, but in fact they are not used because they turn out not to be needed (somewhat to the surprise of the community itself). The formal instruments that are in operation are the General Public License, the internet mailing list, and the assignment of names to contributions. Those appear to suffice for self-organisation. Is this self-organisation really so simple, reliable, and generally valid? A counter-example is the development of Wikipedia, which according to Mateos-Garcia and Steinmueller (2006) was as follows.

Wikipedia was launched in 2001, and started as completely open, with only a small number of essential rules: non-ownership of contributions, i.e., the right for others to create a modified version, use of the Wiki software (for creation and editing pages, and interlinking between pages), the requirement of a neutral, nonpartisan point of view, and several stylistic conventions. However, publicity and success of Wikipedia generated an influx of participants that included a surge of problematic users and abuses of the system, such as vandalism, spam, partisan points of view, pursuit of a political agenda, self-promotion, commercial gain, and lack of respect for expertise. This evoked interventions by the editor to selectively block entry, withhold IP, protection of pages and article tracking. An arbitration committee was instituted to solve ongoing disputes between members. However, these interventions were perceived as going against the anti-authoritarian, egalitarian ideology of the beginning. Yet, the interventions were insufficient to quell complaints about incorrect or libelous content in articles. This led to further but highly contested requirements for users to register and login as a condition to editing articles, and the proposed creation of 'stable/developmental' branches of the project. A new problem appeared where people entered under a false identity to make adjustments to entries that were giving negative information on them. This has been repaired to some extent with a mechanism that can identify and expose such actions by tracing the real identity of authors by their IP-address (with the complication that sometimes such an address is used by multiple users).

Now, there are significant substantive differences between an encyclopedia and professional software. Software development requires a certain minimal knowledge and reasonably circumscribed professionalism, with limited dispersion of knowledge and experience, which yields a self-selection that is more stringent than for people who think they have something to tell an audience that is unlimited in its variety. Software does not lend itself to misuses of partisan interests or politicking as much as articles do. Bugs in software are more objectively identifiable than the 'correctness' of an encyclopedic article, whatever that would mean. Standards of 'elegance' that apply to software are not so readily available for articles, and more subject to variety of style and taste. Motives for contributing to open software may differ greatly from motives to contribute to an encyclopedia. In sum, the lesson is that the extent of self-organisation and openness that is viable depends on the content and the type, extent, and variety of professionalism associated with the system, as well as the corresponding motives.

8.7 INNOVATION IN GOVERNMENT SERVICES

We have sketched the development of new forms of organisation that involve users in co-production of services. Could this become as important in the government sector as in private services? We believe they can and will. In areas like pensions planning or the administration of taxes and subsidies the scope for co-production of government services is immense.

An example is the British 'Directgov' project. This is a portfolio of facilities and services (central web portal, kiosk, digital TV channel) intended to deliver online government services more efficiently and effectively. It is designed to be a central point of reference for citizens to access core government services online. Instead of being based on the way departments are organised, it is intended that users access the services they need according to topics (such as motoring) or groups (such as disabled people). According to the Cabinet Office, it is the largest e-transformation in the UK (NAO 2006).

The new organisational form of platforms yields considerable potential to improve the efficiency and quality of government services. Here we make the usual distinction between innovations that improve efficiency and those that improve quality. A complication is what the meaning is of the concept of quality in public services. Quality in such services not only includes features of user quality such as relevance of content, ease of use, ease of access, and speed of response. It also includes public quality in equity of use and access, due process, and public accountability. In our analysis we employ the well-known distinction between front and back office in service firms, already used in figure 8.1.

In figure 8.2 we illustrate where different kinds of innovation may come from. In business, early innovations in services were mostly oriented towards efficiency, based on technology, especially IT and Web-based changes, and a strong push from firms' managements to cut costs and improve the impacts of given spending. Most efficiency innovations, in the upper row of figure 8.2, have been back-office initiated, achieved by increasing the division of labour inside organisations, pushing through new technologies, outsourcing inputs and exploiting new synergies between activities. Later innovations of user quality, in the middle row of figure 8.2, mostly came, as they should, from the pull of customer demand, the need to introduce new services or provide a faster response in a better location, and in a more differentiated and convenient manner for customers, with the front office as the channel of information between firm and customers. Often there is also a pressure for quality innovations from staff keen to try out new products or new ways of doing things, or to exploit new capabilities. Here the back office plays some role, indicated in the figure with a dotted arrow.

Figure 8.2 Sources of innovation

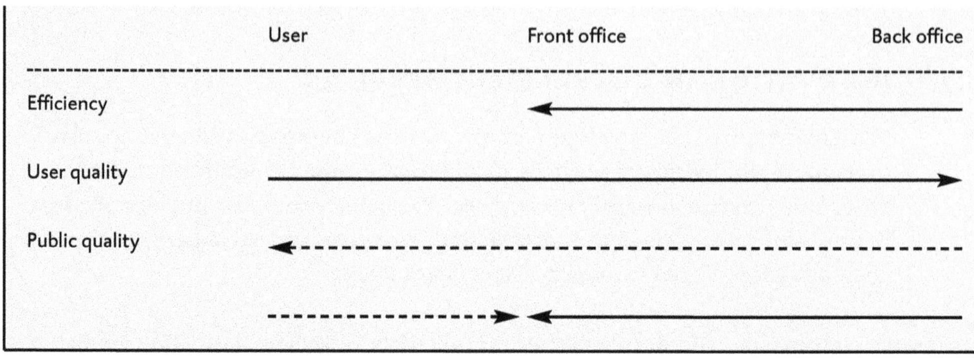

In private firms the importance of public quality, in the bottom row, has increased, as can be seen from the emergence of 'socially responsible management'. While individual customers may be a factor , this dimension of quality is mostly a corporate affair, in corporate communication with the public at large. In figure 8.2 this is expressed with a dotted line from users, and an uninterrupted line from the back office. While the importance of public quality is increasing in business, it is still less prominent than in public services. While in business the pressure for public accountability has increased, equity of access is much less of an issue than in public services, due to the fact that political processes are absent.

When we turn to public services, we expect that the institutional logic of the governmental sector, with a great emphasis on due process and accountability, combined with a risk-averse hierarchy that has to avoid political incidents, will sit in the way of initiatives to try out new approaches or make use of new possibilities. Thinking about government services in terms of figure 8.2, one therefore might predict that the bulk of innovations will be initiated by the back office and will focus on efficiency rather than quality. A report on a British investigation of 125 innovations in central government services (NAO 2006) offers many indications that this is indeed the case, pointing to processes that are very top down, influenced by an over-emphasis on hierarchy and rank. This is difficult to reconcile with the fact that innovation of quality from customer pull requires room for initiatives from the front office, which is lower in rank and hierarchy. The report indeed shows that around two and a half times as many innovations in the data set related to efficiency matters as opposed to quality improvements.

"Journalist Pieter Hilhorst argues that the public sector is also severely troubled by a "not invented here syndrome". Policymakers in one town rarely get "contaminated with good ideas from another town", the public sector has many laboratories but barely any production units that dole out successes. The consequence is that many successful projects and useful experiences end up on the scrapheap. The biggest flaw of the public sector is thus not that it lacks creativity or good plans, but that good ideas are hardly imitated. In the public sector intelligence is not contagious'" (*de Volkskrant*, August 7, 2007).

From above in the political hierarchy processes are imposed through the back office, to ensure political grasp, to safeguard political responsibility, and to protect politicians from political discomfiture. Public accountability is also imposed from the back office. Requirements are restrictive, requiring constant measurement and control and raising challenges of whether we can de-bureaucratize enough to allow front-line staff in government service more space for experimentation and improvisation. The NAO report highlights findings that civil servants are over-cautious and that their influence on provision has in recent years narrowed down to service delivery and internal organisational issues rather than improvement of user quality. Departments and agencies often know of opportunities to make improvements but they can be reluctant to act on these opportunities until they are pushed. These barriers may take time and persistence to counteract.

In chapter 7, we argued that monitoring and control can drive out innovation. In the public sector, that applies with a vengeance. Since the 1980s, when countries found themselves confronted with economic decline and increased international competition, governments have begun to spend more time and money on performance measurement and evaluation in the public sector, where result-based management has become "the talk of the day at all levels" (van Thiel and Leeuw 2002). Following the "New Public Management Hype" (Raad van Economische Adviseurs, [REA] 2007), the Dutch government has copied management-driven organisational practices from the private sector in its initiative for a new budget system to increase accountability. A new system of performance budgeting was thus introduced in the budgetary year 2002, which was christened "From policy budget to policy account" (Van Beleidsbegroting tot Beleidsverantwoording, VBTB). The purpose of this new system was to make it easier for parliament to compare budget plans and annual reports and to thereupon assess government policies. Over the years, VBTB has also developed into an instrument to advance a more efficient government (Tweede Kamer 2004; see also van der Dussen 2006).

There is no denying that performance measurement can be of great value in the public sector, and as the IMF (2006) has noted, "VBTB can, and in some instances has already, had a positive impact on the transparency of budget and final account documents. Importantly, it has improved policy accountability and has the potential to improve allocative decision-making." But it has also become clear that despite all good intentions, the way performance measurement has been implemented has had several unintended negative consequences. The Council of Economic Advisers (REA 2007) argues that the public domain is so different from the private sector that it is often counterproductive to transfer practices from one domain to the other, and emphasises two 'perversions' this has given way to: the autonomy of professionals has been corroded, and management has been elevated and become dominant. We would add that this has been destructive in business as well. Van Thiel and Leeuw (2002: 270) draw a number of more specific conclusions from their review of relevant studies: "First of all, the proliferation of regulators and auditors has increased the monitoring costs of the organisations and the state. Second, within policy-implementing organisations, the increased measure pressure can create dysfunctional effects such as ossification, a lack of innovation, tunnel vision, and sub optimisation. These unintended effects can jeopardize the effectiveness and efficiency of policy implementation. Third, there is some evidence that monitoring has led to symbolic behaviour; that is, monitoring appears to be in place but is in fact not. And finally, in some cases it is unclear what is actually being measured (e.g., the definition of quality)" (see also Aardema 2002).

"(B)ureaucratic performance measurement schemes can inhibit innovation, and lead to *ossification*: organizational paralysis brought about by an excessively rigid system of performance evaluation. The need to choose performance measures and set targets in advance means that new threats and opportunities may be ignored by managers. Moreover,... a predictable system of performance evaluation may offer considerable scope for manipulation on the part of management.

The danger of ossification arises because of the inevitable delay in designing and putting in place a performance evaluation scheme, and the effort required to change it subsequently. Yet... any scheme is likely to be deficient to a greater or lesser extent, and so will need to be regularly reviewed and updated" (Smith 1995: 299-300).

However, accountability by itself is not the only, and perhaps not the most important problem. Related to it are the organisational and mental obstacles to opening innovation up to initiatives from low-rank and low-status front-office staff, as well as citizens via their access to the system through the front office.

8.8 CONCLUSIONS

Although there are signs of a positive relation between organisational innovation – or 'social innovation' – and the success of innovation within Dutch companies, the innovation debate in the Netherlands has so far not paid much attention to the management and organisational features of successful organisation. Some researchers even argue that the decline of the Netherlands in some international rankings on competitiveness and innovativeness can to a large extent be explained by a concentration on technology-related macro-variables. However, more recently social innovation has been put on the agenda others, including trade unions, employers organisations, and the Innovation Platform, resulting in the institution of the Netherlands Centre for Social Innovation (NCSI). Mirroring day-to-day practice, however, social innovation has so far mainly been taken on as 'operational excellence', i.e., improving exploitation. Thus, the development of policy reflects the very problem that is at stake. As in organisations, here also exploitation takes precedence over exploration because it yields improvements in the short term, as opposed to improvements of exploration, and here as elsewhere the short term mostly wins. Furthermore, we tend to focus on exploitation because we understand it better and know better how to improve it. The main challenge for the future thus seems to be to extend projects to stimulate social innovation towards the exploration of new products and applications. There is an enormous potential for further innovation, not least in government services.

It seems inevitable that exploitation and exploration are to some extent separated, in time or space, because they require different organisational and cognitive logics. At the same time, however, they need to be connected in ways that allow them to build upon each other, in some form of 'ambidextrous organisation', within or between organisations. There are several ways to do this, but tensions tend to remain, to a greater or lesser extent. In earlier chapters we noted that while markets can serve exploitation well, intense price competition can erode the resources and the basis for trust needed for exploration. In chapter 9, Anderson and Gasteiger present evidence that time pressure forms a hindrance to creativity in R&D (Amabile 1988). This has implications for policy, in view of possible tensions between competition policy and innovation policy. As argued by Von Hippel (2005) the dynamic of openness and disclosure of innovation raises questions for policy concerning the protection of intellectual property rights, which we will not discuss here.

Since exploration is inspired by novel experience, across a variety of use contexts, and this variety derives from users, including citizens in the case of governmental services, users have an important role to play in innovation. The emergence of ICT and internet facilitates new forms of organisation, that allow much more involvement of users and citizens, yielding a different logic of relations between users and producers, in platform organisations, user-led innovation, and open source communities. A key feature is that with the use of internet-based facilities allowances can be made for asymmetry of information between users and producers, and activities can be located where the source of the information is. In open source communities, users are the producers, in systems that are highly self-organised, as evolutionary systems.

Policy implications concerning the organisation of private firms first of all regard industry rather than government. However, there still is a role for government in putting social innovation on the agenda, and in stimulating and coordinating activities of firms. Furthermore, there are implications also for the organisation of government services: the new organisational form of platforms yields considerable potential to improve the efficiency and quality of government services.

NOTES

1 See www.ncsi.nl

2 Encountered by author during a course given to consultants of Cap Gemini, in 2001.

3 For unto every one that hath shall be given, and he shall have abundance: but from him that hath not shall be taken away even that which he hath (Matthew XXV: 29, KJV).

REFERENCES

Aardema, H. (2002) 'Pas op voor de veertien ongewenste neveneffecten van bedrijfsmatig werken!', B&G, December, 16-19.

Appleyard, M.M., N.W. Hatch and D.C. Mowery (2000) 'Managing the development and transfer of process technologies in the semiconductor manufacturing industry', in: G. Dosi, R.R. Nelson and S.G. Winter (eds.) *The Nature and Dynamics of Organizational Capabilities*, Oxford: Oxford University Press, 183-207.

Archibugi, D. and B.A. Lundvall (2002) 'Introduction: Europe and the Learning Economy', in: D.Archibugi and Lundvall B.A. *The Globalizing Learning Economy*, Oxford: Oxford University Press.

Armbruster, H., A. Bikfalvi, S. Kinkel, and G. Lay (2007) *Organisational Innovation – The Challenge of Measuring Non-Technical Innovation in Large Scale Surveys*, paper under review for publication.

Biermans, M. and J. Poort (2006) *Rolopvatting Innovatie en Arbeidsproductiviteit*, Amsterdam: SEO.

Birkinshaw, J. and C. Gibson (2007) 'The Ambidextrous Organisation', executive briefing, Advanced Institute of Management Research, London

Brusoni, S. (2006) 'The Limits to Specialization: Problem Solving and Coordination in Modular Networks', *Organization Studies* 12(12): 1885-1908.

Dussen, J. van der (2006) 'Gehaktdag exit?', *Overheidsmanagement* 2006, 9: 221-223.

FNV Bondgenoten (2006) *Slimmer werken begint bij jezelf. Experimenteren met sociale innovatie*, Utrecht: FNV Bondgenoten.

Geest, L. van der and L. Heuts (2006) *Parels van de Nederlandse maakindustrie*, Breukelen: NYFER.

Harrison, M.M., N.L. Neff, A.R. Schwall and X. Zhao (2006) *A Meta-analytic Investigation of Individual Creativity and Innovation*, paper presented at the 21st Annual Conference for the Society for Industrial and Organizational Psychology, Dallas, TX.

Hippel, E. von (1999) '"Sticky information" and the locus of problem solving: Implications for innovation', in: A.D. Chandler, P. Hagström and Ö. Sölvell (eds.) *The Dynamic Firm: The Role of Technology, Strategy, Organization and Regions*, Oxford: Oxford University Press, 60-77.

Hippel, E. von (2005) *Democratizing Innovation*, Cambridge, MA. and London: MIT.

International Monetary Fund (IMF) (2006) *Kingdom of the Netherlands – Netherlands: Report on the Observance of Standards and Codes – Fiscal Transparency Module, and the Aide-Mémoire Regarding the Fiscal Framework*, IMF Country Report No. 06/124.

Jansen, J. (2005) 'Ambidextrous Organizations: A Multiple-level Study of Absorptive Capacity, Exploratory and Exploitative Innovation, and Performance', Ph.D. dissertation, Rotterdam: Rotterdam School of Management.

Lakhani, K.R. (2005) 'Why Hackers Do What They Do: Understanding Motivation and Effort in Free/open Software', in: J. Feller, B. Fitzgerald, S. Hissam and K.R. Lakhani (eds.) *Perspectives on Free and Open Software*, MIT Press.

Lam, A. (2005) 'Organizational Innovation', in J. Fagerberg, D. Mowery and R. Nelson, *The Oxford Handbook of Innovation*, Oxford: Oxford University Press.

Lester, R. and M. Piore (2004) *Innovation – The Missing Dimension*, Cambridge and London: Harvard University Press.

Li, J., W. Vanhaverbeke, and W. Schoenmakers (2007) 'Interpretation, Operationalization and Determinants of Exploration and Exploitation in Innovation: Beyond a Literature Review', unpublished paper.

Mateos-Garcia, J. and W.E. Steinmuller (2006) 'Open, But How Much? Growth, Conflict and Institutional Evolution in Wikipedia and Debian', paper presented at DIME conference on Communities of practice, Durham, October 27-28, 2006.

Miron, E., M. Erez and E. Naveh (2004) 'Do Personal Characteristics and Cultural Values that Promote Innovation, Quality and Efficiency Compete or Complement Each Other?' *Journal of Organizational Behaviour* 25: 175-199.

Muller, P. (2006) 'How Does Leadership Support the Activity of Communities of Practice?' Ph.D. dissertation BETA , University of Strasbourg, document de travail 2006-14.

NAO, National Audit Office (2006) *Achieving Innovations in Central Government Organizations*, available at www.governmentontheweb.org.

Nooij, M. de, and J. Poort (2005) *Vooruit met procesinnovatie*, Amsterdam: SEO.

Nooteboom, B. (2000) *Learning and Innovation in Organizations and Economics*, Oxford: Oxford University Press.

Nooteboom, B. (2004) *Inter-firm Collaboration, Learning and Networks: An Integrated Approach*, London: Routledge.

Organisation for Economic Co-operation and Development (OECD) (2002) *Employment Outlook*, Paris: OECD.

Pisano, G. (2000) 'In Search of Dynamic Capabilities: The Origins of R&D Competence in Biopharmaceuticals', in: G. Dosi, R.R. Nelson and S.G. Winter (eds.) *The Nature and Dynamics of Organizational Capabilities*, Oxford: Oxford University Press, 129-154.

Quinn, J.B. (1992) *Intelligent Enterprise*, New York: The Free Press.

Raad van Economische Adviseurs (REA) (2007) *Lof der eenvoud*, Kamerstukken II, vergaderjaar 2006-2007, 30 942, nr. 2.

Smith, A. (1776) *An Inquiry Into the Wealth of Nations.*

Smith, P. (1995) 'On the Unintended Consequences of Publishing Performance Data in the Public Sector', *International Journal of Public Administration*, 18(2&3): 277-310.

Sociaal-Economische Raad (SER) (2006) *Welvaartsgroei door en voor iedereen. Thema sociale innovatie*, The Hague: SER.

Thiel, S. van and F. Leeuw (2002) 'The Performance Paradox in the Public Sector', *Public Performance & Management Review* 25(3): 267-281.

Tweede Kamer (2004) *Evaluatie VBTB*, Brief van de minister van Financiën, Kamerstukken II, vergaderjaar 2004-2005, 29 949, nr. 1.

Visitatiecommissie Emancipatie (VCE) (2005) *Emancipatiebeleid en gender mainstreaming bij het Ministerie van Economische Zaken: Voorlopig Beeld*, The Hague: VCE.

Volberda, H. (2007) 'Het innovatieplatform, een sociale innovatie binnen de Nederlandse polder', in: J.P. van den Toren and S. Akkerman (eds.) *Handboek Innovatieplatform: wetenschappelijke visies op de werkwijze van het innovatieplatform'*, The Hague: Innovatieplatform.

Volberda, H. and F. van den Bosch (2005) 'Rethinking Innovation: Management and Organization Matter Most', paper presented at the Copenhagen Conference on Strategic Management.
Volberda, H., F. van den Bosch and J. Jansen (2006) *Slim managen & innovatief organiseren. Onderzoeksverslag*, Rotterdam: RSM Erasmus Universiteit.
Wendel de Joode, R. van (2005) 'Understanding Open Source Communities', Ph.D. dissertation, Technical University Delft.

9 INNOVATION AND CREATIVITY IN ORGANISATIONS: INDIVIDUAL AND WORK TEAM RESEARCH FINDINGS AND IMPLICATIONS FOR GOVERNMENT POLICY

Neil R. Anderson and Rosina M. Gasteiger

9.1 INTRODUCTORY COMMENTS

Facing rapid technological changes and being challenged, for example by emerging markets, today's organisations have to adapt quickly in order to maintain or, ideally, to increase their effectiveness. Within this scope the development and adoption of innovations have become a critical determinant of organisational productivity, competitiveness, and longevity. Hence, it is not surprising that a major research effort has focused on variables that facilitate or hinder the development and implementation of innovations (Howell and Higgins 1990). The aim and scope of the present chapter is to present an overview of existing research findings into innovation and creativity in the workplace as these relate to potential government policies for innovation facilitation in the Netherlands. By necessity, as many of the applied studies originate from the disciplines of organisational psychology and management sciences, these disciplines constitute the predominant theoretical perspective adopted in this chapter. Given this disciplinary background, this chapter considers in detail two particular 'levels of analysis' with regard to creativity and innovation in workplace settings: (1) Individual creativity and work role innovation and (2) work group creativity and team-level innovation.

One of our main reasons for focusing upon these two levels of analysis was the WRR summary description of this whole project area itself. To quote from the homepage related to the project on 'Innovation: The Need for Renewal':

"The fundamental unit of analysis is people in interaction with other people, within and between businesses, with businesses, macro-conditions and institutions as 'enabling constraints'. The analysis at micro-level provides a basis for recommendations which are largely situated at the institutional level" (http://www.wrr.nl/english/content.jsp?objectid=3949&pid=3947).

Whilst other, more macro levels of analysis undoubtedly exist with regard to organisational innovation (the organisational level of analysis), and the diffusion of innovations across industrial sectors or organisations (the societal level of analysis), the focus of this chapter is intentionally more micro- and meso-analytical in orientation. Creativity in individual work roles can build to transform organisational performance overall. The implementation of new and improved ways of doing things in organisations can, and does, occur primarily at the individual and work-team levels of analysis (Anderson, De Dreu and Nijstad 2004; West, 2002), and so these two primary levels of foci for this chapter are justifiable. In addition, the cumulative effects of numerous individual-level work-role innovations with an ongoing climate for work-group innovation may be argued to be a major source of competitive advantage for Dutch organisations. Indeed, in rapidly changing global markets where innovation and adoptive capacities are key determinants of organisational success, it can be argued that creativity and innovation are crucial factors for organisational survival and longer-term performance.

In this chapter we review important findings and empirical studies as they relate to potential government policy for innovation facilitation in organisations. Although we do not develop these research findings into specific recommendations for government policy, we hint at the need for these findings to be taken into clear account and to be borne in mind with regard to the general formulation of various policies and procedures that may impinge upon the innovativeness of work organisations in the Netherlands.

REMIT

Building upon these opening comments, the agreed remit for this chapter as laid out in our proposal was as follows:

As an integral component of this overall project, to review existing research findings into creativity and innovation in the workplace. Particular attention will be given to studies into individual-level work role creativity and to empirical research into work group-level innovation. Attention will also be focused upon setting forth a general research context in summarising the major study findings upon which potential government policy and procedures can in future be based.

9.2 DEFINING INNOVATION AT WORK

An initial problem faced by all practitioners and researchers is to be able to accurately define the terms 'creativity' and 'innovation' as they pertain to organisational practices. In fact, precise definitions of creativity and innovation have long been a source of dispute and debate amongst academic researchers (West 2002). Fortunately, although there remains no general consensus over precise definitions, the most widely accepted definition of innovation was proposed by West and Farr (1990). They define workplace innovation as:

... the intentional introduction and application within a role, group or organization of ideas, processes, products or procedures, new to the relevant unit of adoption, designed to significantly benefit the individual, the group, the organization or wider society' (West and Farr 1990: 9).
By use of this definition the authors distinguish between creativity and innovation in the crucial regard that the latter involves 'intentional introduction and application' of new and improved ways of doing things, whereas creativity can simply refer to idea generation alone (see also King, 1992). This distinction is essential: Creativity can refer simply to idea generation; unlike creativity the critical aspect of innovation is that this term refers alone to implemented ideas. That is, the

application component is crucially important for the latter. For this reason, the innovation research typically draws distinctions between the early phases of idea generation (so-called 'ideation'), and latter-stage implementation to suggest that successful innovation attempts will require both phases to be successfully navigated. Second, West and Farr (1990) point out that innovation must confer intended benefits at one or more levels-of-analysis: the work role, group, or organisation. Again, this is not necessarily the case for creativity where benefits can be impossible to quantify or assign. In this case, however, it is often advisable to be somewhat flexible as to whether benefit was truly intended before the innovation process was begun. For instance, the 3M product of Post-it actually resulted from errors that were made by one technician in developing a glue for totally another purpose, which in fact was a 'failure' for the purpose actually at hand. The glue was then re-used as an unusual combination of being adhesive enough to stick down but only temporarily on the reverse side of notelets, hence the development of this new product which only turned out subsequently to be of considerable commercial benefit. Third, and finally, innovation implies relative novelty as opposed to absolute novelty (Zaltman, Duncan and Holbeck 1973). That is, any change need only be new to the relevant unit of adoption, not absolutely new allowing for innovations to be adopted and adapted from other organisations or work teams (West and Anderson 1996). This concept of 'relative novelty' is also important in that new and improved ways of doing things need only be new to the section or department where they are implemented. That is, they should be new in the perceptions of those involved with the innovation attempt, rather than being absolutely new when compared with practices in all other organisations of a similar nature. This distinction thus allows for innovations that are in common usage elsewhere to be adopted and adapted within the new unit of adoption, thus still fitting within West and Farr's overall definition of innovation in the workplace. Few would argue that for a new idea or process to be regarded as an innovation requires absolute novelty (Anderson et al. 2004), and so this forces practitioners and researchers to regard innovation policies from the 'eyes of the beholders', that is from the perspective of employees and managers in organisations who are charged with innovation implementation on a daily basis.

These debates over defining innovation are clearly not just matters of semantics or rather ephemeral academic discussions. For any government policy aimed at innovation stimulation and facilitation to be successful, it is crucial for innovation to be viewed from the perspective of those affected by the suite of policies in force at any particular time. Top-down policy that is imposed is unlikely to be stimulative of response innovations by organisations, work teams, or individual workers (Van de Ven, Angle, and Poole 1989). Rather, innovation policies need to be aimed at establishing frameworks and climates that are supportive of creativity and innovation. It is thus vitally important to consider the involvement, where appropriate, of key workers and their representatives in early-stage policy formulation. We return to each of these themes in subsequent sections of this chapter.

9.3 ANTECEDENTS AND FACILITATORS OF INNOVATION

A substantial body of findings across the research literature has firmly established which factors act as either facilitators or inhibitors of creativity and innovation in the workplace. Such is the weight of evidence from the huge number of primary studies, that robust and fairly definitive conclusions can be drawn from this body of empirical research. This stated, one of the most intriguing aspects of these findings is that several factors have been found to be related in a curvilinear manner to innovativeness, rather than in a direct, linear manner. Most common have been research findings that, when combined, reveal an inverted U-shaped curvilinear relation with subsequent innovativeness (e.g., rule independence and general mental ability at the individual level-of-analysis; group cohesiveness at the work team level-of-analysis). Caution is therefore warranted concerning the interpretation of existing findings, but for cases where such curvilinear relations have been found there is a readily understandable explanation. For instance, a work group needs to be high on cohesiveness for innovation attempts to emerge, but not so high that conformity to current methods and practices takes over as the climate for maintenance (so-called 'groupthink', Janis, 1972). Thus, middling levels of group cohesiveness have been found to be most facilitative of work group innovativeness, with low cohesiveness and high cohesiveness being found paradoxically to be related to significantly lower levels of innovation.

Some caution is also warranted in that, as is fairly typical in the social sciences, studies sometimes report contradictory findings with regard to specific predictors of work role or group innovativeness. In the present chapter we have sought to overcome this by reporting the balance of weight of particular findings and by giving priority to review papers published in the top-tier journals in the management sciences where possible (e.g., *Academy of Management Journal, Journal of Applied Psychology, Academy of Management Review, Journal of Management, Journal of Organization Behavior*). Other studies, often published in second- or even third-tier journals have been cited albeit briefly, and with the caveat that we have given precedence of interpretation to the findings appearing in top-tier journals where review and publication procedures are far more stringent and reliable. Of course, it could also be that such contradictory results have come about because single studies have restricted sample sizes, or have calculated a single linear correlation where as mentioned above curvilinear relations may exist, or have been based in particular industries or at particular levels within the organisation hierarchy. Here, we attempt to provide a balanced and reasoned summary of the key research findings, and also to note studies and programs of research of particular noteworthiness and scientific impact.

One final point needs to be made at this introductory stage. Most of the research efforts by work psychologists referred to in the present chapter have focused upon larger organisations, usually on the basis of attempting to maximize sample size once access has been negotiated. Many of the studies cited in our chapter concern individual and team level innovation in larger, often multinational organisations. The generalisability of these findings to smaller organisations, family businesses, and the like therefore can be open to question. Rather, the bulk of the research referred to in this chapter concerns larger numbers of individual employees (e.g., in engineering companies, public sector departments, research and development teams, hospitals, and so forth) and teams working in large- or medium-sized organisations. Much of the research has also been carried out within the US, and so again, some caution is warranted in applying these findings, although where generalisability across countries has been investigated the findings have tended to hold up remarkably well across different countries (and especially from the US to European organisations in general).

Table 9.1, summarises the key findings over factors found to be antecedents to work role innovation and work group innovation in overview.

9.3.1 INDIVIDUAL-LEVEL ANTECEDENT FACTORS: INDIVIDUAL AND JOB CHARACTERISTICS

As table 9.1 sets out, four key factors have been found to be predictive of work role innovation at the individual level-of-analysis:
- Personality;
- Motivation;
- Cognitive ability;
- Mood states.

In addition, job characteristics, for example the autonomy in ones work role, have been shown to be important predictors for individual innovation at the workplace.

In early studies on creativity and innovation scholars thought that certain personality characteristics constitute an individual's potential to be creative and to innovate (Barron and Harrington 1981). Correspondingly, it was attempted to find measures to identify a 'creative personality' (Gough 1979). Contemporary studies analyse the relationship between personality and innovation in more detail. Within that scope the so-called 'big five'[1] or the 'Five-Factor Model' (FFM) of human personality play an important role. In terms of individual personality types found in the applied research to be likely to be more innovative, the profile is not particularly surprising. Individuals with distinct self-confidence, tolerance of ambiguity, unconventionality, and independence were shown to be more likely to be innovative than individuals showing these attributes to a moderate extent. Furthermore, individuals high on openness to experience as well as lacking conscientiousness and agreeableness have been found to have a higher propensity to innovate in their job roles (George and Zhou 2001a, 2001b; Patterson 1999). Openness to experience was shown to be the personality variable with the most consistent relationship with creativity and innovation (Harrison et al. 2006). Given the conceptual similarity between openness and creativity, this result is not astonishing. Individuals high on openness to experience are characterised as being curious and interested in novelties in general; furthermore, they are for example more likely to try out new products and to generate new ideas as well (McCrae 1987). Looking at the relationship of the 'big bive' and creativity in comparison to job performance ratings it becomes obvious that creativity may not be comprised in global ratings of job performance. Conscientiousness, for instance, which is known as the most consistent personality predictor of job

Table 9.1 Innovation research findings in overview: Facilitators of innovation at two levels-of-analysis – individual and work group

Level-of-analysis	Characteristic	Dimension	Key studies/ evidence [1]
Individual and Job 1.2 characteristics	Personality	Tolerance of ambiguity	Barron & Harrington, 1981; Patterson, 1999
		Self-confidence	Barron & Harrington, 1981
		Openness to experience	West 2001; Patterson, 1999; George & Zhou, 2001a; 2001b
		Unconventionality	West & Wallace, 1991
		Originality	West & Wallace,1991; Patterson, 1999
		Rule governed (negative relation)	Simonton, 1991
		Authoritarianism (negative relation)	Simonton, 1991
		Independence	West & Wallace, 1991; Patterson, 1999
	Motivation	Intrinsic (versus extrinsic)	West, 2001
		Determination to succeed	Amabile, 1983
		Personal initiative	Frese & Zapf, 1994
	Cognitive ability	Above average general ('g') intellect	Barron & Harrington, 1981; Patterson, 1999
		Task-specific knowledge	West & Wallace, 1991
		Divergent thinking style	Kirton, 1976; 1978
		Ideational fluency	Barron & Harrington, 1981
	Mood States	Negative moods	George & Zhou, 2002
	Job characteristics	Autonomy	Axtell et al, 2000
		Span of control	Axtell et al, 2000
		Support for innovation	Axtell et all, 2000
		Mentor guidance	Simonton, 1991; Van de Ven, et al., 1989
		Appropriate training	Van de Ven, et al., 1989; West, 2001, George & Zhou, 2001a; 2001b
Work group			
	Team structure	Minority influence	De Dreu, 2006
		Cohesiveness [2]	West & Anderson, 1996
		Longevity [2]	King, 1992; West & Anderson, 1996
	Team climate	Participation	West & Anderson, 1996; De Dreu & West, 2001
		Vision	West & Anderson, 1996; De Dreu & West, 2001
		Norms for innovation	West & Anderson, 1996; De Dreu & West, 2001
		Conflict	De Dreu, 2006
		Constructive controversy	West, 2001; 2002
	Team member characteristics	Heterogeneity of members	Kirton, 1976; 1978
		Education level	Sauer & Anderson, 1992
	Team processes	Reflexivity	West, 2002
		Minority dissent	De Dreu & West, 2001
		Integration skills	Stevens & Campion, 1994
		Decision making style	King & Anderson, 2002
	Leadership style	Democratic style	West, 2001, Krause, 2004
		Participative style	West & Anderson, 1996; De Dreu & West, 2001
		Openness to idea proposals	West & Anderson, 1996

1) Source: Developed, modified and extended from Anderson, et al. (2004)
2) As noted earlier, for several characteristics at different levels-of-analysis curvilinear effects have been proposed, and in a few instances, found. Mostly inverted U-shaped curvilinear effects have been reported in a relatively small number of more complex study designs.

performance, was shown to hold nearly zero-relationships or even slightly negative relationships with creativity and innovation (Harrison et al. 2006). A potential explanation for these findings is that the 'big five' constructs are too broad and heterogeneous for predicting important criteria like, for example, overall job performance or innovation potential (Barron and Harrington 1981). Moreover, these results suggest that personality traits might not be paramount drivers for creativity and innovation on the individual level-of-analysis as it was hypothesised in earlier research.

Research indicates that individuals need to possess a driving force in order to cope with the challenges associated with creativity and innovation at work (Tagger 2002; Harrison et al. 2006). Regarding the motivation underlying creative and innovative behaviour non-material reasons, such as the aspiration to shape one's own professional environment, to ease work, and to search for remedies to defects and risks as well as the possibility of cost-saving, have been proven to be important (Maier et al. 2001). Intrinsic motivation, which refers to the inherent human tendency to seek out novelty and challenges that represent a principal source of enjoyment throughout life (Csikszentmihalyi and Rathunde 1993), is one of the most important driving forces for creative and innovative behaviour (Amabile 1988; West 1987; Frese et al. 1999). Furthermore, psychological empowerment, which refers to the set of cognitions essential for intrinsic motivation (Spreitzer 1995), the determination to succeed (Amabile 1983; Howell and Higgins 1990), an orientation towards learning, self-efficacy[2] footnotes as in previous chapters] and creative self-efficacy (Axtell et al. 2000; Tierney and Farmer 2002), as well as personal initiative[3] (Frese and Zapf 1994) and have been shown to be important predictors for innovative behaviour at work. Additionally, also extrinsic motivation has been shown to have a positive relationship with creativity although on a smaller scope (Harrison et al. 2006).

Moreover, in addition to these personality-related attributes, other studies have found that individuals who are of higher intellect (higher general intelligence, abbreviated as 'g') are typically more innovative. This is simply due to the fact that these individuals possess greater information processing and cognitive capacities, and therefore, they are able to generate more ideas for ways to improve existing ways of doing things at work. Of course, this then requires the personality, motivation, and determination to implement such ideas, which is where other person-related attributes become more important. Following the 'investment theory of creativity' by Sternberg and Lubart (1991) creative individuals are characterised by taking up ideas with high development potential at a time when these ideas are still largely unknown, relatively unpopular or even disrespected by others. They stick to their idea and make an effort to convince others of its usefulness in order to put it into practice eventually. Hence, creative behaviour requires not only specific personality traits and motivation but also intelligence, knowledge, expertise, and specific thinking and working styles on part of the ideator (Amabile 1988; Sternberg and Lubart 1996; Patterson 2002). Amabile's componential theory of creativity (1983) and creativity in the workplace (1996) suggests that domain-relevant skills are a fundamental component of creativity. Domain-relevant skills comprise knowledge and expertise within a certain domain. According to Amabile (1983, 1996) domain-relevant skills are prone to be affected by formal and informal education as well as an individual's perceptual, cognitive, and motor abilities. Through education individuals are exposed to a variety of experiences, viewpoints, and knowledge bases (Perkins 1986). Recent meta-analytical findings by Harrison and colleagues (2006) suggest that general cognitive ability and education have a moderately strong direct effect on creative and innovative performance.

With regard to mood and creativity research findings are less clear. While some articles come to the conclusion that positive moods are positively related to creativity and innovation (Russ 2000) other research has shown that job dissatisfaction and negative mood states can provoke creative and innovative outcomes. An intriguing theme of recent research into rather more negatively connotated factors at the individual and group levels and how these relate to subsequent innovation has been examined and published by Jennifer George and Jing Zhou (2001a, 2001b, 2002). Whilst much previous innovation research has suffered from what has been termed a 'pro-innovation bias' (that is, innovation being assumed to be positive per se regardless of the context or outcomes), George and Zhou (2001a, 2001b) took a contrary stance. They argued, and found empirical support for, the point that individuals will also attempt to change matters at work from the basis of a negative mood state, dissatisfaction, or as an attempt to alleviate the stress experienced in an underperforming situation. These negative mood states (such as dissatisfaction, irritation, and unhappiness with work role processes and outcomes) were found to be far more positively predictive of innovation attempts than had previously been thought. Further research is needed in order to shed light on the interplay of mood states and creativity and innovation. In doing so it is crucial to be aware of the complexity of affect. Affect can be studied either as a trait or as a state variable while the latter can be further differentiated into 'moods' and 'emotions'.

All predictors described above are directly related to the individual. In studying creativity and innovation at work not only person-related but also situational variables have been shown to play a decisive role. Within this connection job characteristics have been widely studied as predictors of creativity and innovation. Amabile (1988) found that challenging, non-routine jobs are conducive to creative and innovative behaviour, presumably due to the fact that they include diverse activities. In addition, research shows that individuals holding down a complex job demonstrate more creative self-efficacy (Tierney and Farmer 2000). Furthermore, autonomy in one's work role, i.e., the freedom to decide how, when, and with whom to work (West and Farr 1990), has been shown to be predictive of creative and innovative behaviour (Krause 2004). In addition, Amabile (1988) argues that time pressure is a frequently reported hindrance to creative behaviour in the field of R&D on the grounds that time pressure results in shallow, restricted thinking, instead of broad thinking. Unlike many other factors that are likely to influence creativity and innovation at work, such as individual personality, time pressure is a variable that may be more easily controlled for. For instance, supervisors and managers may be able to take action to protect time for their employees that enables them to engage in creative thinking for example within the scope of think tanks (Amabile et al. 2002).

In sum, the previously explained findings suggest that jobs can be (re)designed in order to facilitate creative and innovative behaviour at work, for example by increasing work autonomy and complexity as well as protecting time where employees can give free rein to their ideas. For each of these areas of findings there are concomitant implications for the design of jobs in work organisations, human resource management (HRM) practices, employment legislation, and indirectly, for policy formulation at the national level to support and facilitate innovation. It therefore follows that governmental and industry-sector policies can influence such initiatives, maybe not in a direct and immediate manner, but certainly with regard to longer-term outcomes and in relation to establishing the underlying conditions, regulatory structures, and societal conditions facilitative of innovation in the workplace

9.3.2 WORK GROUP LEVEL ANTECEDENT FACTORS

At the level-of-analysis of the work group or team, a further five variables have been found to be associated with generally higher levels of innovativeness. These are:
- Team structure;
- Team climate;
- Team member characteristics;
- Team processes; and
- Leadership style.

Team variables were shown to be momentous for the promotion of creative and innovative work behaviour (West 2002). With regard to the structure or composition of a work group diversity has been suggested to be conducive to creativity whereas the management of diversity is seen to be the critical factor. Following West (2002) the effective management of diversity comprises integrative group processes, which in turn contain reflexivity, the development of intra-group safety, participative decision-making processes and the way of dealing with minority influence. Furthermore, the cohesiveness and the longevity of the group (West and Anderson 1996) are crucial determinants for the creative and innovative potential of a group.

With regard to the team climate for creativity and innovation the following four dimensions can be distinguished: vision, participative safety, objectives and task orientation, and support for innovation (West 1990; Anderson and West 1994). *Vision* refers to the extent to which superior goals are perceived as motivating, clear, coherent, and accomplishable. *Participative safety* relates to the degree to which the participation in decision-making processes is felt to be unbiased, motivating, and rewarding by the employees. *Objectives and task orientation* bear on the extent of the commitment to quality and high achievment by the employees.

Finally, *support for innovation* refers to supporting social norms and expectations that are for example beneficial for the introduction of new practices. These dimensions influence the different phases of the innovation process.

With regard to team member characteristics the educational level of the team members and their heterogenity (e.g., concerning age, gender, ethnic origin, function) have been suggested to be crucial for creativity at work (Amabile 1988).

Furthermore, the way that groups work together has an impact on their innovativeness. With regard to team processes a high degree of participation and autonomy in the work group (Axtell et al. 2000), reflexivity, a good communication structure in terms of problem-oriented discussions and active support, a shared vision (West 2002), and reciprocal feedback (Zhou and George 2001a) have a positive impact on a group's creative.

In addition, leadership has been shown to be an important factor throughout the innovation process. Recent meta-analytical findings show that leader-member-exchange (LMX) in terms of supervisor support, is positively related to creative and innovative behaviour of followers (Harrison et al. 2006). In accordance with the theory on LMX, the relationship between a leader and her or his subordinates emerges from a formal interpersonal relationship to one characterised by mutual trust and respect. Subordinates in high-quality LMX relationships are supposed to have more autonomy in their work roles (Graen and Uhl-Bien 1995) and, thus, are more likely to be creative and innovative in their jobs (Jaussi and Dionne 2003; Krause 2004). Furthermore, transformational leadership has been shown to be positively related to creative and innovative behaviour of followers (Harrison et al. 2006). This may be due to the fact that transformational leaders have a positive impact on followers' motivation along with their support for ideas and social support (Jaussi and Dionne 2003). Research conducted by Zhou and George (2003) provide evidence for the important role of leaders throughout the innovation process, i.e., from idea generation to implementation, in particular in terms of the supervisor's emotional intelligence. Different leadership styles will lead to specific climates and cultures primarily, that are either conducive or detrimental to creativity and innovation at the workplace. Taken all together, the findings described above underscore how important leadership is as the major lever to gearing innovation (Zhou and George 2003).

9.4 THE PROCESS OF INNOVATION

One of the most commonly misunderstood aspects relating to workplace innovation is the process or stages through which any innovation progresses from its earliest instigation to its routinisation of usage or 'exnovation' (the intentional termination of a former innovation) by a work group or individual employee (Van de Ven et al. 1989). This misunderstanding, it can be argued, is most likely a result of poorly researched but strikingly popular management texts that almost universally suggest that innovations proceed though neat, linear stages that can be managed or directed by those at the top level in an organisational hierarchy. Rather, several robustly conducted processual studies now firmly illustrate that innovation attempts proceed through a rather messy, multi-phase 'star-burst' trajectory where processes are cyclical, iterative, and will often involve "two steps forwards and then one step backwards" before an innovation progresses from initiation to final implementation (King and Anderson 2002).

Further international evidence over the stages involved in different types of innovations in organisations is provided by Van de Ven et al. (1989, 1999) in their major programme of research in North America, usually referred to as the Minnesota Innovation Research Program (MIRP).

Building upon earlier work, figure 9.1 presents a summary of potential negative antecedents, processes, and outcomes from innovation attempts. Counterposing these three stages against the three levels of analysis of the individual, group, and organisation, this model thus generates a nine-cell representation of how innovation attempts may be driven by, go through phases of, or result in, rather less than positive elements within organisations. The following nine cells have been derived.

Figure 9.1 **Generic Model of Negative Innovation Attributes at Three Levels of Analysis**

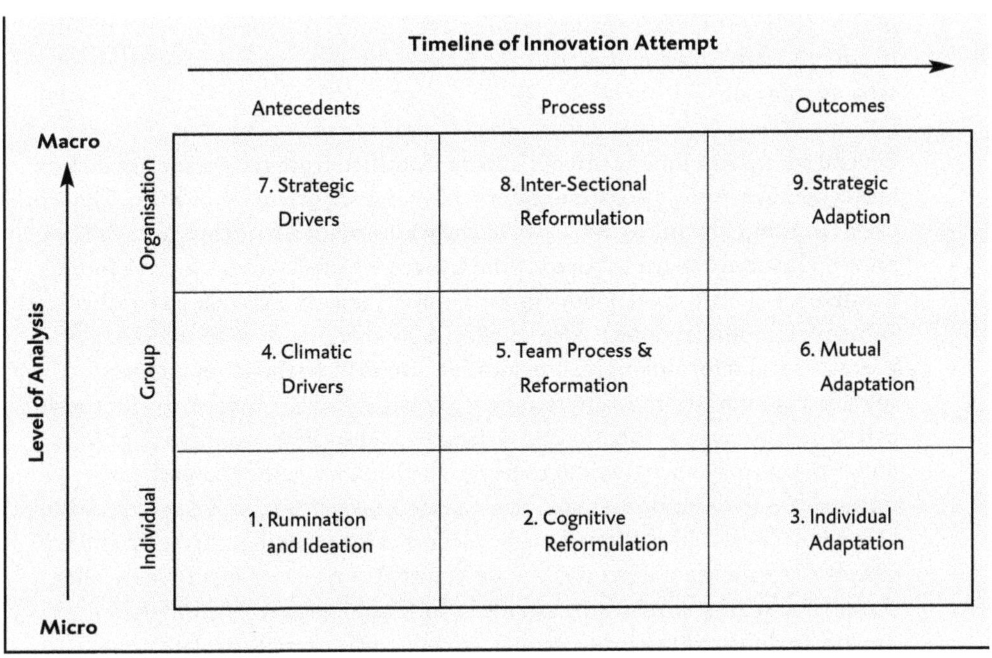

Figure 9.1 thus summarises much of the very recent research that suggests that innovation attempts, processes, and outcomes commonly have negative as well as positive attributes (Anderson and Gasteiger 2007). Considering each level of analysis in turn, at the individual and job characteristics level, some studies have already shown that, for instance, a negative rather than a positive mood state induces greater innovativeness (Zhou and George 2001a, 2001b). As negative attitudes such as job dissatisfaction, lack of adherence to rule structures, and so on, these attributes can be grouped under the heading of 'rumination and edeation'. Further studies strongly indicate that the process an individual worker goes through to be innovative is also far from smooth sailing, with numerous backward steps, reformulations of the original proposal, and notable resistance from work colleagues or one's immediate supervisor (Amabile 1988, 1996). Thus, we coin the term 'cognitive reformulation' to refer to the way in which individual employees will need to re-think and reformulate their original ideas in order to move them toward implemented work role innovations. Finally, at the individual level of analysis, some research hints at the range of adaptations the innovator will need to go through once their new idea or proposal is accepted and implemented toward routinised practice, even if just in their own job role (Bunce and West 1996). Innovation attempts are likely to cause stress for the initiator, along with other challenging feelings and imperatives to update one's cognition toward one's working environment. We therefore term this third cell 'individual adaptation' and it is meant to refer to any change the individual innovator experiences directly as a result of their innovation efforts. A final point of caution is warranted. Of course, not all attempts at being creative will result in a 'successful' innovation for employees. One of the potential negative outcomes can often be that the innovation attempt failed to result in modified work role procedures or outcomes. It is therefore wise to bear in mind that one outcome of innovation may be feelings of being unsuccessful or even ineffective for employees involved in attempted innovations that did not make it through to implementation for whatever reason.

Several so-called 'climatic drivers' can be identified from the extant research as antecedents carrying a negative connotation of work group innovation. The work of de Dreu and his colleagues is particularly informative in this regard (de Dreu 2006). This research team, based at the University of Amsterdam, have found positive relations between work group conflict, at least task-related conflict, and subsequent innovativeness. Also categorisable within our cell titled 'team processes and reformulation', it is sensible also to view the group process involved in innovation attempts as being one that can be problematic for those concerned. Some in a team may be resistant to innovation attempts (e.g., West, 2001), others may wish the idea to be pursued down a particular path only, whereas groups extremely high upon maintenance needs may wish to avoid any hint whatsoever of the disagreement or conflict inherent in innovation and try simply to retain the status quo (Van de Ven et al. 1999). Not surprisingly, then, innovation within work teams can result in several negatively connotated outcomes alongside the innovation itself, including changes in the group

Figure 9.2 Summary of innovation typologies

Author(s)	Damanpour (1987) Damanpour & Evans (1984)	Zaltman, Duncan & Holbek (1973)	Peters & Waterman (1982) Kanter (1983)	Anderson (1990)	West & Anderson (1996)
Description of Typology	Sociotechnical systems typology of innovation	Interaction typology of innovation characteristics	Radicalness of innovation typology	Sources of innovation typology	Innovation characteristics and outcome effects of typology (Health Care Organisations)
Innovation types proposed	*Technical;* Changes in production methods or products manufactured *Administrative;* Changes in social relationships at work *Ancillary;* Innovations which cross organisations-environment boundaries	*Programmed-Non programmed;* innovations vary in the extent of pre-planning, scheduling and organisational programming untertaken *Instrumental-Ultimate;* Instrumental innovations facilitate larger-scale ultimate innovations which can be self-contained change processes *Radicalness;* (a) novelty-degree of newness of the innovation (b) risk-degree of risk involved in implementation	*Evolutionary-Revolutionary continuum;* Evolutionary innovations as minor improvements on existing designs/systems Revolutionary innovations as paradigm-breaking redesigns of existing designs/systems *See also* Kirton (1976, 1978) at the individual level of analysis who proposes an adaption-innovation continuum of creativity styles	*Emergent innovations;* Novel, unproven ideas and proposals developed and implemented unique to a particular organisation or organisational subunit *Adopted innovations;* Systems/procedures already in use within comparator organisations which are replicated and adopted by other organisations *Imposed innovations;* Where changes in environmental contingencies force the organisation to develop innovative responses	*Magnitude;* Size of innovation and the extent of likely consequences *Novelty;* Relative newness of the innovation *Radicalness;* Extent of likely change to the status quo *Effectiveness-patient care;* Likely benefit to patient care *Effectiveness-staff well-being;* Likely benefit to staff well-being *Effectiveness-hospital administration;* Likely benefit to hospital administrative efficiency

Source: *Internat. Review of Industrial and Organisational Psychology*, (1993) 8, 1-34, John Wiley & Sons, Ltd. Reproduced with permission

processes, changes in interaction patterns and interpersonal relationships, and patterns of ongoing working (West and Anderson 1996). This, we term 'mutual adaptation' in our model presented in figure 9.1.

At the wider organisational level of analysis, the same three effects can be postulated to occur during the antecedents, process, and outcomes of innovation. Less empirical research is available to support these propositions than at the individual or group levels, however, so some caution needs to be emphasised with regard to our model being applied to overall organisational innovation. This stated, Damanpour (1987), for instance, in his meta-analysis of organisational factors associated with company innovativeness reports some borderline negative factors such as loose structure and non-directive leadership styles as being conducive to company innovativeness. Such 'strategic drivers' as we term these variables will also include aspects such as the organisation's structure (Daft 1978), the number of levels in the hierarchy (Kanter 1983, 1988), organisational resources and time devoted to support innovation attempts especially within research and development (Kimberly 1981), and the industrial sector within which the organisation needs to survive by being either more or less innovative (Kanter 1988). Many such organisation-wide innovations will result in unavoidable changes to the old organisation structure, a facet we term 'inter-sectional reformulation' in the table. Finally, the introduction of new products, services, or working practices will often result in necessary 'strategic adaptation' for any organisation. By definition, these more substantial forms of innovation will require an organisation to restructure itself in some manner and to engage in a period of reflection and review of its own strategic plans.

9.5 CONCLUSIONS

In this chapter we have reviewed key findings into innovation primarily in job roles and work teams. There is now sufficient mass of consistent findings to at least offer firm bases upon which to formulate national policies and procedures which will be supportive of creativity and innovation within organisations in the Netherlands. Clearly, cultural and work practice differences will exist between different organisations and between different industrial sectors, so it is unwise to make too specific recommendations at such an early stage. However, this stated, it is apparent that the growing body of research into creativity and innovation at the level of analysis of the individual and the work team holds considerable sway upon the rational formulation of policies and procedures designed to stimulate innovative capacity in Dutch industry and service organisations. Although the level of analysis of work psychologists has been more micro-analytical in nature, the weight of empirical findings from research now published holds clear implications for the wider formulation of policy and procedure. Whether such policies should be aimed at maximising innovative capacity is a moot point; indeed, we have cautioned against this in this chapter. A more balanced approach is to recognise the benefits of workplace creativity and innovation on the one hand, and on the other to bear in mind the necessarily disruptive aspects of innovating as the process emerges in individual work roles and teams in organisations. Truly, not necessarily more innovation is unconditionally better, but strategically stimulated innovation in key areas that is effectively managed can produce huge benefits to competitiveness and productivity. It is this balance that is the dilemma inherent in innovation management and one that is at the core of the complexity of this process even at more micro-levels of analysis.

Appendix I Key research findings at different levels of analysis

Author(s)	1.2.1 Research Object	1.2.2 Dimensions	Level of Analysis 1.2.3			Key Findings or Proposed Outcomes of Innovation	
			Individual	Work Team	Organisation	Positive Outcomes	Negative Outcomes
Livingstone, Nelson, & Barr (1997)	Creativity and person-environment fit; examining supply value and demand ability versions of fit	Creativity, person-environment fit, strain, job satisfaction, performance	×		×	A fit between demands for creativity and abilities for creativity was related to lower strain and higher job satisfaction.	Discontentment and low performance as consequence of a lack of fit between creativity demands, individual skills and organisational conditions.
James, Clark, & Cropanzano (1999)	Antecedents and outcomes of positive and negative creativity in organisations	Positive and negative creativity	×			Positive creativity may result in e.g., job improvement, reduced health costs, adaptability to change in environment, product creation, and marketing ideas.	Negative creativity may result in e.g., theft, sabotage, harmful behaviour to other employees, and undermining of organisational goals, and policies.
Shalley, Gilson, & Blum (2000)	Work environment characteristics facilitating or inhibiting creativity at work; effects on employee satisfaction and intentions to leave	Creativity, work environment, satisfaction, intention to leave	×			Individuals were more satisfied and reported lower intentions to leave when their work environments complemented the creativity requirements of their jobs. Job-required creativity, job-environment (J-E) fit or complementarity had a significant effect on employee's affective outcomes.	
Zhou & George (2001a)	Creativity as consequence of job dissatisfaction depending on conditions conducive to the expression of voice	Creativity, job dissatisfaction, continuance, commitment	×			If new ideas proposed by the employees are accepted and subsequently implemented by the organisation, it is likely that employees' job dissatisfaction decreases.	If new ideas proposed by the employees are not accepted and implemented by the organisation, employees' may become even more dissatisfied with their job.
Janssen (2003)	Conflict and less satisfactory relations with co-workers as consequence of innovative behaviour and job involvment	Conflict, job involvement, satisfaction with co-worker relations	×	×		A worker's innovative initiatives may contribute to organisational effectiveness.	A worker's innovative behaviour interacts with job involvement in providing conflict and less satisfactory relations with co-workers.
Janssen, Van de Vliert, & West (2004)	The bright and the dark sides of individual and group innovation	Conflict, performance, success/failure, job attitudes, well-being	×			Constructive conflict, innovation success, performance improvement, positive job attitudes, well-being.	Destructive conflict, innovation failure, lowered performance, negative job attitudes, stress.
		Success/failure, group cohesion and potency, clarity of objectives and leadership		×		Successful innovation, group cohesion, group potency, clear objectives, clear leadership, group effectiveness, receptivity to future innovation.	Failure, lowered group cohesion, and potency, unclear objectives, lack of clarity over leadership, group ineffectiveness, resistance to future innovation.

Appendix I Key research findings at different levels of analysis (continued)

Author(s)	1.2.1 Research Object	1.2.2 Dimensions	Level of Analysis 1.2.3 Individual	Work Team	Organisation	Key Findings or Proposed Outcomes of Innovation: Positive Outcomes	Negative Outcomes
Miron, Erez, & Naveh (2004)	Personal and organisational factors that enhance or hinder innovation; factors that contribute to quality and efficiency	Cognitive styles, initiative, innovative culture	×		×	Creativity positively affected innovation at the implementation stage given high initiative and an organisatioanl culture that supports innovation.	Creativity was found to have a significant negative effect on performance quality when the task required accuracy, and adherence to rules.
Townsend, DeMarie, & Hendrickson (2004)	Implementation of virtual team work	Stress, trust, and cohesion issues, and structural resistance	×	×	×	Synergy of teamwork and use of information and communication technology.	Employees' being assigned to more teams, working in a more complex environment may experience more stress and are more likely to suffer from burnout. Free flow of team members' communication may be inhibited by concerns about privacy and system security. Organisational restructuring may cause resistance by employees.
De Dreu (2006)	Relationship between task conflict and innovation in teams; effect of innovation on team performance	Task conflict, team innovation		×		Task conflict in teams can be positively related to innovation; potential positive impact on performance parameters like e.g., learning, innovativeness.	Too much or too little conflict was detrimental to team innovation, i.e., teams were more innovative when the level of task conflict was moderate. While teams may benefit from innovations in the long run, immediate goal attainment may be reduced due to lowered efficiency of work processes.

Source: N. Anderson & R. M. Gasteiger, 2007

NOTES

1 Openness to experience (referring to individual differences in the tendency to pursue new experiences in life), conscientiousness (differences in diligence and reliability in performing assigned tasks), extraversion (differences in assertiveness and dominance in social settings), agreeableness (differences in sensitivity in interpersonal relationships) and neuroticism (referring to individual differences in the tendency to experience distress and anxiety). These five dimensions are held to be a complete description of human personality.

2 Perceived self-efficacy is defined as people's beliefs about their capabilities to produce designated levels of performance that exercise influence over events that affect their lives (Bandura, 1994).

3 Personal initiative refers to an active approach characterised by a persons self-starting and proactive nature to overcome difficulties that arise in the pursuit of a goal.

REFERENCES

Amabile, T.M. (1983) *The Social Psychology of Creativity*, New York: Springer.

Amabile, T.M. (1988) 'A Model of Creativity and Innovation in Organizations', in: B.M. Staw and L.L. Cummings (eds.) *Research in Organizational Behavior,* 10: 123-167, Greenwich, CT: JAI Press.

Amabile, T.M. (1996) *Creativity in Context*, Boulder, CO: Westview Press.

Amabile, T.J., J.S. Mueller, W.B. Simpson, C.N. Hadley, S.J. Kramer, and L. Fleming (2002) 'Time Pressure and Creativity in Organizations: A Longitudinal Field Study', available at http://www.hbs.edu/research/facpubs/workingpapers/papers2/0102/02-073.pdf (retrieved on March 30, 2007).

Amabile, T.M., S.G. Barsade, J.S. Mueller and B.M. Staw (2005) 'Affect and Creativity at Work', *Administrative Science Quarterly,* 50: 367-403.

Anderson, N.R. (1990) 'Innovation in Work Groups: Current Research Concerns and Future Directions', paper presented to the British Psychological Society Annual Conference, Bowness-on-Windermere, January.

Anderson, N. and N. King (1993) 'Innovation in Organizations', in: C. L. Cooper & I. T. Robertson (eds.) *International Review of Industrial and Organizational Psychology* 8: 1-34, Chicester: John Wiley & Sons.

Anderson, N. and R.M. Gasteiger (2007) 'Helping Creativity and Innovation Thrive in Organizations: Functional and Dysfunctional Perspectives', in: J. Langan-Fox, C.L. Cooper, and R.J. Klimoski (eds.) *Research Companion to the Dysfunctional Workplace: Management Challenges and Symptoms*, Chichester: Wiley.

Anderson, N., C.K.W. de Dreu and B.A. Nijstad (2004) 'The Routinization of Innovation Research: A Constructively Critical Review of the State-of-the-Science', *Journal of Organizational Behavior,* 25: 147-173.

Axtell, C.M., D.J. Holman, K.L. Unsworth, T.D. Wall and P.E. Waterson (2000) 'Shopfloor Innovation: Faciliating the Suggestion and Implementation of Ideas', *Journal of Occupational and Organizational Psychology,* 73: 265-285.

Baer, M. and M. Frese (2003) 'Innovation is Not Enough: Climates for Initiative and Psychological Safety, Process Innovations, and Firm Performance', *Journal of Organizational Behavior,* 24: 45-68.

Bandura, A. (1994) 'Self-efficacy', in: V.S. Ramachaudran (ed.) *Encyclopedia of Human Behavior,* , New York: Academic Press, 71-81

Barron. F. and D.W. Harrington (1981) 'Creativity, Intelligence and Personality', *Annual Review of Psychology,* 32: 439-176.

Bunce, D. and M.A. West (1996) 'Stress Management and Innovation Interventions at Work', *Human Relations,* 49: 209-232.

Csikszentmihalyi, M. and K. Rathunde (1993) 'The Measurement of Flow in Everyday Life: Toward a Theory of Emergent Motivation', in: J.E. Jacobs (ed.) *Developmental Perspectives on Motivation*, Lincoln: University of Nebraska Press, 57-97.

Daft, R.L. (1978) 'A Dual-core Model of Organizational Innovation', *Academy of Management Journal,* 21: 193-210.

Damanpour, F. (1990) 'Innovation Effectiveness, Adoption and Organizational Performance', in: M.A. West and J.L. Farr (eds.) *Innovation and Creativity at Work:*

Psychological and Organizational Strategies, Chichester: Wiley, 125-141.
Damanpour, F. (1987) 'The Adoption of Technological, Administrative, and Ancillary Innovations: Impacts of Organizational Factors', *Journal of Management* 13(4): 675-688.
Damanpour, F. and W.M. Evans (1984) 'Organizational Innovation and Performance: The Problem of 'Organizational Lag'', *Administrative Science Quarterly*, 29: 392-409.
Dreu, C.K.W. de (2006) 'When Too Little or Too Much Hurts: Evidence for a Curvilinear Relationship between Task Conflict and Innovation in Teams', *Journal of Management*, 32: 83-107.
Dreu , C.K.W. de and L.R. Weingart (2003) 'Task versus Relationship Conflict, Team Performance, and Team Member Satisfaction: A Meta-analysis', *Journal of Applied Psychology*, 88: 741-749.
Dreu, C.K.W. de and M.A. West (2001) 'Minority Dissent and Team Innovation: The Importance of Participation in Decision Making', *Journal of Applied Psychology*, 86: 1191-1201.
Drucker, P.F. (1985) 'The Discipline of Innovation', *Harvard Business Review*, 85: 67-72.
Frese, M. and D. Fay (2001) 'Personal Initiative: An Active Performance Concept for Work in the 21st Century', in: B.M. Staw and R.M. Sutton (eds.) *Research in Organisational Behavior*, Amsterdam: Elsevier Science, 133-187
Frese, M., E. Teng and C.J.D. Wijnen (1999) 'Helping to Improve Suggestion Systems: Predictors of Making Suggestions in Companies', *Journal of Organizational Behavior*, 20: 1139-1155.
Frese, M. and D. Zapf (1994) 'Action as the Core of Work Psychology: A German Approach', in: H.C. Triandis, M.D. Dunette, and L.M. Hough (eds.) *Handbook of Industrial and Organizational Psychology*, Palo Alto, CA: Consulting Psychologists Press, 271-340.
Gebert, D., S. Boerner and R. Lanwehr (2003) 'The Risks of Autonomy: Empirical Evidence for the Necessity of a Balance Management in Promoting Organizational Innovativeness', *Creativity and Innovation Management*, 12: 41-49.
George, J.M and J. Zhou (2001a) 'When Job Dissatisfaction Leads to Creativity: Encouraging the Expression of Voice', *Academy of Management Journal*, 44: 682-696.
George, J.M. and J. Zhou (2001b) 'When Openness to Experience and Conscientiousness are Related to Creative Behavior: An Interactional Approach', *Journal of Applied Psychology*, 86: 513-524.
George, J.M. and J. Zhou (2002) 'Understanding when Bad Moods Foster Creativity and Good Ones Don't: The Role of Context and Clarity of Feelings', *Journal of Applied Psychology*, 87: 687-697.
Gough, H.G. (1979) 'A Creative Personality Scale for the Adjective Check List', *Journal of Personality and Social Psychology*, 37: 1398-1405.
Graen, G.B. and M. Uhl-Bien (1995) 'Relationship-based Approach to Leadership: Development of Leader-member Exchange (LMX) Theory of Leadership Over 25 Years: Applying a Multi-level Multi-domain Perspective', *Leadership Quarterly*, 6(2): 219-247.
Harrison, M.M., N.L. Neff, A.R. Schwall and X. Zhao (2006) 'A Meta-analytic Investigation of Individual Creativity and Innovation', paper presented at the 21st Annual Conference for the Society for Industrial and Organizational Psychology, Dallas, TX.

Harmon, R.L. (1992) *Reinventing the Factory*, New York: MacMillan.

Hart, P.M. and C.L. Cooper (2001) 'Occupational Stress: Toward a More Integrated Framework', in: N. Anderson, D.S. Ones, H.K. Sinangil, and C. Viswesvaran (eds.) *Handbook of Industrial, Work, and Organizational Psychology*, Thousand Oaks, CA: Sage, 93-114

Herscovitch, L. and J.P. Meyer (2002) 'Commitment to Organizational Change: Extension of a Three-component Model', *Journal of Applied Psychology*, 87: 474-487.

Hippel, E. von (1988) *The Sources of Innovation*, Oxford: Oxford University Press.

Hippel, E. von (2005) *Democratizing Innovation*, Cambridge, MA: MIT-Press.

Howard, A. (ed.) (1995). *The Changing Nature of Work*, San Francisco: Jossey-Bass.

Howell, J.M. and C.A. Higgins (1990) 'Champions of Change: Identifying, Understanding, and Supporting Champions of Technological Innovations', *Organizational Dynamics*, 19: 40-54.

Jackson, S.E. (1996) 'The Consequences of Diversity in Multidisciplinary Work Teams', in: M.A. West (ed.) *Handbook of Group Psychology*, Chichester: Wiley, 53-75

James, K., K. Clark and R. Cropanzano (1999) 'Positive and Negative Creativity in Groups, Institutions, and Organizations: A Model and Theoretical Extension', *Creativity Research Journal*, 12: 211-226.

Janis, I.L. (1972) *Victims of Groupthink*, New York: Houghton Mifflin.

Janssen, O., E. van de Vliert and M. West (2004) 'The Bright and Dark Sides of Individual and Group Innovation: A Special Issue Introduction', *Journal of Organizational Behavior*, 25: 129-145.

Janssen, O. (2004) 'How Fairness Perceptions Make Innovative Behavior More or Less Stressful', *Journal of Organizational Behavior*, 25: 201-215.

Janssen, O. (2003) 'Innovative Behaviour and Job Involvement at the Price of Conflict and Less Satisfactory Relations with Co-workers', *Journal of Occupational and Organizational Psychology*, 76: 347-364.

Jaussi, K.S. and S.D. Dionne (2003) 'Leading for Creativity: The Role of Unconventional Leader Behavior', *Leadership Quarterly*, 14: 475-498.

Jehn, K.A. (1995) 'A Multimethod Examination of the Benefit and Detriments of Intragroup Conflict', *Administrative Science Quarterly*, 40: 256-282.

Kanter, R.M. (1988) 'When a Thousand Flowers Bloom: Structural, Collective, and Social Conditions for Innovation in Organization', *Research in Organizational Behavior*, 10: 169-211.

Kanter, R.M. (1983) *The Change Master: Corporate Entrepreneurs at Work*, New York: Simon & Schuster.

Kaschube, J. (2006) *Eigenverantwortung – Eine neue berufliche Leistung. Chance oder Bedrohung für Organisationen? [Self-responsibility – A new occupational performance. Opportunity or risk for organizations?]*, Göttingen: Vandenhoeck & Ruprecht.

King, N. (1992) 'Modeling the Innovation Process: An Empirical Comparison of Approaches', *Journal of Occupational and Organizational Psychology*, 65: 89-100.

King, N. and N. Anderson (2002) *Managing Innovation and Change: A Critical Guide for Organizations*, London: Thompson.

Kimberly, J.R. (1981) 'Managerial Innovation', in: P.C. Nystrom and W.H. Starbuck (eds.) *Handbook of Organizational Design*, Oxford: Oxford University Press, 84-104

Kirton, M.J. (1976) 'Adaptors and Innovators: A Description and Measure', *Journal of Applied Psychology*, 61(5): 622-629.
Kirton, M.J. (1978) 'Hava Adaptors and Innovators Equal Levels of Creativity', *Psychological Reports*, 42: 695-698.
Krause, D.E. (2004) 'Influence-based Leadership as a Determinant of the Inclination to Innovate and of Innovation Related Behaviors: An Empirical Investigation', *Leadership Quarterly*, 15: 79-102.
Livingstone, L.P., D.L. Nelson and S.H. Barr (1997) 'Person-environment Fit and Creativity: An Examination of Supply-value and Demand-ability Versions of Fit', *Journal of Management*, 23: 119-146.
Maier, G.W., D. Frey, S. Schulz-Hardt and F.C. Brodbeck (2001) 'Innovation', in: G. Wenninger (eds.) *Lexikon der Psychologie [Encyclopedia of psychology]*, Heidelberg: Spektrum, 264-267
McCrae, R.R. (1987) 'Creativity, Divergent Thinking, and Openness to Experience', *Journal of Personality and Social Psychology*, 52: 1258-1265.
Miron, E., M. Erez and E. Naveh (2004) 'Do Personal Characteristics and Cultural Values that Promote Innovation, Quality, and Efficiency Compete or Complement Each Other?', *Journal of Organizational Behavior*, 25: 175-199.
Patterson, F. (2002) 'Great Minds Don't Think Alike? Person-level Predictors of Innovations at Work', *International Review of Industrial and Organizational Psychology* 17: 115-144, Chichester: Wiley.
Patterson, F. (1999) *Innovation Potential Predictor*, Oxford: Oxford Psyehologists Press.
Peters, T.J. and R.H. Waterman (1982) *In Search of Excellence: Lessons From America's Best Run Companies*, New York: Harper & Row.
Pelled, L.H. (1996) 'Demographic Diversity, Conflict, and Work Group Outcomes: An Intervening Process Theory', *Organization Science*, 7: 615-631.
Perkins, D.N. (1986) 'Thinking Frames', *Educational Leadership*, 43: 41-10.
Rank, J., V.L. Pace and M. Frese (2004) 'Three Avenues for Future Research on Creativity, Innovation, and Initiative', *Applied Psychology: An International Review*, 53: 518-528.
Russ, S.W. (2000) 'Primary-process Thinking and Creativity: Affect and Cognition', *Creativity Research Journal*. 13: 27-35.
Sackett, P.R. and C.J. DeVore (2001) 'Counterproductive Behaviors at Work', in: N. Anderson, D.S. Ones, H.K. Sinangil and C. Viswesvaran (eds.) *Handbook of Industrial, Work, and Organizational Psychology*, 2: 145-199, London/New York: Sage.
Sauer, J. and N. Anderson (1992) 'Have We Misread the Psychology of Innovation? A Case Study from Two NHS Hospitals', *Leadership and Organizational Development Journal*, 13: 17-21.
Shalley, C.E., L.L. Gilson and T.C. Blum (2000) 'Matching Creativity Requirements and the Work Environment: Effects on Satisfaction and Intentions to Leave', *Academy of Management Journal*, 43: 215-223.
Simons, T. and R. Peterson (2000) 'Task Conflict and Relationship Conflict in Top Management Teams: The Pivotal Role of Intragroup Trust', *Journal of Applied Psychology*, 68: 102-111.

Simonton, D.K. (1991) 'Latent-variable Models of Posthumous Reputation: A Quest for Galton's G', *Journal of Personality and Social Psychology*, 60: 607-619.

Spector, P.E. and S.M. Jex (1998) 'Development of Four Self-report Measures of Job Stressors and Strain: Interpersonal Conflict at Work scale, Organizational Constraints Scale, Quantitative Workload Inventory, and Physical Symptoms Inventory', *Journal of Occupational Health Psychology*, 3: 356-367.

Spreitzer, G.M. (1995) 'Psychological Empowerment in the Workplace: Dimensions, Measurement, and Validation', *Academy of Management Journal* 38, 5: 1442-1465.

Staw, B.M. and R.D. Boettger (1990) 'Task Revision: A Neglected Form of Work Performance', *Academy of Management Journal*, 33: 534-559.

Sternberg, R.J. and T.I. Lubart (1991) 'An Investment Theory of Creativity and its Development', *Human Development*, 34(1): 1-32.

Sternberg, R.J. and T.I. Lubart (1996) 'Investing in Creativity', *American Psychologist*, 51: 677-688.

Stevens, M.J. and M.A. Campion (1994) 'The Knowledge, Skill, and Ability Requirements for Teamwork: Implications for Human Resource Management', *Journal of Management*, 20: 503-530.

Taggar, S. (2002) 'Individual Creativity and Group Ability to Utilize Creative Resources: A Multilevel Model', *Academy of Management Journal*, 45: 315-330.

Tierney, P. and S.M. Farmer (2002) 'Creative Self-efficacy: Its Potential Antecedents and Relationship to Creative Performance', *Academy of Management Journal*, 45: 1137-1148.

Townsend, A., S. deMarie and A.R. Hendrickson (2004) 'Virtual Teams: Technology and the Workplace of the Future', in: R. Katz (eds.) *The Human Side of Managing Technological Innovation: A Collection of Readings*, New York: Oxford University Press, 269-282

Ven, A. van de, H.L. Angle and M. Poole (eds.) (1989) *Research on the Management of Innovation: The Minnesota Studies*, New York: Harper & Row.

Ven, A.H. van de, D.E. Polley, R. Garud and S. Venkataraman (1999) *The Innovation Journey*, New York: Oxford University Press.

West, M.A. (2001) 'The Human Team: Basic Motivations and Innovations', in: N. Anderson, D.S. Ones, H.K. Sinangil and C. Viswesvaran (eds.) *Handbook of Industrial, Work, and Organizational Psychology*, 2: 270-288, London/New York: Sage.

West, M.A. (2002) 'Sparkling Fountains or Stagnant Ponds: An Integrative Model of Creativity and Innovation Implementation in Work Groups', *Applied Psychology: An International Review*, 51(3): 355-424.

West, M.A. and N.R. Anderson (1996) 'Innovation in Top Management Teams', *Journal of Applied Psychology*, 81: 680-693.

West, M.A. and N.R. Anderson (1992) 'Innovation, Cultural Values and the Management of Change in British Hospital Management', *Work and Stress*, 16: 293-310.

West, M.A. and J.L. Farr (1990) 'Innovation at Work', in: M.A. West and J.L. Farr (eds.) *Innovation and Creativity at Work*, Chichester: Wiley, 3-13

West, M.A. and M. Wallace (1991) 'Innovation in Health Care Teams', *European Journal of Social Psychology*, 21: 301-315.

Zaltman, G., R. Duncan and J. Holbek (1973) *Innovations and Organizations*, New York: Wiley.
Zhou, J. and J.M. George (2001) 'When Job Dissatisfaction Leads to Creativity: Encouraging the Expression of Voice', *Academy of Management Journal*, 44: 682-696.
Zhou, J. and J.M. George (2003) 'Awakening Employee Creativity: The Role of Leader Emotional Intelligence', *The Leadership Quarterly*, 14: 545-568.

10 INTER-ORGANISATIONAL NETWORKS AND INNOVATION

Marius T.H. Meeus, Leon A.G. Oerlemans, and Patrick Kenis

10.1 INTRODUCTION

In different scientific fields (e.g., organisation and business studies, economics), and in policy practice, there is growing attention for the importance of inter-organisational relationships (IORs) and networks (IONs). Central issues are the functioning and performance of the business community in general, and the relationship between IORs and IONs for the innovative behaviour of organisations in particular. It is believed that having IORs and being part of IONs is beneficial to innovation. However, reviews that have been done so far, predominantly focus on the effects of so-called dyads on innovation (i.e., the IORs between two organisations) (Pittaway et al. 2004), whereas only Provan et al. (2007) reviewed generic effects of complete networks on all kinds of organisational outcomes. Our main aim with this chapter is to identify features of so-called egocentric and whole (or complete) networks and their impact on the innovation of individual firms. Moreover, the findings on this relationship are used to evaluate the extent to which current Dutch innovation policy portfolio is geared towards IONs.

This chapter is organised as follows. Section 2 briefly introduces the relevance of the topic to the Dutch economy by showing to what extent innovation and collaboration are empirically associated. Section 3 introduces the research model and the main arguments supporting it. Subsequently section 4 reports on a review of the literature pertaining to the features of inter-organisational networks and their potential impact on innovation The fifth section shows to what extent current innovation policies impact on networks to achieve their goals. In the final section we draw some conclusions.

10.2 THE IMPORTANCE OF COOPERATION FOR INNOVATION: SOME EMPIRICAL UNDERPINNINGS

Each ION starts out with cooperation between two legally independent organisational entities, which eventually leads to additional relations. To set the stage for this chapter it is relevant to know whether innovation and collaboration are related. Table 1 displays a review of eight surveys covering a period of more than ten years, on collaboration in the context of R&D, and the type of actors with whom a focal firm has R&D relations.

It is the diversity of samples, their varying composition, and spatial range, which helps us to identify patterns in R&D collaboration. Only the CBS (column 7) and the Poot and Brouwer (column 8) samples covered the Dutch population of

Table 10.1 **An overview of R&D partnering in surveys by Meeus and Oerlemans (1992/1993, 1996, 1997, 1998), Meeus, Smits en Stuurman (1996), and Meeus, Faber, Moors en Hekkert (2002), Dutch Bureau of Statistics (1998), Poot and Brouwer (2001): Source (Meeus, 2003)**

Types of R&D cooperation	1 % yes	2 % yes	3 % yes	4 % yes	5 % yes	6 % yes	7 % yes	8 % yes
R&D with suppliers	42.6	46.9	17.6	54.3	50.0	42.4	46.9	66.0
R&D with buyers	47.6	46.9	55.1	25.7	36.8	36.5	43.7	25.7
R&D with rivals	4.5	7.6	4.0	0	7.9	4.8	34.7	8.4
R&D engineering or consultancy firms	18.3	22.3	10.8	17.1	26.3	12.9	20.2	--
R&D with public R&D organisations:								
university/academic Hospital	11.9	17.4	11.4	17.1	60.5	5.9	21.4*	10.7
technical university								30.9
large technologial inst.								39.5
polytechnics								14.4

1 MINT 1992/1993, (n = 689), R&D in 420 companies (South-East Brabant, mainly industry)
2 CINT 1996, (n = 363), R&D in 224 companies (The Netherlands, mainly industry)
3 Standardisation 1996 (n = 886), R&D in 535 companies (The Netherlands, only industry)
4 IOP Image Processing 1997 (n = 65), R&D in 42 companies (The Netherlands, full sample of one specific technological field)
5 IOP Man Machine Interaction 1998 (n = 50), R&D in 38 companies (idem as in 5)
6 CISU 2002 (n = 144), R&D in 85 companies (Province of Utrecht, mainly industry)
7 CISII CBS 1998 (n = 17.193), R&D in 13.178 companies (representative sample of the Dutch population of companies)
8 Knowledge acquisition in partnerships. Poot & Brouwer 2001, (n = 2117) (idem as in 7)
* average percentage of cooperation with polytechnic and university

organisations. All other surveys had a more or less biased sample, with an emphasis on industry and low numbers of service companies.

Notwithstanding this sampling bias and temporal variation of data acquisition, it is remarkable how consistent the findings are over time. The largest survey – the Community Innovation Survey II of 2002 – has some findings that deviate in terms of the size of percentages of R&D with rivals and R&D with public R&D organisations such as universities, although the ranking of the percentages of types of R&D partners is the same as in the other surveys.

The main inferences from table 1 are: (a) R&D collaboration is a pervasive phenomenon among innovative organisations in the Netherlands, (b) cooperation is mainly positioned in the value chain; (c) R&D with rivals is rather infrequent; and (d) R&D collaboration with actors in the knowledge infrastructure is even more unusual. These observations indicate that Dutch industrial companies already in the early phases of the innovation process integrate inputs and requirements of buyers and suppliers. However, these findings also indicate that a larger distance of partners from the day-to-day business seems to be associated with a lower significance for R&D collaboration.

Findings displayed in table 1 confirm that, given the prevalence of R&D collaboration with a variety of partners, there must be an impact of collaboration on innovation. The sheer fact that R&D collaboration is associated with innovation is only an indication, not an empirical proof, of the causal effects of being engaged in an ION on innovation.

In this chapter, two related issues concerning the causal effects of participation in an ION on innovation are addressed. The first is: which aspects of the innovation process are especially affected by or related to IONs? The second issue focuses on Dutch innovation policy measures. The questions asked are twofold: To what extent does the current Dutch innovation policy portfolio consist of policy measures geared toward facilitating specific mechanisms that affect the network effect on innovation?

10.3 A FRAMEWORK LINKING INNOVATION POLICY TO EFFECTS OF NETWORKS ON INNOVATION

The hybrid nature of this paper is due to its unusual combination of a literature review study, a policy review, and a policy evaluation study. This requires an explanation as to how these elements are linked to each other. The linkages are specified below in 5 equations

Consider a generic model of a firm's performance at time t:

(1) $FP_t = f(FP_{t-1}, A_{t-1}, C_{t-1}, ED_{t-1})$

FP_t = financial performance at time t
FP_{t-1} = firm performance in previous period
A_{t-1} = activities performed at t_{-1}
C_{t-1} = set of explorative and exploitative capabilities at t_{-1}
ED_{t-1} = environmental dynamics at t_{-1}

The financial performance of a firm is considered a function of its financial performance at $_{t-1,}$ its activities (A_{t-1}), and its capabilities and environmental dynamics. The main part of A consists of producing its existing product portfolio and the associated processes (AE_{t-1}). A part of A at t_{-1} consists of innovative activities (IA), which means a radical or incremental renewal of a firm's products, processes, or organisation.

(2) $A_{t-1} = f(AE_{t-1}, IA_{t-1})$

In which:
AE_{t-1} = production of existing product portfolio and related production processes at t_{-1}
IA_{t-1} = innovative activities at t_{-1}

The innovative activities are often induced by environmental dynamics (ED_{t-1}) such as for example product differentiation processes of competitors that are anticipated or reacted upon. Also the announcement of breakthrough technologies can induce innovation, as well as major regulatory changes (e.g., liberalisation of former monopolistic markets).

To be able to anticipate or react to environmental dynamics firms must develop their capabilities, which is mostly done by innovative activities organised as formal and informal (after office hours) research and development. Such innovative activities are not without cost, quite the contrary, they require large and often risky investments (I_{t-1}) with an unknown trade-off of costs and benefits.

(3) $IA_{t-1} = f(It_{-1}, RN_{t-1})$

In which:
I_{t-1} = investments in innovation at t_{-1}
RN_{t-1} = knowledge flows through inter-organisational networks at t_{-1}

Innovation related investments (I_{t-1}) are the addition of internal investments at t_{-1}, and another parameter of external money available for innovation activities of companies ($EF\ t_{-1}$). The financial resources that have to be invested in innovative activities are the point where innovation policies become relevant because most measures in the policy portfolio offer additional funding options to make the investments less costly and risky, and therefore more attractive.

(4) $I_{t-1} = f(II_{t-1}, EF_{t-1})$

In which:
II_{t-1} = internal investment in innovation at t_{-1}
EF_{t-1} = external governmental funding of innovation at t_{-1}

In order to clarify RN_{t-1} (eq 3) we go back to table 1. Table 1 shows that knowledge flows (only R&D in table 1) are associated with inter-organisational relationships and networks (RN), mainly in the value chain, and eventually in the broader innovation system to actors such as universities, large technological institutes (LTIs), or consultancy firms, and brokers. The relationships with these partners have several functions: to make these innovative activities aligned with user needs and capabilities, as well as with supplier capabilities, to develop joint projects that enable risk sharing and resource pooling (financial and knowledge). The deployment of these functions relies on three levels of networks: the so-called node and relational features (NV_{t-1}). For instance one can have strong and weak partners in the network, which have a distinct impact on the innovation activities. The characteristics of the whole network at $_{t-1}$, such as cohesiveness or centralisation of a network, determine knowledge flows in networks and also impinge upon the impact of networks on innovation. Finally NV_{t-1}, the structural configuration of a whole network differentiates the network impact on innovation. Now we know that many innovation policy measures favour cooperation, and even in several instruments cooperation is a requirement. Since we do not know for now, we put a specific type policy measure ($EFCC_{t-1}$) in this equation that makes innovation funding conditional on cooperation. This kind of measure will have its own effect on the prevalence of innovative collaboration, for example as the Bay-Dohle Act did in the US in the 1980s. Whether this is a valid operation will become clear later in our evaluation of current innovation policy measures.

(5) $RN_{t-1} = f(NV1_{t-1},, NV2_{t-1}, NV3_{t-1}, EFCC_{t-1})$

In which:
$NV1_{t-1}$ = network node and relational features at t_{-1}
$NV2_{t-1}$ = characteristics of the whole network at t_{-1}
$NV3_{t-1}$ = the structural configuration of ties in the whole network at t_{-1}
$EFCC_{t-1}$ = the external governmental funding of innovation conditional on cooperation available at t_{-1}.

The first contribution of this chapter consists of trying to unravel whether there is an empirically tractable impact of *networks of* relationships on innovation, which means that we are analysing the impact of the whole set of direct relationships of all network actors, and also the whole set of indirect relationships. In other words, in our literature review we asked: is innovation (IA) indeed a function of whole networks (RN) (eq. 3)?

As was explained above, most research in the field focuses on the dyad level and investigates the influence of IORs, which can be defined as "the relatively enduring transactions, flows, and linkages that occur among and between an organisation and an organisation in its environment" (Oliver 1990: 241), on innovation by organisations. In this study, however, we focus predominantly on so-called 'whole networks'. A whole network consists of multiple organisations linked through multilateral ties (Provan et al. 2007: 482). Below, we will further elaborate on this distinction.

The second contribution of this paper consists of answering the question regarding the extent to which the funding and in-kind facilities provided for innovative activities by Dutch innovation policy measures draw upon the effect of whole networks on innovation (eq. 5). In fact it means that we want to know whether Dutch innovation policies integrate the network effect in their policy design and if so, how this affects the network effect?

For instance, policy measures can provide sheer financial inputs without any further conditions as to the ways in which innovation projects are organized. In contrast, policy design can condition the approval of certain grants on ways of organizing e.g. in networks with specific partners. If the external funding is simply adding to internal investment without any organizational conditions, this would simply imply additional monetary inputs. If the external funding is conditioned upon organizational forms such as involving parts of networks or even whole networks (consortia) as in the current Innovation Programmes, this would imply that an additional set of mechanisms impact upon the relation between policy inputs and their policy outcomes.

Of course this has a number of implications, for example that firms with strong exploratory capabilities and existing networks are favoured above those with less exploration capabilities and weakly developed networks. Eventually this might induce exploratory networks with science myopia and a technology bias, who prefer the production of state-of-the-art technology over commercial product development.

The general reasoning is presented in figure 10.1, which will be explained systematically below.

The dependent variable

Innovation at firm level is the dependent variable in figure 10.1. Innovation consists of all the behaviours that are directly related to performing innovation and can be defined at the macro-level of the total population of firms in a nation state; at the meso-level of sectors, networks of organisations, or product categories; and finally at the micro level of individual firms or business units within companies. Innovation can be distinguished in the adoption and the generation of new technology, products, services, and/or processes. Innovation – adoption and generation – comprises R&D investments and a portfolio of R&D and implementation projects. Generation of innovation and technology means that a company itself performs research and development. This can take the shape of either informal or formalised R&D organised in R&D departments, and specialised R&D personnel, both technical and organisational. Adoption implies that a firm buys new technologies of manufacturers, which are implemented in the buyer's company.

The independent variables

An interaction between two actors – the focal unit or ego (**E**), and alter/partner (either between ego and a partner (**P**) or between two Ps) is defined here as a voluntary and enduring arrangement between legally independent organisations to have any type of relationship. Figure 10.1 lists a number of variables in the centre area showing distinct levels of analysis of these interactions. Node features such as size, sector of partners, or types of partners (NV1) can have strong impacts on innovation of the focal firm E, but they are different from effects of the characteristics from the whole network (NV2). It is important to note that not relations or networks per se contribute to the innovation but certain attributes of the structural patterns of interactions in which the organisation is embedded have such an effect (NV3).

For the sake of completeness figure 10.1 also contains firm performance (FP) and individual firm features (IF), and collective efforts of network members (CE). Our literature review is limited to the network variables to the extent that there is empirical research available.

The primary effect of engaging in an ION implies that the partner's partner, as well as the indirect partner's partner, is also considered to be part of E's network. Consequently it is in fact almost impossible to foresee what kind of network one enters when a relationship starts. This effect determines the initial complexity of engaging in relationships. It is mainly the attraction of P that determines the linkage formation.

Another factor that complicates matters considerably is – and that is what we are focussing on in this chapter – if collaboration pursues innovation. Because innovation always means either doing new things or doing existing things in a new way, innovation has multiple impacts. Innovation not only erodes market positions of competitors, but it potentially destroys internal firm capabilities as well as the main capabilities of customers, suppliers, and knowledge producers. Consider the situation that customers, suppliers, knowledge brokers, and producers of company X were involved in producing innovation at t_{-1} and company X takes up an initiative to adopt or develop breakthrough technologies. This decision will have a tremendous impact upon the capabilities of company X's direct partners, as well as on indirect links in the network. Despite the critical role of company X's partners in its innovation at t_{-1}, this does not necessarily imply that the innovation at t_0 renders related effects for company X's partners, and the wider network. The main message being this: whereas at time t_{-1} company X depended upon its partners' capabilities to innovate, this is not necessarily the case at time t_0. Former collaborative relationships clearly contributing to a company's competitive edge can eventually become detrimental to this later on (Afuah 2000), and hence collective efforts in networks are not constant over time. This simple consequence also raises the intricate strategic issue of staying with or switching suppliers, customers, or other innovation partners that cannot keep up with the pace of innovations. A lock-in situation is always looming ahead when innovative companies do not dare considering leaving old partners. So it is not trivial question whether the impact of complete networks on innovation will be as positive as often claimed. A network is not just an aggregate of the strengths of its members, but also of its weaknesses at certain points in time. From a dynamic perspective both strengths and weaknesses are re-constructed with each innovative step made by members of the network.

To avoid redundancy, illustrations of how attributes of the structural patterns of interactions potentially impact on innovation are described and explained in the literature review itself (for further details on NV 1-3 see appendix A).

Innovation policy measures

On the left-hand side of figure 10.1 there is a set of innovation policy measures 1 to N that can either have targeted or untargeted impacts, direct or indirect effects on network variables, or the collective effort invested in innovation networks.

Table 10.2 Effects of innovation policies on features of networks

	Direct	Indirect
Targeted	Cooperation required and evaluated at funding application	Cooperation promoted but not required in funding application
Untargeted	NWO* fundamental research asking for team composition and former cooperation.	Measures that distribute R&D grants Measures that promote start-up founding

* NWO is the Dutch science foundation.

Most of the functions of IONs for innovation have also been recognised by policy-makers. From an innovation policy point of view, any form of collaboration with regard to innovation is very often considered to be 'easy money', inducing a low-cost multiplier effect. This applies because resources are pooled, knowledge deficiencies turn into complementarities, and the networks offer the conduits that let the life blood of innovation and knowledge flow, while the partnering provides the prisms with which the value of knowledge is tested and improved in situ. Do policy makers in their current set of instruments utilise the leverage that networks potentially have for innovation (eq. 5) or not? And if so, does the design of policy instruments and the policy portfolio draw upon network effects directly or indirectly, targeted or untargeted?

10.4 THE LITERATURE REVIEW

10.4.1 METHOD OF LITERATURE REVIEW

In an effort to identify recent empirical work on the relationship between (whole) inter-organisational network characteristics and innovation, an extensive review of the literature was undertaken. First, a search for journal articles utilising Web of Science, ProQuest ABI/Inform, and Google Scholar was conducted. Consistent with the broad range of definitions of both networks and innovation in the literature, our search terms included: networks, inter-organisational networks, centrality, small worlds, density, innovation, innovative performance, innovation outcomes, as well as combinations of these search terms. The search was limited to the 27-year period from 1980 to 2007 because most empirical work on networks and innovation has been done in recent years. The results were then analysed with respect to their relevance for organisation studies. Because many network studies exist in a variety of fields, those articles falling outside of studies of networks and organisations were discarded. For example, many articles in computer science address computer networks, and many in the health field deal with neural networks. After the initial analysis, the abstracts of the articles that remained were investigated. Often, the abstracts provided ample information about the methods and unit of analysis. Based on a reading of these abstracts, many articles that focused on individuals and their social networks could be eliminated. Once relevant articles and potentially relevant articles were identified, the complete articles were studied to make sure that each fit the requirements of this study.

Second, each article that fulfilled the requirement of focusing on the relationship between inter-organisational network features and aspects of innovative behaviour of organisations was then indexed. A summary was produced for each article, which listed the following topics:

1. Source: Reference to the empirical study reviewed;
2. Network measure: Description of the network measure(s) applied in the study;
3. Type of tie: Description of the type of tie studied in the paper (e.g., information sharing or joint product development);
4. Boundary specification: Description of the type of actors that are part of the inter-organisational network;
5. Innovation outcome measure: Description of measure applied to indicate the innovative performance of organisations;

Figure 10.1 The basic research model

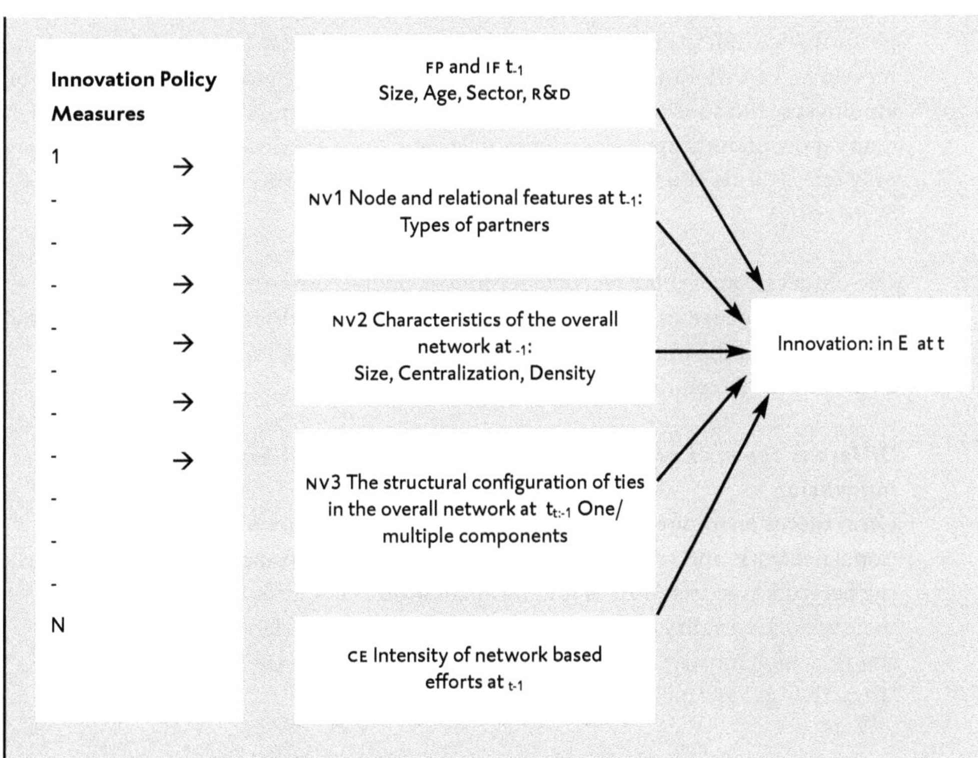

6. Key findings: Description of the key empirical findings of a study as far as the relationship between network characteristics and innovative behaviour of organisations is concerned.

The article summaries provided us with easily identifiable markers for comparisons of the research being conducted in the field.

10.4.2 FINDINGS: NETWORKS OF INTER-ORGANISATIONAL RELATIONS AND INNOVATION

This section reports on a literature review of empirical studies using measures indicating network characteristics on the one hand, and innovation on the organisational level on the other. The results of this review are presented in table 3.

Innovation: multiple measures

The majority of the studies listed – ten out of 25 – used patent-related indicators as innovation outcome measure: patent intensity, counts of patent applications, number of granted patents, successful patenting, and received patent applications. In addition to this, there are a number of studies that used self-reports of innovative performance (Oerlemans, Meeus, Boekema 1998; Ebadi and Utterback 1994; Ruef 2002; Zollo, Reuer and Singh 2002). Some studies focus on the market introduction of innovations using the number of (new) products introduced on the market (Zahra and Nielsen 2002; Fey and Birkinshaw 2005; Capaldo, 2007) or innovative sales (Faems, Van Looy and Debackere 2005) as performance indicators. Other studies apply rather unique innovation outcome measures. Powell, Koput, and Smith-Doerr (1996) apply growth of the number of employees, and subsequent R&D alliances, whereas Baum, Calabrese, and Silverman report on R&D spending and growth, and R&D employees growth. There is only one study that uses creativity as innovation outcome measure (Uzzi and Spiro 2005).

The observation of this diversity in innovation measures reveals that network research: (a) focuses mainly on innovation generation, (b) is more oriented to the early phases of the innovation process (exploration, R&D) than to the commercialisation and exploitation stage.

Different types of centrality in inter-organisational networks and innovation

Centrality is an important property of an individual actor within an organisational network, and is a structural measure of the importance of a given player in the network (see appendix A for an explanation of commonly used centrality measures). Centrality derives from being the object of relations from other actors, which mostly[1] as in earlier chapters] implies that the central actor is 'attractive' as a partner in the eyes of other actors.

Figure 10.2 Degree, Closeness and Betweenness Centrality

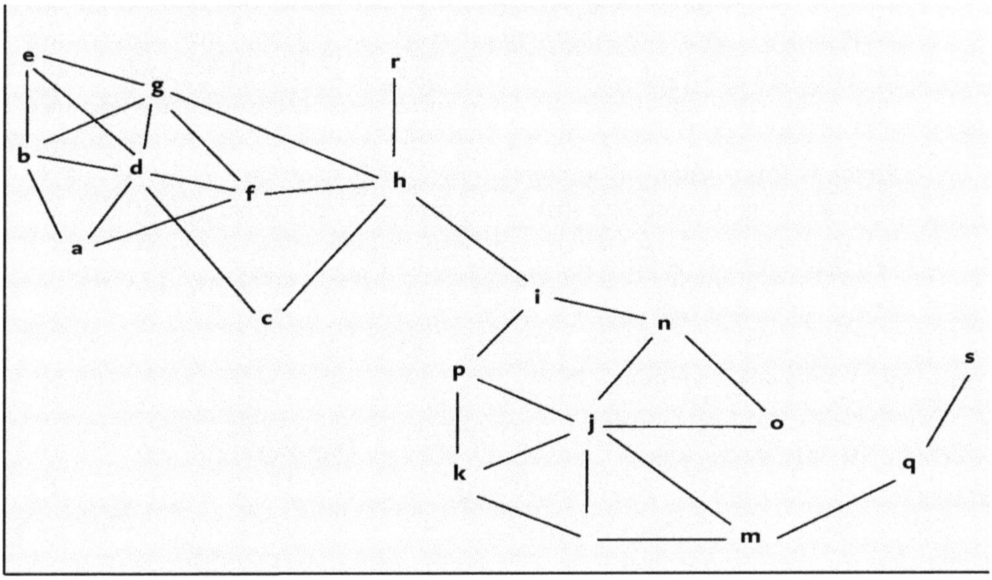

Notes: **j** is the actor with the highest degree centrality
i is the actor with the highest betweenness centrality
h is the actor with the highest closeness centrality

Most scholars hypothesize a positive relationship between levels of centrality and innovation. In literature, three theoretical mechanisms are proposed to substantiate this hypothesis. First, there is a resource-based argument: if centrality is taken as a proxy for the quantity and quality of critical resources available to an actor, it can be argued that actors that are more central are more likely to have 'slack' resources that facilitate experimentation and innovation. Second, there is a knowledge-based argument: Innovation is more likely to occur in a rich and complex knowledge environment, as organisations are exposed to a wide variety of cues that stimulate innovation. A more central actor is at the crossroads of a larger number of knowledge and information sources, and it is therefore better positioned to innovate. Furthermore, a more central actor is more likely to receive earlier innovation-related information, knowledge and influence as compared to less central organisations in the same network. Third, there is a status-based argument: a more central actor is unlikely to imitate already widespread practices, which are in use by 'followers'. Rather, the former will either innovate or imitate other highly central peers. Put differently: an imitator may not be selected as an attractive partner as much as an innovator would be. Such an argument is consistent with the proposition that first movers must continue to innovate in order to maintain a central position.

Extensive empirical support confirms that centrality is positively related to innovation. Most studies (see table 3) use a combination of various centrality measures (degree, betweenness, closeness) and patent-related performance measures. All studies report that higher levels of centrality are positively related to innovative performance. Studies using other outcome measures (Powell et al. 1996) arrive at a similar conclusion. Therefore, one can conclude that (a) the empirical evidence about the relationship between centrality and innovative performance of organisations is rather robust, and (b) that as far as the evidence applies both innovation in the exploration and exploitation stage benefits from E's centrality.

Range in inter-organisational networks and innovation

Network range is the second structural property of the inter-organisational network of the focal actor that is assumed to influence innovative behaviour. Network range implies that the organisations that are part of the network are dissimilar in some way.

Figure 10.3 The Range of Networks

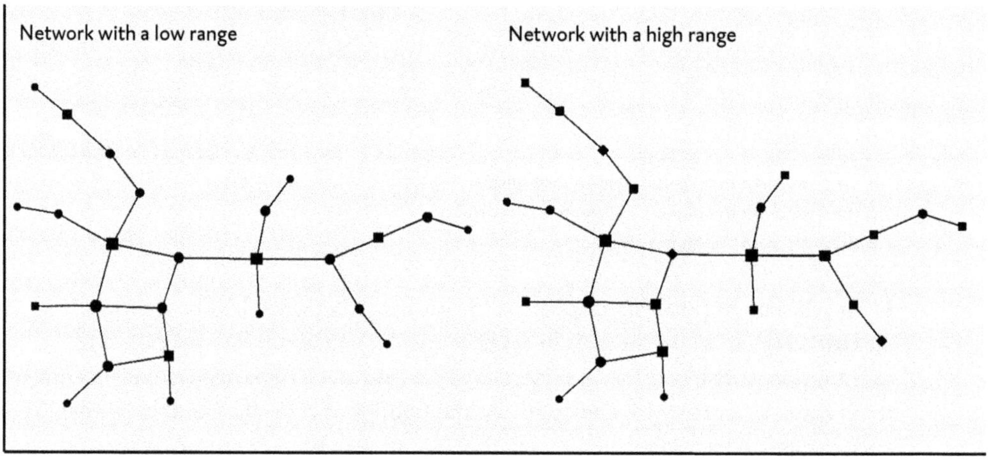

Note: Different shapes and colors of nodes indicate differences between the actors in the network

Two theoretical arguments can be found in the literature that ground the relationship between network range and innovation. First, if a focal actor would rely on inter-organisational relationships with actors of the same type (a so-called homogeneous set of ties; Ruef, 2002), there is a risk that there are no mechanisms for iterative and diverse learning feedback with respect to an innovation. This implies that innovators depending exclusively on these ties will tend to adapt novelties to their own circumstances and needs without necessarily engaging in further innovation. Focal actors related to a diverse set of actors may subject their innovations to further modifications, as they receive diverse feedback from multiple actors. In sum, in this argument network range or diversity functions as a sounding board for the innovating focal actor.

Secondly, having inter-organisational relationships with a diverse set of actors might imply access to complementary assets needed to turn inventions into successful new or improved products on the market. Moreover, interacting with a more diverse set of actors might encourage the transfer of knowledge and information, which, when combined with internally available knowledge resources, could lead to the creation and development of processes and products that would otherwise be difficult to mobilise and to develop. An organisation with higher network range will, therefore, have access to a more diverse and unique set of (knowledge and information) resources that could lead to competitive advantages. Moreover, this could mean that an actor within such a diverse network has a more efficient-effective network as it has access to resources that are not duplicated or redundant.

The most commonly found hypothesis on the relationship between network range and innovation is a positive one: higher levels of range are beneficial to innovation.

In table 3, four studies report on the empirical relationship between range and innovation (Powell et al. 1999; Baum et al. 2000; Ruef 2002; Faems et al. 2005). Although these studies use different criterion variables, they all arrive at the same finding: higher levels of network range (diversity of Ps) are positively related to organisational innovative performance. Faems et. al. (2005) confirm the effect of network range for the exploitation stage, whereas the other three studies (Powell et al. 1999; Baum et al. 2000; Ruef 2002) do the same for the exploration phase.

Type of actor and innovation

A number of scholars research the relationship between the type of actor(s) an organisation interacts with on the one hand and innovative behaviour on the other. It is assumed that a specific type of actor is a source of specific knowledge and information. For example, universities or research labs are assumed to be sources of basic research knowledge that enables the innovative performance of the organisations that are on the receiving end of the transfer (Rothmaermel and Deeds 2004). In most cases, it is hypothesized that partnering with a specific type of partner has a positive impact on (parts of) the innovative process or performance of an organisation. A notable exception is a tie with a rival or competitor. We start our discussion with this type of actor.

Forming a relationship with a competitor can be beneficial to an organisation as this competitor possesses complementary assets or can give access to new markets, which could enhance organisational performance. However, for a number of reasons allying with a competitor can be a "risky business" (Park and Russo 1996):

- Collaborations between competitors have a higher risk of uncontrolled information disclosure;
- Competitors have, due to comparable knowledge bases and competences, a greater capability to appropriate disclosed knowledge and information (high relational absorptive capacity).

Combined, this results in higher probabilities of (goal) conflict and lower levels of willingness to share knowledge, which in turn might lead to tie instability or even failure (Park and Russo, 1996), or lower performance levels. In sum, the literature assumes a negative impact of collaborating with rivals on innovative performance. This hypothesis is confirmed in the empirical literature (Baum, Calabrese and Silverman 2000; Mowery, Oxley and Silverman 1996).

In other studies other actor types, varying from universities with specific R&D expertise (Fey and Birkinshaw 2005) to links with institutional partners and providers of financial resources (Goes and Park 1997), are introduced as being conducive for the innovative behavior and performance of organisations. It can be derived from table 3, that in most cases (with the exception of administrative links) a positive impact is empirically found irrespective of the way innovation outcomes are measured.

Some scholars take a more configurational approach in which groups of types of network actors are assumed to have similar (knowledge) resources, which have effects on specific parts of the innovation process and its outcomes. A good example of this can be found in the paper by Rothaermel and Deeds (2004), who argue that so-called exploration alliances, in which it is the aim to jointly discover something new, are of importance to product development in firms. These exploration alliances refer to ties with organisations conducting basic research, drug discovery or development. Exploitation alliances, that consist of collaborations with actors competent on improvement and refinement of existing capabilities (e.g., actors conducting clinical trials and are specialized in marketing and sales of products), are thought to be conducive for the successful market introduction of product innovation.

Both Rothmaermel and Deeds (2005) and Faems et al. (2005) empirically explore the effects of exploration and exploitation on innovation outcomes of organisa-

Figure 10.5 The Density of Networks

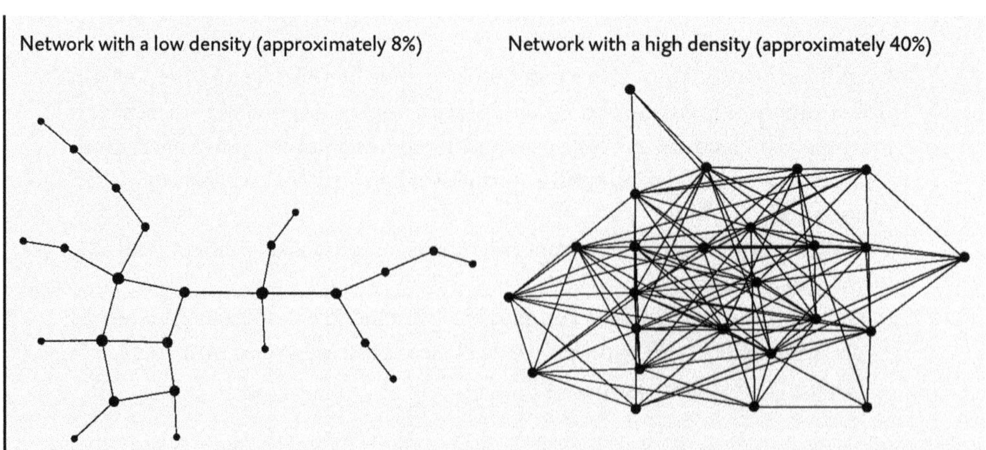

tions in which a combination of type of partner and degree centrality are used as independent variables. They conclude that exploitation ties are beneficial for the development or sales of new products, whereas exploration ties tend to have a positive impact on market introduction and sales of product innovations.

Network density and innovation

Theoretical literature suggests that network density will be positively related to the innovation .

There are two main arguments for the proposed positive effect between the density of a network and innovation:

- An information argument: high network density indicates higher levels of communication in the network, which increases the probability that actors will be exposed to novelties, information, and knowledge. These external inputs stimulate innovation;
- A socialisation argument: high density networks function as 'cliques' creating strong behavioural pressures that facilitate the development of trust and shared values and norms, and lower the chance of opportunistic behaviour and its related negative reputation effects. As a result, high density networks have favourable conditions for (especially tacit) knowledge transfer.

The empirical studies reporting on the relationship between network density levels and innovative outcomes produce mixed findings. Ahuja (2000) shows, for firms in the chemical sector, that a higher density impacts positively on innovation outcomes. McEvily and Zaheer (1999), however, report an opposite effect: lower density levels in advice networks are positively related to the adoption of new organisational practices. Obviously, these different effects of network density on innovation outcomes can be caused by sectoral differences and different ways of measuring network density and innovation outcomes.

Gilsing and Nooteboom (2005) find indications that density levels perform differently for different networks. In networks that are focused on exploration, higher density levels are important because it is not clear yet which knowledge will be important and consequently, having ties with many actors that produce different types of new knowledge, keeps options open. Moreover, having redundancy in ties can be beneficial due to the fact that network entry and exit is high. Third, under the condition that the level of newness of knowledge is high in exploration networks, one may need third parties to supplement one's absorptive capacity.

Tie strength and innovation

In the literature, there is an ongoing debate on the influence of the strength of inter-organisational ties in networks on innovation in general and on innovation outcomes in particular (Gilsing and Nooteboom 2005). One stream of research argues that the stronger the tie with a partner, the higher the actor's commitment is to the collaboration, and the more likely the tie is a factor in its behaviour. Uzzi

(1997) proposed that strong ties facilitate the exchange of fine-grained information, the development of trust, and complex adaptation to environmental changes. Moreover, strong ties are associated with strong behavioural pressures to conform. These are the result of the desire to keep a relationship going and the wish not to endanger it by behaving in an opportunistic way. Thus, the argument is that strong ties favour the exchange of (often tacit) and more complex knowledge in networks, which is beneficial to the innovative performance of organisations.

Another stream of research (Granovetter 1973; Burt 1992) has proposed that weak ties are more beneficial to innovation. (Many) strong ties in networks are redundant as they lead to actors that possess the same or similar knowledge. This is even more the case in densely knit networks. As a result, this knowledge is available to many, and few innovative and competitive advantages can be derived from it. Moreover, establishing and maintaining strong ties is costly and organisations with strong ties might become myopic and not open to new knowledge and insights. Strong ties might also lead to routinisation of behaviour, which hampers creativity and innovation. Therefore, from an innovation perspective it is more effective to maintain weak ties.

Although the concept of tie strength is widely accepted in the field of network studies, there is far less agreement on how the concept should be measured. Operationalisations of tie strength vary from the frequency or intensity of knowledge or information transfer to indicators of the level of commitment (e.g., dedicated investments or the level of reciprocity) in an inter-organisational relationship. By implication, it can be concluded that there is no consensus on the measurement of this important concept.

Despite these measurement differences, a majority of the studies in our literature sample reports a positive relationship between tie strength in networks and innovation outcomes (Oerlemans, Meeus and Boekema 1998; Rindfleish and Moorman 2001; Zollo, Reuer and Singh 2002; Fey and Birkinshaw 2005; Capaldo 2007). Some studies, however, found the opposite effect (Ruef, 2002) or no statistically confirmed relationship (McEvily and Zaheer 1999).

An interesting result is presented by Capaldo (2007) in his case study on the innovative performance of three design-intensive furnishing manufacturers. Although the strong tie argument is supported in this study, he also finds that so-called dual networks, which are not dominated by weak nor by strong ties, are most conducive to innovative performance. Put differently, in network structures dominated by strong ties innovative performance slows down as compared to dual network structures.

Small world networks and innovation

Small world networks are network structures which are characterised by high local clustering, while at the same time having short path length.

Figure 10.6 The Smallworldiness of Networks

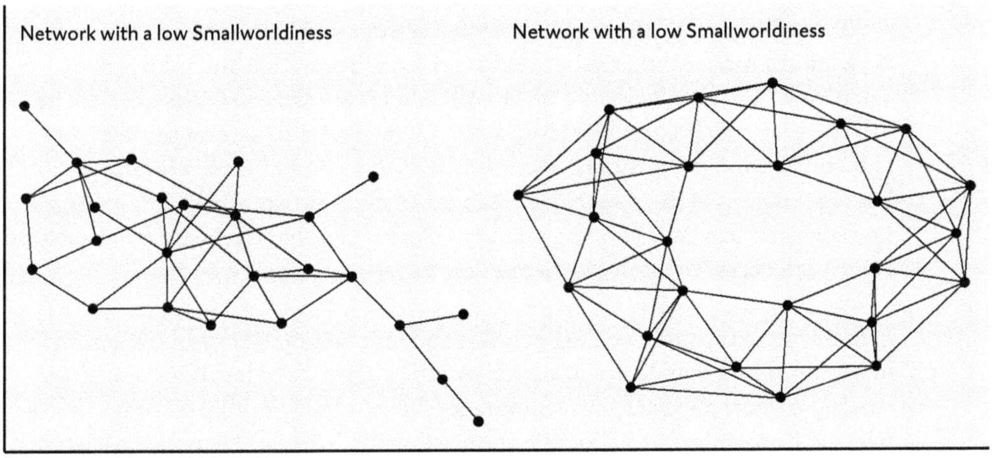

High local clustering means that a high proportion of an actor's collaborators are also collaborators with one another, whereas short path length indicates that there is a relatively low number of intermediaries between all pairs of actors in the network. A small network influences behaviour through a structural and a relational mechanism. The more a network becomes 'small worldly', the more the links between network components/clusters increase in frequency, which potentially enables the knowledge available within network components to be distributed throughout the global network (structural mechanism). Moreover, the more a network becomes small worldly, the more the links between the network components/clusters are made up of repeated ties and third-party ties, which potentially increased the level of cohesion in the global networks.

However, the theoretical literature proposes two hypotheses on how these potential effects materialise in actual (innovation) performance of organisations: a positive and a curvilinear relationship. As to the former, it is argued that as the network's connectedness and cohesiveness increases, the flow of creative ideas and promising collaborations between network components are facilitated. Moreover, the greater the level of repeated ties and third-party links, the greater the risk-sharing and the development of trust. These lines of thought suggest that an increase in a network's small world character can boost the performance of the global network and its participants (depending, however, on their network position) by making the exchange of conventions, as well as risk taking, more likely.

The latter argument suggests that both low and high levels of connectivity and cohesion in a network can be a liability for innovative performance. In case of low levels, there are few links between network components and links are not disproportionately formed through credible third-party or repeated ties, which isolates creative knowledge in separate clusters in the global network. As levels of connectivity and cohesion increase, separate network components become more interlinked and linked by actors who know and trust each other. These processes facilitate the flow of creative knowledge and help to build cohesive network components that support more risky collaboration and innovative behaviour. However, beyond a certain threshold, the same processes can create liabilities for collaboration and its outcomes. Increased structural connectivity reduces the innovative distinctiveness of network components, which leads to a homogenisation of the pool of knowledge. At the same time, problems of excessive cohesion (e.g., overlooking relevant information that is deviant from existing views) can occur. Both processes produce lower performance levels.

In their empirical study on small worlds and innovation, Flemming, King, and Juda (2004) find that decreasing path length and increased local clustering (the two ingredients needed to determine the level of small worldliness) have a positive impact on innovation outcomes (measured as successful patent applications). However, their small world measure turned out to be statistically insignificant in all models investigated. Flemming et al. argue that this lack of results might be due to a lack of diversity within a network component or a tendency that over time pressures of conformity result in a culture of risk aversion.

Uzzi and Spiro (2005) studied networks of teams of creative artists producing musicals. They hypothesize and find an inverted U-shape relationship between small world network structures and levels of creativity of the production teams.

Actor characteristics and innovation

Under this heading, two partner characteristics and their relationship with the innovative outcomes of organisations are discussed: alliance experience and partner characteristics.

There is an emerging literature on alliance capabilities of organisations (Gulati 1999; Kale, Dyer and Singh 2002; Ireland, Hitt and Vaidyanath 2002; Rothaermel and Deeds 2006). An important argument in this literature is that the relevance of features of inter-organisational relationships and networks has been overemphasised, whereas the capabilities of organisations to manage these relationships and networks often have been neglected. Amongst others, this literature proposes that the more experienced an organisation is in collaboration, the higher its alliance success will be. An important way of developing alliance capabilities is through prior or repeated collaborative ties (prior ties with the same partner). Prior and repeated ties do not only signal commitment and produce trust between partners (cf. the tie strength argument), but also enable an organisation to learn from its past experiences. Thus, the more prior ties an organisation has, the higher the probability that its collaboration will be successful (i.e., will generate higher innovation outcomes).

While reviewing this literature, an interesting pattern emerged: although the number of prior ties seems to be beneficial for the success of collaboration in general, this is not the case when it is related to the innovative performance of organisations. Both in the Stuart (2000) paper and in the paper by Zollo, Reuer, and Singh (2002), it was found that the number of prior ties (even in similar technological fields) did not impact on innovation outcomes. The number of prior ties with the same partner (repeated ties) does, however, have an impact on innovative performance, which also might be interpreted as the effects of the development of a stronger tie over time.

The innovativeness of a direct or indirect partner is a partner characteristic that influences the innovative outcomes of a focal organisation. Stuart (2000) shows that more innovative direct partners are beneficial for the ego, whereas Ahuja (2000) presents similar findings for indirect partners using weighted path length as an indicator (i.e., path length to an indirect actor times the innovativeness of this actor).

10.4.3 SUMMARY: MAIN FINDINGS IN THE LITERATURE

Table 4 summarises the main findings of this literature review. The table depicts the network characteristics applied, the hypothesized relationship between a network measure and the innovative outcomes of an organisation, and the empirical findings in relation to these hypotheses (confirmed, mixed findings, rejected).

From the findings reported in table 4, it can be deducted that in a majority of cases the hypothesized relationship between a network characteristic and the innovative outcomes of organisations is confirmed. This seems to be true for 'network centrality', 'network range', 'type of actor', and 'repeated ties', often irrespective of the way in which the network characteristic or innovative performance actually is measured. For these cases one can conclude that despite the relatively small number of papers, empirical findings tend to be rather robust and consistent. In other cases ('density', 'tie strength', 'small worldiness', and 'prior ties'), this cannot be concluded. Explanations for these latter results are the use of differing measures for these network features and innovative performance, sample differences, or the possible influence of diverse characteristics of innovation (e.g., incremental versus radical) on empirical findings.

The large majority of the papers looked at the generation of innovation (22), two had adoption as a topic. A real comparison with network effects on adoption processes is not doable. As to the stages of innovation it turned out that most of the papers were limited to the exploration stage (ten papers), six focused on the exploitation stage, and six papers used criterion variables combining exploration and exploitation stages. There were no systematic differences in effects of distinct network measures on innovation in the exploration and exploitation stages.

Table 10.3 **Inter-organisational relationship and network characteristics and their impact on the innovative behaviour of organisations**

Source	Network measure	Type of tie	Boundary specification	Innovation outcome measure	Main findings
Hagedoorn & Schakenraad (1994)	Weighted degree centrality (number of alliances per unit of sales)	Strategic technology alliances (formal R&D linkages)	IORs among firms in information technologies, electronics, mechanical engineering and process industries	Patent intensity	Higher levels of centrality are positively related to patent intensity
Powell, Koput & Smith-Doerr (1996)	(1) Centrality (degree and closeness) (2) Centrality (closeness)	Formal R&D alliance; outside investor; clinical trials; manufacturing; marketing/licensing; idea transfer; supply/distribution; JV; multiplex	Dedicated biotechnology firms; pharmaceuticals, research institutes, universities; venture capital firms; chemical firms.	(1) Growth employees (2) Subsequent R&D alliances)	(1) Higher levels of centrality are positively related to growth of firm size (2) Higher levels of centrality are positively related to new R&D tie formation
Owen-Smith & Powell (2004)	Betweenness centrality	Formal ties: R&D alliances, information transfer, transfer of funds	Dedicated biotechnology firms, public research organisations, venture capital firms, government agencies, pharmaceutical, chemical and health care organisations	Count of patent applications	Higher levels of centrality are positively related to innovation output, but only in geographically dispersed networks
Powell, Koput, Smith-Doerr, Owen-Smith (1999)	Own value centrality (true local centrality) network diversity (number of different ties = range)	Formal R&D alliance; outside investor; clinical trials; manufacturing; marketing/licensing; lidea transfer; supply/distribution; JV; multiplex	Dedicated biotechnology firms, public research organisations, venture capital firms, Government agencies, pharmaceutical, chemical and health care organisations	Count of patent applications	Higher levels of centrality are positively related to innovation outcome. Higher levels of network diversity are positively related to innovation outcome
Shan, Walker, Kogut (1994)	Degree centrality	(1) Formal commercial agreements (of bio tech start-ups) (2) Formal research agreements	Private sector firms, universities, government research agencies, research labs	Number of patents granted	Higher levels of centrality (of commercial ties) are positively related to patents granted
Baum, Calabrese, Silverman (2000)	(1) Network size (number of alliances = degree?) (2) Ties with rivals (3) Network efficiency (= range)	Formal alliances	Alliances with: nonrival bio-tech firms; potential rival BF; pharmaceutical firms; chemical firms; universities; research institutes; government labs; industry associations; marketing firms	R&D spending growth R&D employees growth Patenting rate	(1) Higher network size is positively related to start-up performance (2) Ties with rivals decreased start-up performance (3) Higher levels of network efficiency are positively related to R&D employee growth and patenting rate

Table 10.3 Inter-organisational relationship and network characteristics and their impact on the innovative behaviour of organisations (Continued)

Source	Network measure	Type of tie	Boundary specification	Innovation outcome measure	Main findings
Oerlemans, Meeus and Boekema (1998)	Tie strength	Intensity of knowledge transfer	(technical) universities, colleges for professional and vocational education; trade organisations, TNO, and private consultants; buyers, suppliers.	The extent to which process or product innovations resulted in (1) reduction of the cost price; (2) quality improvement of (a) processes and (b) products; (3) increase in production capacity; (4) improvement in delivery time; (5) increase in sales; (6) increase in profits	The higher the contributions to innovation by networks actors, the higher the innovation outcomes (except for universities)
Ahuja (2000)	(1) Degree centrality (2) Weighted path length (3) Number of structural holes	Technological collaboration	Firms in the chemical sector	Number of patents received by an organisation	(1) The higher the number of direct ties, the higher innovation output (2) The higher the number of indirect ties, the higher innovation output (2a) Impact of indirect ties is weakened by the number of direct ties (3) Higher levels of density (less structural holes) impact positively on innovation outcomes
Ebadi and Utterback (1984)	Network (inter-project) cohesiveness	Communication ties	Principle investigators of other projects	Self-reported success of innovation projects	Higher levels of inter-project cohesiveness are positively related to the success of innovation projects
Uzzi and Spiro (2005)	Small world Q	Relationships among creative artists producing musicals	Creative artists (excluding performers)	Level of creativity (measured as critics evaluation of musicals)	There is an inverted U-shaped relationship between small world network structure and level of creativity

Table 10.3 Inter-organisational relationship and network characteristics and their impact on the innovative behaviour of organisations (Continued)

Source	Network measure	Type of tie	Boundary specification	Innovation outcome measure	Main findings
Flemming, King, Juda (2004)	(1) Path length (2) Local clustering (3) Small world measure (ratio of (1) and (2))	Co-author links in patents	Inventors	Number of successful patent applications	(1) Decreasing path length, increases innovation output (2) Increased local clustering are positively related to higher innovation output (3) No effects
Ruef (2002)	(1) Tie strength (2) Network diversity (= range)	Information transfer	Family members, friends, business associates, competitors, advisors	Objective measure: Number of patent and trademark applications by entrepreneurial start-ups Subjective measure: nine items, selfreports on e.g. attempts to introduce product, process and service innovations, new market entry by entrepreneurial start-ups	(1) Weak ties are more conducive for innovation outcomes as compared strong ties (2) More diverse networks are positively related to higher innovation outcomes. Results only hold for subjective outcome measure
Zollo, Reuer, Singh (2002)	(1) Number of prior ties (general) (2) Number of prior ties (in similar technology fields) (3) Number of prior ties (with same partner) (4) Alliance importance (in terms of dedicated investments) (5) Interdependence (labor division)	Any cooperative agreement of any form aimed at the development, manufacture and/or distribution of new products	Firms in biotech and pharmaceuticals	Alliance performance with two relevant dimensions: (1) Level of satisfaction with knowledge accumulation due to alliance participation (2) Evaluation of the extent to which new opportunities were created due to alliance participation	(1) No result (2) No result (3) The higher the number of prior ties with the same partner, the higher alliance performance (4) The higher the alliance importance, the higher alliance performance (5) Higher levels of interdependence are positively related to alliance performance

Table 10.3 Inter-organisational relationship and network characteristics and their impact on the innovative behaviour of organisations (Continued)

Source	Network measure	Type of tie	Boundary specification	Innovation outcome measure	Main findings
Zahra and Nielsen (2002)	(X1) Use of alliances and joint ventures in manufacturing (X2) Use of outsourcing and licensing	Knowledge transfer In- and out-sourcing of technology	Not specified	Technology commercialisation (TC), 4 indicators: (1) Number of new products (2) Radical new products (3) Number of patents (4) TC speed	Number of new products: (X1) = +; (X2) = + Radical new products: (X1) = +; (X2) = + Number of patents: (X1) = no result; (X2) = + TC speed: (X1) = +; (X2) = +
Fey and Birkinshaw (2005)	(X1) Importance of universities as sources of R&D expertise (X2) Importance of alliances as sources of R&D expertise (X3) Level of technology in-sourcing	Intensity of transfer of R&D expertise	Universities Alliance partners (not specified)	Relative R&D performance (timely market introduction of new products; coming up with technological breakthroughs)	(X1) = positive (X2) = no result (X3) = negative
Stuart (2000)	(X1) Number of technology alliances formed in a five-year period (X2) Innovativeness of partners	Joint product development agreements; joint ventures; technology exchanges; licensing; marketing agreements	Horizontal (intra-industry) alliances	Rate of innovation of firms in the semi-conductor industry (measured as the number of time firm's patents are cited)	(X1) = no result (X2) = positive
Rothaermel and Deeds (2004)	(X1) Number of alliances on basic research, drug discovery and development (degree) (X2) Number of alliances on clinical trials, marketing and sales (degree)	Exploration alliances (jointly discovering something new) Exploitation alliances (jointly improving and refining existing capabilities)	Horizontal (intra-industry) alliances	(Y1) = Number of products in development of bio-tech firm (Y2) = Number of products introduced on the market	(X1) and (Y1) = positive (X2) and (Y2) = positive Both relationships are negatively moderated by firm size
Faems, Van Looy and Debackere (2005)	(X1) Range (diversity of IOR partners) (X2) Number of exploitation-oriented collaborations (X3) Number of exploration-oriented collaborations	Innovation related collaboration	IORs with organisations: in same group; competitors; customers; consultants; suppliers; universities; research institutes.	(Y1) = Innovative sales due to technologically new or improved products (Y2) Innovative sales due to technologically improved products (Y3) Innovative sales due to technologically new products	(X1) and (Y1) = positive (X2) and (Y2) = positive (X3) and (Y3) = positive

Table 10.3 Inter-organisational relationship and network characteristics and their impact on the innovative behaviour of organisations (Continued)

Source	Network measure	Type of tie	Boundary specification	Innovation outcome measure	Main findings
Goes and Park (1997)	(X1) Presence of structural links (X2) Presence of administrative links (X3) Presence of institutional links (X4) Resource links	Information sharing; managerial experience; legitimacy; financial resources	Multi hospital system; other hospitals; institutional associations	Relative adoption rate of service innovations	(X1) = positive (X2) = negative (X3) = positive (X4) = positive
Capaldo (2007)	Tie strength (duration, frequency and intensity of collaboration) Overall network structure (of strong and weak ties = dual network structure)	Design agreements	Design firms	Innovative performance of three design-intensive furnishings manufactures (new product on the market; time of product in the market; critic's ratings)	Stronger ties produce mutual understanding, trust, reciprocity all supporting innovative performance. In network structures dominated by strong ties, innovative performance slows down as compared to dual network structures with both strong and diverse weak ties
Salman and Saives (2005)	(X1) = degree centrality (X2) = closeness centrality (X3) = betweenness centrality (X4) = own value centrality	Strategic partnerships and collaborations	Universities, venture capitalists, manufacturing firms, public and private labs, consultants, equipment suppliers, distributors	Number of patents of biotech firms	All centrality measures impact positively on innovation outcomes (in separate models)
Gilsing and Nooteboom (2005)	(X1) = network centrality (X2) = network density (X3) = network stability (X4) = tie strength	Knowledge transfer	Universities, suppliers, customers, research institutes, dedicated biotech firms	Exploration and exploitation networks in multimedia and biopharma	Exploration = high on density; high on frequency; high on trust. Exploitation = high on stability; high on centrality; high on control; narrow on scope.
McEvily and Zaheer (1999)	(X1) = densiity (non-redundancy in ego network) (X2) = frequency of interaction	Advice	Advisors	Adoption of organisational practices	(X1) Non-redundancy is positively related to adoption (X2) Not supported

Table 10.3 Inter-organisational relationship and network characteristics and their impact on the innovative behaviour of organisations (Continued)

Source	Network measure	Type of tie	Boundary specification	Innovation outcome measure	Main findings
Rindfleisch and Moorman (2001)	(X1) Tie strength	Collaboration	Competitors, suppliers, customers	(Y1) New product-related information acquired (Y2) New product/process creativity (Y3) New product development speed	(X1) and (Y1) = positive (X1) and (Y2) = positive (X1) and (Y3) = positive

10.5 INNOVATION POLICY AND NETWORKS: THE POLICY SECTION

In this section we ask two questions:

1. To what extent is the current Dutch innovation policy portfolio geared toward creating and facilitating IONs?
2. To what extent does the current Dutch innovation policy portfolio (un-) targeted, (in-) directly draw upon network mechanisms to foster innovation?

Table 10.4 Overview of findings

Network characteristic	Hypothesized relationship between network characteristic and innovative outcomes	Main findings
Centrality (different measures)	Positive	Confirmed
Network range	Positive	Confirmed
Type of actor: competitor	Negative	Confirmed
Type of actor: other actors	Positive	Confirmed
Network density	Positive	Mixed findings
Tie strength	Positive	Mixed findings
	Negative	Mixed findings
Small worldiness	Positive	Rejected
	Inverted U-shaped	Confirmed
Alliance experience:		
Prior ties	Positive	Rejected
Repeated ties	Positive	Confirmed

Diagnosis and policy initiatives

In the 'Annual Innovation Policy Trends and Appraisal Report' (EC 2006: 26) an important element relates to this topic. The reviewers infer: "There is limited interaction among the actors in the National Innovation System; inadequate exploitation of research results". This observation, which was characterised as the 'knowledge paradox', has been translated into a number of challenges for Dutch innovation policy, one of which is to promote collaboration among companies in general, as well as more specifically the promotion of collaboration between industry and the public knowledge infrastructure.

Box 10.1 The status of the Dutch National Innovation System (NIS)

Based on the figures presented above, the following conclusions can be drawn concerning the status of the Dutch NIS:

- The innovative performance of the Netherlands as measured by different indicators (input, throughput, and output) can be regarded as good: the quality of scientific research output is high, there is a high level of patenting, a high share of financing of public research by industry, and there are high rates of usage of ICT and access to ICT applications.
- The NIS, however is also characterised by specific features and (structural) problems that weaken the strong innovative performance of the Netherlands, while countries with less-favourable innovation performance seem to be catching up (the Netherlands is 'losing momentum' according to the European Innovation Scoreboard):
  - Total financial efforts in R&D expenditure are stagnating, business expenditure on R&D lags behind, compared to main competitors.
  - Increasing shortage of highly educated people, especially in science and technology; mismatch between outflow from education and demand by industry.
  - Limited interaction between the actors of the NIS; inadequate exploitation of research results.
  - Limited innovation entrepreneurial activity.
  - Problems concerning financing of (early stages of) innovation.

The strengths and weaknesses of the Dutch innovation system are in line with those identified within the framework of the analyses, which forms the basis for the 'Innovation Letter' of 2003. This innovation policy document of the Dutch Ministry of Economic Affairs describes the integrated (interdepartmental) approach towards industry-oriented innovation policy, and forms the basis/framework for the current instruments and their changes in the last years (see section 2.1.3).

The main innovation indicators on human resources, expenditure, employment, and IPR, do not yet indicate any changes compared to the year when the new policy was launched. This has to do with the fact that the current policy mix is under review, based on the results of the Innovation Letter. Most importantly, measures taken in the area of R&D and innovation show a delayed impact on the innovation system in general, and on economic performance in particular. The Dutch economy itself, therefore, seems unchanged in the last years, with some indications of recovery in the first quarter of 2006. This is partly caused by the budget restrictions imposed by the current cabinet, as well as by the further rationalisation of production by the industry in order to cut costs. A more structural economic growth, based on increased labour productivity levels, is not predicted.

Source: Based on EC, 2006: 33

It is interesting to see how this issue has drawn the attention of policymakers. In figure 3 one can see the number of recent initiatives that address this challenge.

Figure 10.7 Innovation challenges and policy responses

Exhibit 11: Innovation challenges and plicy responses

Key challenge	Measures responding to the challenge
1. Total financial efforts in R&D expenditure are stagnating	• (intensifying) WBSO (NL_5 – a generic fiscal scheme to stimulate investments R&D by entrepreneurs • Innovation Subsidy for Collaborative Projects (NL_45) – stimulates national and international technological collaboration between companies and the Ministry of Economic Affairs to invest more in innovation.
2. Increasing shortage of HRST.	• Delta Plan Science and Technology • Casimir scheme (NL_48) – promoting public-private mobility of researchers
3. Limited interaction between the actors of the NIS; inadequate exploitation of research results	• Innovation omnibus (NL_52) – a legal framework for innovation programmes in selected key areas. Many current innovation measures will be brought under the omnibus regulation as 'modules'. A pilot innovation programme is currently developed in three sectors (Food & Flowers, Water, and Hightech Systems & Materials) • IOP-LTI Module (NL_54) – uniting 'old' measures Leading Technological Institutes (NL_19) and Innovation oriented Research Programmes (NL_18) • Challengers module (NL_55) • Smart-Mix (NL_53) – to stimulate public-private consortia performing excellent and relevant research • Casimir (NL_48) to stimulate public-private mobility • TechnoPartner Knowledge Exploitation Subsidy Arrangement (NL_43) • Investment Grants for Knowledge infrastructure (NL_29) • Programme-basid funding of Large Technological Institutes and TNO • Funding of excellent relevant scientific-technological university research via the Technology Foundation STW • Valorisation grant (NL_49) – to promote the creation of spin-offs • Innovation vouchers (NL_47) to stimulate interaction and exchange between SMEs and the knowledge infrastructure • Regional Attention and Action for Knowledge circulation (RAAK; NL_51) – to improve interaction and exchange between SMEs and vocational higher education institutes
4. Problems concerning financing of (early stages of innovation.	TechnoPartner Action Programme (NL_43), including: • Seed Facility – to improve and mobilize the Dutch venture capital market • Pre-seed Facility – to help potential technostarters • Patent Facility – to help knowledge institutes to professionalise patent policies • TechnoPartner Label – to help techno-starters obtain a bank loan

Source: Annual Innovation Policy Trends and Appraisal Report, 2006: 37

Besides the fact that the third key challenge – expanding collaboration – is covered with the largest number of policy measures, it turns out that all other key challenges (1, 2, and 4), are part of measures mentioned under key challenge collaboration. This is the first strong indication that the Dutch innovation portfolio is drawing upon generic network mechanisms. In order to understand how these policy measures do utilise network mechanisms each policy measure is briefly explained.

The measure Innovation Subsidy for Collaborative Projects (NL-45) stimulates, as one of the exceptions, *international* technological collaboration between companies and/or public knowledge institutes. In terms of budget, the scheme is a substantial measure (approximately 100 million euros per year). In 2007, this measure became integrated into the new innovation omnibus legislation (NL 52), with part of it being transformed into a credit facility rather than a grant (EC 2006: 45). In the Innovation Performance Contracts there is also the opportunity to make part of the agreement with the Ministry of Economic Affairs a network effort. The Casimir scheme is based on the assumption that there are exchange relations that exist between organisations; otherwise exchange of personnel is an empty proposition.

In the Innovation Oriented Research Programme and the Large Technological Institutes module (IOP-LTI NL 54) the target is to align applied scientific research to the needs of the business community, and to involve the industry in this type of research. As one can see in table 1, there are large differences in the prevalence of R&D collaboration between the two IOPs. In the Man-Machine Interaction IOP some 60 percent of the IOP participants reported relations with the knowledge infrastructure, whereas the Image Processing IOP reports about 17 percent of R&D cooperation with knowledge institutes.

In the Challenger module (NL 55) the aim is to create spin-offs from knowledge institutes. Because this is very often associated with the maintenance of relations with the parent institute it indeed contributes to the web of relations adding to the valorisation frontiers of scientific knowledge. The start-up company usually initiates its activity with a search for commercialisation options in the business community, and there the linkage formation starts.

The Smart Mix module (NL 53) is the most prominent programme in this series of measures. € 100 million euros are annually available to stimulate public-private consortia on collaborative research of users, or buyers of scientific knowledge, and knowledge producers like university and large public research institutes. It is important that this programme requires that funding recepients organise a network of 'excellent' partners out of the business and science community.

The Technopartner programme (NL 43) also focuses on the foundation of high-tech start-ups. In terms of providing support for the creation of collaborative networks, this programme follows the same rationale as the Challenger module.

The so-called Large Technological Institutes have as a specific target the valorisation of scientific research findings. Of course this means that finding utilisation partners is the priority of this measure. There are several LTIs in the Netherlands that actively collaborate with industry and government. For instance the Waterloopkundig Laboratory has a strong interaction with the Rijkswaterstaat (authority for infrastructure and water).

The innovation vouchers (NL 47) concentrate entirely on getting scientific knowledge from university to industry. Organisations that are in urgent need of specialist knowledge can apply for an innovation voucher in order to engage university researchers on a consultancy basis on requested topics.

Finally the Regional Attention and Action for Knowledge circulation (RAAK; NL 51) has the explicit aim of stimulating the exchange of knowledge at the regional level.

In addition there are many regional programmes that provide an impetus to collaborate locally. Also there is STW (Foundation for technical sciences) which organises collaborative research projects in the way it is done in the IOPs; applicants for research funding have to organise a utilisers/buyers network, which has to confirm their contribution to the project, either in time or with in-kind investments.

The Dutch innovation policy portfolio: An assessment from a network perspective

The remaining question is to what extent do these policy measures implicitly or explicitly target the kind of network features that we found to have more or less robust – but significant – effects upon innovative behaviour? It is clear that none of the measures has an explicit target of affecting one specific network feature. For that reason it is better to look at the *portfolio of innovation measures* and see to what extent their impacts cumulate or counteract, and to what extent their network impacts are more or less explicit. Subsequently, we will answer the following questions:

- How is network size affected?
- How is density affected?
- How is centrality affected?
- How is tie strength affected?
- How are structural holes, i.e., brokerage affected?
- How are small world effects affected ?

In table 3, *network range* turned out to be beneficial for innovation if the ties are more diverse. There are at least two policy measures that explicitly have an impact on this network feature: Technopartner and Challenger. The functions of the Technopartner and additional measures such as Challenger are very important because they foster the founding of start-ups, and these new entrants add new and diverse nodes to the population rendering *potential* new relations: Technopartner opts for backing innovative technology-driven start ups, though related to the mainstream technologies. In Challenger start-ups, or programmes promoting the odd-one-out technologies, are favoured, which broadens the technological opportunity set. This latter category is in general considered to be one of the ways in which the crowding out of R&D investments is prevented. Both measures induce new R&D trajectories, which are signalled by start-ups founded in the Challenger module.

Except for these main effects of creating new entrants in a sector, additional nodes have several other effects. First, new entrants enlarge the partnering opportunities, which is a continuous issue in sectoral vitality. Funding start-ups has signalling value for technology opportunities, and simultaneously legitimises partnering options. Of course one should take into account the relatively high proportion of tie failures (Oerlemans, Gössling and Jansen 2007, for a review; Powell et al. 2005: 1165, for figures in the life sciences industry; Van der Valk and Faber 2007, for figures in Dutch dedicated life sciences firms). However, there is a positive side to tie failure, because it most certainly adds to the partnering capabilities. Given the positive impacts of networks on innovation, partnering capabilities are a key part of the innovation process. The more competent partners are selected out. A decisive criterion to continue or to stop a partnership is the contributions of partners to the innovation project – and hence the innovative performance of collaborators guides network expansion over time although in an indirect way.

As we saw in table 3, there is a tendency that *strong ties* (Rindfleisch and Moorman 2001; Gilsing and Nooteboom 2005; Oerlemans, Meeus and Boekema 1998), and *weak ties* (Ruef 2002), respectively configurations of strong and weak ties (small world) contribute to creativity (Capaldo 2007; Uzzi and Sprio 2005). New entrants and especially start-ups serve in inter-organisational perspective – not from a social network perspective – as weak ties, they create non-redundancy in a network. Weak ties are in general considered to have a bridging function, which is often pivotal to solve problems in innovation projects, and even more to generate innovation and creativity, so there is a genuine need for weak ties. Start-ups add to the number of weak ties, even despite their relatively low absorptive capacity, and hence create additional options to achieve satisfactory innovative performance.

Because start-ups are in general small and young, especially in case of university spin-offs, but also with entrepreneurial company founding, they have to cope with the liability of newness, and one way to do so is by means of partnering. Because of the inclination to search locally, many start-ups initially prefer partnerships with their old departments, or the company they spun out from. This leads to what is called a rewiring of the network, in fact existing relations are repeated although now with another governance form. Former colleagues e.g., in STW-collaborative projects become allies and this adds to the network size and its diversity, and yields additional means to carry out R&D and innovation projects. The point is that expanding the number of partnering oriented nodes – therefore particularly small high-tech start-ups – also has an impact upon density of a network. As density is defined as the proportion of total nodes in a population of organisations that is actually engaged in a relationship, this would mean that density increases only if each node added initiates more than one partnership. In high-tech sectors that is likely.

Enlarging the *size of the network* by adding nodes with a high likelihood of partnering can also be achieved by means of active business venturing of large companies such as DSM or Shell. Shell LiveWIRE and DSM business venturing, or Rijkswaterstaat's Future Centre, because all promote the founding of new companies by (former) employees. It is very interesting to see that large numbers of start-ups realised by DSM venturing are located in the US. Each player in the Dutch banking sector has this kind of business venturing activities as well: Fortis, Rabo Bank, ING, and ABN AMRO.

As to the current portfolio of policy measures, there is none that explicitly draws upon *network centrality*. Despite the positive effects of centrality measures on innovative behaviour it is difficult to imagine how one can affect this *directly* with policy measures. The only agent who can work on centrality is the focal organisation (E) in the network. By means of partnering E can expand the number of ties. However, it is pretty well known that for instance in the biotech sector, in health research, or in food research there is a central role for universities and LTIs, and a very small number of MNOs, with large R&D laboratories. Policy measures, for example SMART MIX, STW, and IOP, put those players in leading positions because they have the absorptive capacity, the financial means, and the partnering capabilities to compete for larger R&D projects that have to be prepared with consortia of excellent partners.

For *tie strength* our assessment is similar to centrality: it is E in particular who decides about the intensity and frequency of contacts. However, here the criteria for funding developed in application protocols do define base-line tie strength. In the STW, EET,[2] IOP, LTI one can find specifications as to the kind of exchange, the manner of resourcing a partnership (in kind, or additional funding), which in fact transforms relations into literally valued graphs. For instance in the SMART MIX module one can see that per participant fundamental research is funded for 75 percent at maximum, applied research is funded at maximum for 50 percent, whereas for pre-competitive research only 25 percent is funded. E is also asked to indicate its own investment in the project. This implies that partners who cannot deliver their negotiated share of the matching are dropped.

How can *brokerage* be affected? We think that the Innovation Relay Centres have a strong brokerage effect in the Dutch innovation community. These intermediaries have contacts in many different sectors, and can combine this into creative solutions. Also E can organise brokerage more easily by the recruitment of people who have experience across industrial/service sectors and in that sense can link distinct knowledge and technological trajectories.

With respect to *small world effects*, indicated with a configuration of locally strong ties, and globally weak ties, the aforementioned arguments with respect to tie strength apply. Yet there is one relevant specification here, which is that global dispersion of ties is facilitated more at the EU level in the framework programme (which is left out of this story), than at the national level. On the contrary even, in general there is a restrictive approach to the funding of international collaboration. A conclusion could be that in the Dutch innovation policy portfolio there is little room, and virtually no explicit attention to, the small world effect on innovative performance. The international dimension needs further attention. For instance in almost 25 years of STW research involving hundreds of projects there have been less than a handful of international collaborations. Recently Peeks in the Delta paid some attention to regional innovation policies and the need of the northern regions to develop international collaboration with Eastern Europe. Unfortunately those policy schemes are the exceptions, which emphasise even more the sort of introversion of Dutch innovation policy as far as small worlds are concerned.

Finally, it is important to emphasise that the WBSO scheme plays an important – though generic – role in creating attractive partners. According to Ahuja (2000) two distinct forces drive partnering: inducements and opportunities. Inducements urge a company to collaborate, not being able to go it alone. But a company can also be invited to collaborate because of its attractivity for another company. Winning WBSO-grants can easily be such an attractor, and many companies use those grants in this way, because it proves that they are doing state-of-the-art R&D. In that sense the WBSO scheme plays an essential role, because it improves actor characteristics fostering inclusion in innovation networks.

10.6 SOME CONCLUDING REMARKS

The critical part of our innovation policy measures assessment derives from two observations: (1) creation of too few start-ups, and (2) not targeting small world effects. It is well known that the number of start-ups in the Netherlands is relatively low, which implies that the Dutch knowledge economy could create more diversity and new partnering opportunities if this policy would work better. Another rather delicate issue is that although the world of science and innovation is international and not national, there seems to be a strong inclination to spend public money on national collaborative innovation activities. Of course one can say that there is too little research to assess the small world effect on innovation – which we fully agree with – yet the logic of local dense networks and strong ties, and globally sparse networks and weak ties has a strong appeal within the science and innovation community, however this has yet to be translated into appropriate policy measures.

Table 10.5 Overview of network characteristics used in innovation policy measures

Network characteristic	Hypothesized relationship between network characteristic and innovative performance	Main findings	Utilization in policy measures
Centrality (different measures)	Positive	Confirmed	Not directly, but indirectly in LTI,IOP, and STW module
Network range	Positive	Confirmed	Yes, indirect, by creating start-ups, by Technopartner, and Challenger module
Type of actor: competitor	Negative	Confirmed	No
Type of actor: other actors	Positive	Confirmed	Yes, getting the knowledge infrastructure more actively linked to the industry: SMART MIX, IOP_LTI, Innovation Vouchers
Network density	Positive	Mixed findings	Indirectly through start-ups
Tie strength	Positive	Mixed findings	Indirectly because of thresholds
	Negative	Mixed findings	baseline R&D efforts created for new entrants in STW, EET, IOP_LTI, and Smart Mix. This implies that a very small population of organisations rewires their network over time
Small worldness	Positive	Rejected	No
	Inverted U-shaped	Confirmed	
Alliance experience:			Indirectly, see tie strength
Prior ties	Positive	Rejected	
Repeated ties	Positive	Confirmed	

APPENDIX A

An overview of the network characteristics influencing the innovative behaviour of organizations

There are such an abundance of network measures that we cannot pretend to present a complete overview of existing network measures (for such overviews see Wasserman and Faust, 1997; Brandes and Erlebach, 2005). For this reason here we present those network features that have, according to the existing academic literature, been shown to have an influence on the innovative behaviour of organisations (see part 4).

Consequently, we need to specify and define those network attributes or variables which have been proven to relate to the level of innovative behaviour of organisations.

From the perspective of an individual network actor, the following attributes are potentially relevant when it comes to explaining innovative behaviour of an organisation, E:

1. The type of actors E interacts with or does not interact with;
2. The type of relationships or ties E has or does not have with other actors;
3. The position of E in the overall network;
4. Characteristics of the overall network E is a part of: centralisation, size, density, cohesiveness, stability;
5. The structural configuration of ties in the overall network.

Attribute 1. The type of actors E does or does not interact with
Here the idea is that the innovative behaviour of E is related to the type of other actors E has relationships with (denoted as Ps in what follows). The following factors can play a role in this:
- The resources the Ps possess (new knowledge, new experience, legitimating power, financial resources, etc.).
- The position the Ps have in the overall network (centrality, status, brokerage position, access to network resources, etc.).
- Whether E has previously interacted with P.
- Whether E interacts with rivals or not.
- The level of innovativeness of the actor E interacts with.

Ad 2. The type of relationships or ties E does or does not have with other actors.
Central to a relational or network approach is the type of relationship (also called relations, lines, or edges) between different actors. These relationships are defined by certain content. Substantive relationships can range from friendships and social contacts to formal contracts, working relationships, giving and/or receiving advice, interlocking directorates.. Although the type of actor and the

type of relationship will often be related (e.g., a contact between a public agency being existent because of the provision of financial resources) often this is not straightforward. It turns out that especially the fact that different types of relationships (multiplexity) exist between two organisations or within a network in which E is embedded makes the patterns of interactions particularly valuable.

The following factors can be at play here:
- The type of the relationship:
  - Informal
  - Formal
  - Exchange of expertise
  - Exploitation oriented
  - Exploration oriented
  - Administrative links
  - Institutional links
  - Resource links
  - Structural links
- Having had prior ties (with P, with actors similar to P, or in general).
- The degree of multiplexity of relationships.

- The age of the relationships.
- The frequency of interaction.
- The strength of the relationship. Network ties are said to be strong when they are exercised frequently, have been in existence for some time, and when there is an emotional component to the link. Tie frequency refers to the number of times a contact is made with the other actor during a stated time period. Tie duration represents the amount of time that has elapsed since the tie was first established. Emotional intensity captures the closeness of the relationships that exist between the E and other actors (e.g., a friendship tie would be closer than a business-only relationship).

Ad 3. The position of E in the overall network

Given the fact that networks differ in their structures (comparable to structures in organisations) it is relevant for an actor's functioning where the actor is located in that structure, in other words what its position is in the network. In what follows we introduce a number of such positions which have been shown to be relevant.

The actor centrality of E

The centrality of an actor is traditionally considered important since it conditions the degree to which an actor E does or does not have access to resources throughout the network.

The following different types of centrality are commonly distinguished:
- *Degree centrality*: the number of direct relationships an actor has in a network (but also the number of relationships an actor does not have in a network). The degree centrality measures the expansivity and the popularity of an actor. An

organisation's direct ties can for example provide both resource-sharing and knowledge spillover benefits.
- *Closeness centrality*: Those actors that have the shortest path to all other members of the network (inverse distance to all other members of the network).
- *Betweenness centrality* captures the role of 'Brokers' or 'Bridges'; those that have the most indirect ties and can connect and disconnect parts in the network.
- *Own-value centrality* starts from the idea that it is important to have many links to other network members. However, it also takes into account the centrality of the actors to which links exists and hypothesizes that it is more important to have links to other central actors.

The indirect ties of E

Indirect ties of E (count, distance weighted count, or distance and information weighted count). The position of E in the inter-organisational network. E's linkages provide it with access not just to the knowledge held by its partners (direct linkages) but also to the knowledge held by its partner's partners. The position of E in the network of inter-firm linkages can serve as a node in a flow of information, with each firm in network being both a recipient and a transmitter of information. Strategic positioning of E: is E a 'boundary spanner', 'hub', or 'connector'?

Attribute 4. Characteristics of the overall network E is part of: centralisation, size, density, cohesiveness, stability

Here a number of network level variables are introduced. The idea being that the broader network context in which E operates has a direct influence on its innovative behaviour. For example, whereas actor centrality is an attribute of E influencing its behaviour, network centralisation is an attribute of the (entire) network E is a part of. Whether E is part of a highly centralised network or a network with a low centrality will create different constraints and possibilities.

Centralisation is computed by taking the average sum of the individual actor's centrality. It shows the tendency of a single firm to be more central than all others.

The *size* of a network is the number of actors in the network. In order to determine the size of a network it is important to have a precise understanding of what the boundaries of the network are. Network size is important as it represents the potential of the actor to utilise connections to gather resources.

The *range* of a network refers to the diversity of the contacts represented in the network. Although larger networks tend to have greater range, this is not always the case. An actor may have a great number of ties to one type of actor but few ties to other types of actors. Another actor may have relatively few ties, however, they may be connected to a greater diversity of other actors. Range is important because it reflects the actor's potential access to alternative resources.

Density captures how tightly knit a group or subgroup is. It is a proportion that indicates the number of actual ties present in the group relative to the number of possible ties in the group (i.e., if everyone had a relationship with everyone else in the group). When calculating the density of E we look at how closely connected the Ps are to each other.

There are several measures of *cohesion*, including density. However, one common measure is the average number of ties it takes for an actor in the network to 'reach' another actor in the network. If E is connected to P_1 who is connected to P_2, then E is at a distance of 2 from P_2. The average path distance for the group gives an indication of the group's cohesion.

Stability of the overall network: the degree to which actors enter and exit a network.

Attribute 5. The structural configuration of ties in the overall network

- *Closed social networks*: Members of closed networks are connected to each other.
- *Disconnections* (or *structural holes*) between Ps: ties are redundant to the degree that they lead to the same actors. Structural holes are gaps in information flows between Ps linked to the same E but not linked to each other. A structural hole indicates that the Ps on either side of the hole have access to different flows of information. Networks rich in structural holes imply access to mutually unconnected partners and, consequently, to many distinct information flows. Thus, maximizing the structural holes spanned or minimising redundancy between partners is an important aspect of constructing an efficient, information-rich network (Burt, 1992).
- The *inter-subgroup cohesiveness* in a network points to the interactions across subgroups (e.g., projects).
- *Small world*' properties of networks: Small world is a network structure that is both highly locally clustered and has a short path length (short global separation and high local clustering). In a small world network, the clusters can be linked by actors who are members of multiple clusters, making it possible for even large networks that are made up of many separate clusters to be connected and cohesive. The higher the small world degree of a network the more separate clusters become more interlinked and linked by actors who know each other. These processes distribute creative information among subgroups and help to build a cohesive social organisation within subgroups that support risky collaboration around good ideas.
- *Strong tie network*, i.e., a network in which strong ties predominate.
- *Dual network*, i.e., a network architecture wherein a small core of strong ties is integrated with a larger periphery of weak ties.
- A *structural autonomy network*: Structural autonomy is putting brokerage (= valuable but risky trust) and closure (= making it safe to trust) together. A structurally autonomous group consists of actors strongly connected to one another, with extensive bridge relations beyond the group.

NOTES

1 This particularly applies to degree centrality but not necessarily to other centrality measures. For example an E with two direct links in a large network – and for that reason not attractive at all – can have a high betweenness centrality score.

2 EET stands for Ecology, Economy and Technology programme.

REFERENCES

Afuah, A. (2000) 'How Much Do Your Competitors' Capabilities Matter in the Face of Technological Change?', *Strategic Management Journal*, 21(4): 387-404.

Ahuja, G. (2000) 'Collaboration Networks, Structural Holes, and Innovation: A Longitudinal Study', *Administrative Science Quarterly*, 45(3): 425-55.

Baum, J.A.C., T. Calabrese and B.S. Silverman (2000) 'Don't Go it Alone: Alliance Network Composition and Startups' Performance in Canadian Biotechnology', *Strategic Management Journal*, 21(3): 267-94.

Brandes, U. and T. Erlebach (eds.) (2005) *Network Analysis: Methodological Foundations*, Berlin: Springer.

Burt, R.S. (1980) 'Models of Network Structure', *Annual Review of Sociology*, 6: 79-141.

Burt, R.S. (1987) 'Social Contagion and Innovation: Cohesion versus Structural Equivalence', *The American Journal of Sociology*, 92(6): 1287-1335.

Burt, R.S. (1992) *Structural Holes: The Social Structure of Competition*, Cambridge, MA.: Harvard University Press.

Capaldo, A. (2007) 'Network Structure and Innovation: The Leveraging of a Dual Network as a Distinctive Relational Capability', *Strategic Management Journal*, 28(6): 585-608.

Cohen, W.M. and S. Klepper (1996) 'Firm Size and the Nature of Innovation within Industries: The Case of Process and Product R&D', *The Review of Economics and Statistics*, 78(2): 232-243.

Ebadi, Y.M. and J.M. Utterback (1994) 'The Effects of Communication on Technological Innovation', *Management Science*, 30: 572-85.

European Community (2006) 'Annual Innovation Policy Trends and Appraisal Report'.

Faems, D., B. van Looy and K. Debackere (2005) 'Interorganizational Collaboration and Innovation: Toward a Portfolio Approach', *Journal of Product Innovation Management*, 22(3): 238-50.

Fey, C.F. and J. Birkinshaw (2005) 'External Sources of Knowledge, Governance Mode, and R&D Performance', *Journal of Management* 31(4): 597-621.

Flemming, L., C. King and A. Juda (2003) 'Small Worlds and Innovation', working paper, Harvard Business School.

Freeman, C. and L. Soete (1997) *The Economics of Industrial Innovation*, London: Pinter.

Gilsing, V. and B. Nooteboom (2005) 'Density and Strength of Ties in Innovation Networks: An Analysis of Multimedia and Biotechnology', *European Management Review*, 2: 179-97.

Goes, J.B. and S.H. Park (1997) 'Interorganizational Links and Innovation: The Case of Hospital Services', *Academy of Management Journal*, 40(3): 673-96.

Granovetter, M. (1973) 'The Strength of Weak Ties', *American Journal of Sociology*, 78: 1360-80.

Gulati, R. (1999) 'Network Location and Learning: The Influence of Network Resources and Firm Capabilities on Alliance Formation', *Strategic Management Journal*, 20(5): 397-420.

Hagedoorn, J. and J. Schakenraad (1994) 'The Effect of Strategic Technology Partnering on Company Performance', *Strategic Management Journal*, 15(4): 291-309.

Hippel, E. von (1976) 'The Dominant Role of Users in the Scientific Instrument Innovation', *Research Policy*, 5(3): 212-39.

Hippel, E. von (1986) 'Lead Users: A Source of Novel Product Concepts', *Management Science*, 32: 791-805.

Ireland, R.D., M.A. Hitt and D. Vaidyanath (2002) 'Alliance Management as a Source of Competitive Advantage', *Journal of Management*, 28(3): 413-46.

Johnson, J.C. (1986) 'Social Networks and Innovation Adoption: A Look at Burt's Use of Structural Equivalence', *Social Networks*, 8: 343-364.

Johnson, B. (1992) 'Institutional Learning', in: B.A. Lundvall (ed.) *National Systems of Innovation. Towards a Theory of Innovation and Interactive Learning*, London: Pinter Publishers, 23-44.

Kale, P., J.H. Dyer and H. Singh (2002) 'Alliance Capability, Stock Market Response, and Long-term Alliance Success: The Role of the Alliance Function', *Strategic Management Journal*, 23(8): 747-67.

Lundvall, B.A. (1992) *National Systems of Innovation: Towards a Theory of Innovation and Interactive Learning*, London: Pinter.

McEvily, B. and A. Zaheer (1999) 'Bridging Ties: A Source of Firm Heterogeneity in Competitive Capabilities', *Strategic Management Journal*, 20(12): 1133-56.

Meeus, M.T.H. (2003) 'Innovation and Organization: On the Explanation of (In-)compatible Entities', inaugural address, Utrecht University.

Meeus, M.T.H. and J. Faber (2006) 'Interorganizational Relations and Innovation: A Review and a Theoretical Extension', in: J. Hage and M.T.H. Meeus (eds.) *Innovation, Science, and Institutional Change: A Research Handbook*, New York: Oxford University Press, 67-87.

Midgley, D.F., P.D. Morrison and J.H. Roberts (1992) 'The Effect of Network Structure in Industrial Diffusion-processes', *Research Policy*, 21(6): 533-52.

Nelson, R. (1982) 'The Role of Knowledge in R&D Efficiency', *Quarterly Journal of Economics*, 97: 453-70.

Mowery, D.C., J.E. Oxley and B.S. Silverman (1996) 'Strategic Alliances and Interfirm Knowledge Transfer', *Strategic Management Journal*, 17: 77-91.

Oerlemans, L.A.G., M.T.H. Meeus and F.M.W. Boekema (1998) 'Do Networks Matter for Innovation? The Usefulness of the Economic Network Approach in Analysing Innovation', *Journal of Economic and Social Geography*, 89(3): 298-309.

Oerlemans, L.A.G., T. Goessling and R.J.G. Jansen (2007) 'Tie Failure: A Literature Review', in: T. Goessling, L.A.G. Oerlemans and R.J.G. Jansen (eds.) *Inside Networks*, Cheltenham: Edward Elgar, 195-221.

Owen-Smith, J. and W.W. Powell (2004) 'Knowledge Networks as Channels and Conduits: The Effects of Spillovers in the Boston Biotechnology Community', *Organization Science*, 15(1): 5-21.

Park, S.H. and M.V. Russo (1996) 'When Competition Eclipses Cooperation: An Event History Analysis of Joint Venture Failure', *Management Science*, 42(6): 875-90.

Pavitt, K. (1984) 'Sectoral Patterns of Technical Change', *Research Policy*, 13(6): 343-73.

Powell, W.W., K.W. Koput and L. Smith-Doerr (1996) 'Interorganizational Collaboration and the Locus of Innovation: Networks of Learning in Biotechnology', *Administrative Science Quarterly*, 41(1): 116-45.

Powell, W.W., K.W. Koput, L. Smith-Doerr and J. Owen-Smith (1999) 'Network Position and Firm Performance: Organizational Returns to Collaboration in the Biotechnology Industry', *American Journal of Sociology*, 110: 1132-1205.

Powell, W.W., D.R. White, K.W. Koput and J. Owen-Smith (2005) 'Network Dynamics and

Field Evolution: The Growth of Interorganizational Collaboration in the Life Sciences', *American Journal of Sociology*, 110: 1132-1205.
Provan, K.G., A. Fish, J. Sydow (2007) 'Interorganizational Networks at the Network Level: A Review of the Empirical Literature on Whole Networks', *Journal of Management*, 33(3): 479-516.
Rindfleisch, A. and C. Moorman (2001) 'The Acquisition and Utilization of Information in New Product Alliances: A Strength-of-ties Perspective', *Journal of Marketing*, 65(2): 1-18.
Rogers, M. (2004) 'Networks, Firm Size and Innovation', *Small Business Economics*, 22(2): 141-153.
Rothaermel, F.T. and T.L. Deeds (2004) 'Exploration and Exploitation Alliances in Biotechnology: A System of New Product Development', *Strategic Management Journal*, 25(3): 201-221.
Ruef, M. (2002) 'Strong Ties, Weak Ties and Islands: Structural and Cultural Predictors of Organizational Innovation', *Industrial and Corporate Change* 11(3): 427-449.
Salman, N. and A.L. Saives (2005) 'Indirect Networks: An Intangible Resource for Biotechnology Innovation', *R&D Management*, 35(2): 203-15.
Shan, W., G. Walker and B. Kogut (1994) 'Interfirm Cooperation and Startup Innovation in the Biotechnology Industry', *Strategic Management Journal*, 15(5): 387-94.
Stuart, T.E. (2000) 'Interorganizational Alliances and the Performance of Firms: A Study of Growth and Innovation Rates in a High-technology Industry', *Strategic Management Journal*, 21(8): 791-811.

Teubal, M. (1976) 'On User Needs and Need Determination: Aspects of the Theory of Technological Innovation', *Research Policy*, 5: 266-89.
Teubal, M. (1976a) 'Performance in Innovation in the Israeli Electronics Industry: A Case Study of Biomedical Electronics Instrumentation', *Research Policy*, 5(4): 354-79.
Uzzi, B. (1997) 'Social Structure and Competition in Interfirm Networks: The Paradox of Embeddedness', *Administrative Science Quarterly*, 42(1): 35-67.
Uzzi, B. and J. Spiro (2005) 'Collaboration and Creativity: Big Differences from Small World Networks', *American Journal of Sociology*, 111: 447-504.
Valk, T. van der and J. Faber (2007) 'Development of Structure in the Emerging Network of Collaborative Relations of Dutch Dedicated Life Sciences Firms from 2002 to 2004'. In: Valk, Tessa van der (2007) 'Technology dynamics, network dynamics and partnering. The Case of Dutch dedicated life sciences firms, p. 55-70. Netherlands Geographical Studies/ Copernicus Institute for Sustainable Development and Innovation, Utrecht.
Valk, T. van der (2007) 'Technologfy Dynamics, Network Dynamics and Partnering: The Case of Dutch Dedicated Life Sciences Firms', Netherlands Geographical Studies/ Copernicus Institute for Sustainable Development and Innovation, Utrecht.
Ven, A. van de, D.E. Polley, R. Garud and S. Venkataraman (1999) *The Innovation Journey*, New York: Oxford University Press.
Wasserman, S. and K. Faust (1997) *Social Network Analysis: Methods and Applications*, Cambridge: Cambridge University Press.
Zahra, S.A. and A.P. Nielsen (2002) 'Sources of Capabilities, Integration and Technology Commercialization', *Strategic Management Journal*, 23(5): 377-98.
Zollo, M., J.J. Reuer and H. Singh (2003) 'Interorganizational Routines and Performance in Strategic Alliances', *Organization Science*, 13(6): 701-13.

11 REGIONAL INNOVATION POLICY

Ron Boschma

11.1 INTRODUCTION

Why some regions grow more than other regions is a key question in economic geography. Up until the late 1980s, neo-classical economic approaches argued that technology is a key determinant of regional growth. However, these approaches treated technology as an exogenous factor, leaving the geography of innovation unexplained. Inspired by Schumpeter, economic geographers took the lead in criticising this view. Since the early 1980s, they have focused attention on the explanation of the geography of innovation: some regions are more capable of developing and implementing innovations, and region-specific characteristics (including institutions) may be underlying forces. This led to the claim that regions are drivers of innovation and growth. During the last decades, new concepts like industrial districts (Becattini 1987), clusters (Porter 1990), innovative milieux (Camagni 1991), regional innovation systems (Cooke 2001), and learning regions (Asheim 1996) have been launched to incorporate this view.

Many of these concepts have drawn inspiration from evolutionary economics (Nelson and Winter 1982; Dosi et al. 1988; Boschma et al. 2002). In a nutshell, an evolutionary approach argues that "the explanation to why something exists intimately rests on how it became what it is" (Dosi 1997: 1531). The objective of this chapter is to outline how evolutionary economics may provide inputs for regional innovation policy. This is not an easy task, since distinctive strands of thought in evolutionary economics hold opposing views on policy. For example, the neo-Schumpeterian approach (associated with Nelson and Winter, among others) advocates an active role for policy makers, while the Austrian approach (such as Hayek) does not (Wegner and Pelikan 2003). Complexity thinking in evolutionary economics takes a policy view that is again very different. Notwithstanding these different views, we will outline some policy recommendations that incorporate recent thinking in evolutionary economic geography (Boschma and Lambooy 1999).

This chapter is structured as follows. A brief and selective literature review is given in section 2, providing a theoretical and empirical background for the remaining part, which addresses policy implications. In section 3, we claim that system failures should be taken as the point of departure to underpin regional innovation policy. In section 4, we discuss how history should be taken seriously in regional innovation policy. What is essential to recognise is that history determines not only the policy options that are at hand in regions, but also the probable outcomes of regional innovation policy. Building on these insights, in section 5 we sketch some policy options that may direct regional economies into new directions while building on related variety. In particular, we direct attention to various mechanisms through which knowledge transfer may be encouraged at the regional level. Section 6 draws the main conclusions.

11.2 VARIETY AND REGIONAL DEVELOPMENT

Our starting point is a fundamental departure from how conventional neo-classical economics treats knowledge. Knowledge is not a public good that is characterised by diminishing returns to scale. On the contrary, knowledge evolves: it is not reduced when it is used, but it accumulates through processes of learning-by-doing (Arrow 1962). This cumulative and irreversible nature of knowledge development is embodied in individuals (skills) and in firms (competences): they develop different cognitive capacities over time. Due to its tacit and cumulative nature, knowledge is actor-specific and difficult, if not impossible, to copy or imitate by other actors. As a result, variety in an economy is the rule, and knowledge accumulation at the level of individuals and firms is its prime mover.

Variety in economic space

Consequently, an economy consists of numerous pieces of knowledge that are formed as time goes by. To start with, an evolutionary approach to economic geography focuses on the question of how these pieces of knowledge are spatially distributed over time. If one observes the world, it is undeniable that knowledge, knowledge creation, and innovation (i.e., the economic exploitation of new knowledge, as embodied in new products, machines, and organisation techniques) are unevenly distributed over space. This is shown in multiple ways.

To start with, one can observe a high degree of variety in the most urbanised regions. Following Adam Smith, a huge market size enables firms to specialise in activities they can do best, enhancing their productivity levels. As a result, the economies of urbanised regions are characterised by a sharp division of labour between specialised firms, which sustains urban growth (Pred 1966). Another reason has been proposed by Jacobs (1969) who claimed that diversified urban economies trigger new ideas and innovations. Co-location of many different individuals, firms, and sectors enhances knowledge exchange and the recombination of different pieces of knowledge in novel ways, generating even more variety in major cities.

In addition to such intra-regional variety, knowledge creation tends to concentrate in space, leading to interregional variety of knowledge. Research and development is extremely spatially concentrated, favouring only a small set of regions in the world, and empirical studies show this pattern is quite stable over time (Feldman and Audretsch 1999). Studies have found strong relationships between regional stocks of knowledge (as embodied in university research and private R&D) and performance indicators, such as patent intensity and productivity levels (Anselin, Varga and Acs 2000). However, it is not necessarily the case that places of knowledge creation and places of innovation overlap. When there is little overlap, one speaks of a knowledge paradox. The European paradox is a prime example: while Europe excels in (basic) research (i.e., R&D levels and patenting activity in Europe are quite high), Europe is incapable of exploiting this knowledge economically, turning it into innovations. In fact, the geography of knowledge and innovation may be characterised by a strong spatial division of labour, with some places specialising in knowledge creation, other places turning this new knowledge into innovations (such as new products), and again other places focusing on the manufacturing of the new products.

There are countless examples of regions and countries that specialise in a particular knowledge field, and which continue to do so for a long time. Industries often tend to concentrate in space, as shown by the film industry in Hollywood, the financial sector in the city of London, the American car industry in Detroit, and the cut flower sector in the Dutch Westland region.[1] However, spatial variety also occurs within one industry. We already mentioned the fact that the geography of knowledge creation (R&D) does not necessarily overlap with the geography of production of new products within the same industry. For instance, R&D in the Dutch electronics industry is heavily concentrated in the Eindhoven region, while the outputs of the R&D (i.e., new electronic products) are produced elsewhere. In addition, firms operating in the same industry may look very different in different places (Essletzbichler and Rigby 2005). For example, the French, American, and Indian film industries are very different in terms of organisation, state involvement, and market focus (Lorenzen 2007).

Spatial variety and geographical proximity

So, variety in space is paramount. These observations lend support to the fact that knowledge tends to accumulate at the regional level. We stated previously that knowledge tends to accumulate in individuals and firms. So why in regions? The main reason is that knowledge will spill over to other firms now and then, despite the fact that knowledge is actor-specific and difficult to copy. Knowledge is a non-rival good: its use by one firm does not preclude its use by other firms. This means that not only the firm itself, but other firms may benefit from the accumulation of knowledge and human capital. This may result in increasing returns to scale that is external to the firm (Shaw 1992). Empirical studies show that knowledge spillover effects are often geographically localised. That is, they spill over to neighbouring regions at the most, and spillover effects become weaker the higher the distance from the source of knowledge (Audretsch and Feldman 1996).

This suggests that geographical proximity is a prerequisite for knowledge diffusion and innovation. However, there is reason to believe that this position should be reconsidered. In fact, it could be argued that geographical proximity is neither a necessary nor a sufficient condition for interactive learning and innovation (Boschma 2005a). This happens only when other barriers of knowledge diffusion are overcome, such as cognitive, social, and institutional distance. These other forms of proximity need to be secured between actors in order to make them connected, and to enable effective knowledge transfer. Other forms of proximity may act as a substitute for geographical proximity, because they can help to provide the necessary trust to exchange knowledge without the need for geographical proximity. For instance, social proximity may provide a vehicle to connect agents and enable flows of knowledge over large distances, because these agents share a past as former schoolmates or as former colleagues working for the same organisation (Agarwal et al. 2006). However, having said that, effective knowledge transfer may still often be geographically localised because geographical proximity indirectly impacts on the establishment of the other forms of proximity. In fact, geographical proximity may encourage the creation of trust-based relationships or other institutions that facilitate effective knowledge transfer between local agents (Maskell 2001).

The cognitive dimension has attracted most attention in this respect. There is ample evidence that local access to information (e.g., through the provision of ICT infrastructure) is not sufficient. Due to the tacit nature of knowledge, firms can only understand, absorb, and implement external knowledge that is close to their own knowledge base (Cohen and Levinthal 1990). Effective transfer of knowledge requires absorptive capacity of firms and cognitive proximity, that is, firms need to share similar knowledge and expertise to enable effective communication (Nooteboom 2000). In combination with geographical proximity, the need for cognitive proximity may well explain the spatial concentration of tacit knowledge.[2] Once a region specialises in a particular knowledge and competence base, this will act as an incentive, offering opportunities to local firms for further improvements in familiar fields of knowledge on the one hand, and as a selection mechanism, discouraging knowledge creation that does not fit into the regional knowledge base on the other hand (Boschma 2004). As a result, the regional accumulation of tacit knowledge provides an intangible asset for local firms that is hard to grasp for non-local firms, because spatial distance forms an insurmountable barrier to the transfer of tacit knowledge (Gertler 2003).

Empirical studies demonstrate that the key mechanisms through which knowledge between organisations is transferred encourages knowledge accumulation at the regional level. One such mechanism is the spinoff process, through which knowledge diffuses effectively between firms. A spinoff firm is a firm that has been established by a founder that was a former employee of an incumbent firm in the same or a related sector. Crucial is that these new entrants do not start from scratch: these new entrepreneurs have acquired relevant knowledge and skills in a incumbent firm which they can exploit further in their new company. Empirical studies systematically show that this type of entrant performs best, that is, they demonstrate the highest survival rates. Quite a number of sectors are characterised by a high degree of spinoff dynamics during their years of formation, and the most successful firms in those emerging industries tend to be spinoff companies (Klepper 2002). Because most spinoffs locate in the immediate vicinity of their parent organisation, this knowledge transfer mechanism contributes to geographically localised knowledge formation.
Another mechanism through which knowledge diffuses is labour mobility. Since labour is the main carrier of knowledge, employees moving from one firm to the

other will contribute to the exchange of knowledge. Since labour mobility takes place largely at the local level, this implies this type of knowledge transfer contributes to knowledge formation at the regional level. This is especially true for labour markets that have similar or related economic activities: clusters are characterised by local labour mobility that is higher than elsewhere in an economy (Lindgren and Eriksson 2007). In addition, labour mobility creates linkages between firms through social ties between former colleagues. These social relationships facilitate knowledge flows between organisations (Breschi and Lissoni 2003). Since most of the job moves are intra-regional, these social networks are formed locally, enhancing further knowledge accumulation at the regional level.

Thus, a network is a mechanism of knowledge transfer that favors localised learning between firms. Knowledge effectively circulates in networks, as happens in technological alliances and epistemological communities to an increasing extent (see chapters 7 and 10). Basically, networks are a-spatial constructs. Social connectedness is often considered crucial to explain network configurations, as the formation of inventors' networks on the basis of co-patenting shows (Breschi and Lissoni 2003). But because social proximity is enhanced by geographical proximity, networks are often geographically localised, and so is the process of knowledge creation and diffusion.

Networks may be especially beneficial for activities of exploitation, but they may be less suited to exploration (Nooteboom et al. 2007). As chapter 10 showed, network ties may become too close and inward looking, leading to a reduced awareness of developments outside the network (Uzzi 1997). Firms that are involved in embedded relationships may feel morally obligated to stay loyal to their partners, and thus end up choosing less efficient ways of production. If firms do not connect to new groups of firms now and then, and if their own network is not accessible for new partners, it is difficult to break this situation of cognitive lock-in . This over-embeddedness argument has a geographical connotation. Grabher (1993) argues that firms that focus too much on local relationships become less aware of technological and market-related developments outside their region. The establishment of non-local relationships are considered crucial, because they bring new variety into the region (Asheim and Isaksen 2002; Bathelt et al. 2004). However, non-local relations as such do not guarantee effective knowledge transfer either: one needs a certain level of social and cognitive proximity to make effective connections over large distances. So, firms with relationships that are too tight or focus too much on their own region may find it harder to adapt to external changes (Boschma 2005a).

Regional dynamics

There is, however, more in capitalist economies than just knowledge accumulation taking place at the regional level, and consolidating spatial variety. An evolutionary approach to economic geography focuses on the dynamics of urban and spatial systems in the long run. It does so in terms of what Schumpeter (1942) described as creative destruction. What drives a regional economy is the introduction of new variety (as embodied in for example new products, new firms, and new sectors) in the economic system through entrepreneurial activity, because it must offset the decline in other parts of the economy (Saviotti 1996). So, regional growth is about qualitative change, not quantitative change.

This process of creative destruction does not keep the spatial system in balance. On the contrary, regional dynamics is the rule. New basic variety challenges the core-periphery structure of the spatial system (Boschma and Lambooy 1999). New industries require new knowledge, new types of skills, new institutions, among other things, and the existing spatial structure cannot provide these. This makes it unpredictable where new industries emerge in space, although this may differ from industry to industry. Newly emerging sectors do not necessarily favour leading regions, and they provide opportunities to backward regions to a considerable degree. Economic history bears witness to dramatic changes in the spatial system both at the international and national level (Hall and Preston 1988). In the last two centuries, techno-industrial leadership has shifted from Great Britain to the United States and Germany, and some countries in South-East Asia have recently joined the ranks of leading industrial countries. Countries are subject to similar dynamics: in Great Britain, Belgium, and Germany, the leading industrial regions of the nineteenth century have almost been overrun by a set of newly emerging regions in the south east of England, Flanders, and the south of Germany.

As a consequence, the long-term development of regions depends on their ability to create new variety through entrepreneurship and innovation, in order to compensate for the loss of variety through exits and relocations in other parts of their economy. In other words, it is essential for regions to transform and renew their economic base (Pasinetti 1981; Saviotti and Pyka 2004). One reason for this is that (tacit) knowledge not only accumulates in regions, but it may also become standardised (i.e., explicit and codified) in the long run. Since this codification process encourages knowledge diffusion between regions, the regional knowledge base may lose its unique value to local firms (Maskell and Malmberg 1999).

Related variety as source for regional innovation

Regions may have several options to restructure their economies (Martin and Sunley 2006). A key option is to diversify into new fields while building on existing regional assets. There is increasing awareness that the long-term development of regions depends on their ability to diversify into new economic applications and new sectors while building on their current knowledge base. It means that regional economies that branch into new directions may be more stable in the long run than those that start from scratch.

One way to establish this is to develop major innovations that are triggered by knowledge spillovers between different sectors in a region (Henderson et al. 1995). However, knowledge will only be exchanged effectively when the cognitive distance between sectors is not too large. In other words, sectors need to be related or complementary in terms of competences to enable effective knowledge transfer. To be more precise, some degree of cognitive proximity is required to ensure that effective communication and interactive learning take place, but not too much cognitive proximity, to avoid cognitive lock-in (Nooteboom 2000). It is neither regional diversity (which involves too large cognitive distance) nor regional specialisation per se (resulting in too much cognitive proximity), but regional specialisation in related variety that enhances real innovations. The idea of innovations based on related variety comes close to the Schumpeterian definition of innovation as the recombination of pieces of knowledge in entirely new ways (Levinthal 1998). So, major innovations are more likely to occur when knowledge spills over between sectors, rather than within one sector, but only as long as the sectors are related.

Related variety is linked to the concept of technology system that accounts for strong technological interdependencies across industries (Carlsson and Stankiewicz 1991). Economic history has repeatedly given evidence of a high degree of exchange and feedback of technology across a particular set of industries during a particular period (Boschma 1999). For instance, sectors may be technically connected because they originate from a common technology. The discovery of the technological principles behind synthetic dyestuffs in the nineteenth century is an example: this laid the foundations of a range of new chemical sectors, such as synthetic colours, pharmaceutics, explosives, photography, plastics, and synthetic fibres. Major innovations also depend on complementary advances in technology in other industries before they can be fully exploited (Rosenberg 1982). These examples give insights into how related variety enhances knowledge spillovers and sparks of radical innovations, how new growth sectors come into being, and how regional economies branch in new directions.

Another example of how related variety contributes to economic renewal is the post-war experience of the Emilia Romagna region in Italy. For many decades, Emilia Romagna has been endowed with a diffuse knowledge base in engineering. After the Second World War, many new sectors, such as the packaging industry, ceramic tiles, and robotics, emerged out of this pervasive and generic knowledge base one after the other. These new economic applications made the regional economy diversify into new directions. These new sectors not only built and expanded on this extensive regional knowledge base, they also renewed and extended it, further broadening the regional economy of Emilia Romagna.

The economic significance of related variety is also shown through the emergence of new sectors that grow out of old sectors, such as the television industry, which branched out of the radio sector (Klepper and Simons 2000). Because this branching process concerns old sectors giving birth to new sectors, it increases the probability of survival of the new industry. Klepper (2002) demonstrated empirically that prior experience in related industries (like coach and cycle making) increased the life chances of new entrants in the new US automobile sector. This confirms the observation made earlier that the spin-off process is a powerful mechanism that effectively transfers knowledge from one firm to the other. Boschma and Wenting (2007) found evidence that new automobile firms in the UK had a higher survival rate during the first stage of the life cycle of the new industry when the entrepreneur had a background in these related sectors, and when the firm had been founded in a region that was well endowed with these related sectors. So, when diversifying into the new automobile sector, these types of entrants could exploit and benefit from related competences and skills, which improved their life chances significantly, as compared to start-ups lacking those related competences/skills.

Frenken et al. (2007) have more quantitatively assessed the impact of related variety on regional growth in the Netherlands. Making use of the standard sectoral classification, sectors at the five-digit level were defined as related when they shared the same category at a lower level. An outcome was that regions with a high degree of related variety showed the highest employment growth rates in the Netherlands in the period 1996-2002, suggesting the importance of knowledge spillovers across related sectors at the regional level. At the same time, a broad range of unrelated sectors in a region may also be beneficial for regional growth, because unrelated variety spreads risks. When a sector-specific shock occurs, it is unlikely to harm other industries and disturb the regional economy when sectors are unrelated. So, unrelated variety may stabilise regional economies (Essletzbichler 2005).

Before, we pointed out that non-local relationships may be crucial because new variety may be brought into the region through linkages with other regions (Boschma 2004). However, a study on regional growth in Italy, making use of trade data, demonstrated that the inflow of variety of knowledge per se does not affect economic growth in regions: it is not sufficient to attract large flows of extra-regional knowledge (Boschma and Iammarino 2007). The same is true when the extra-regional knowledge is similar to the knowledge base of the region: there is not much to be learned from inflow of knowledge that the region is already familiar with: it does not add to the existing knowledge base of the region, and therefore, does not lead to real innovations and regional growth. However, a crucial finding of this study was that the more related the knowledge base of the region and its import profile was, the more it contributed to growth in the region. This finding suggests that related variety in extra-regional connections ensures that external knowledge sparks of learning and innovation in situ. Thus, a region benefits especially from extra-regional knowledge when it originates from sectors that are related or close, but not quite similar to the sectors present in the region. In those circumstances, cognitive proximity between the extra-regional knowledge and the knowledge base of the region is not too small (avoiding the learning process of being more of the same), but also not too large (enabling the absorption of the extra-regional knowledge).

The need for dynamic innovation systems

However, inter-firm knowledge transfers based on proximity and related variety alone will not lead to innovations. Since the 1990s, the innovation system literature claims that the innovation process should be seen as the outcome of interaction between actors within firms, between firms, and between firms and other organisations such as universities, educational facilities, financing organisations, and government agencies (Freeman 1987). So, being innovative is not just a matter of having access to related variety or to local or non-local knowledge, but whether interaction takes place at all these levels.

According to this literature, a number of organisations (such as research institutes, educational facilities, and financial organisations) provides complementary inputs essential to the innovation process (Edquist 1997). In addition to absorptive capacity, a firm can exploit its innovation only when it is for example able to get access to (venture) capital, when it is able to hire workers with the required new skills, or when it can find a new market. In other words, firms need the presence of a critical mass of organisations that can provide these needs. In many peripheral regions, this critical mass is missing, resulting in low innovative performance.

Besides a critical mass, it is crucial that these organisations are connected and form a system. The innovation process requires organisations to connect in order to enable flows of knowledge, capital, and labour. The key issue is that this is far from self-evident in practice, even quite exceptional (Boschma 2004). Capital suppliers are almost by definition reluctant to invest in innovative projects: (radical and more complex) innovations are a risky business with uncertain outcomes, and financial organisations have built up routines in established markets and technologies. Although the number of inter-firm technology alliances is on the increase (Nooteboom et al. 2007), firms tend to be reluctant to share their core competences with others, because there is a serious risk that knowledge will leak to competitors. Public research institutes such as universities often have difficulties in meeting the demands of innovative firms, because of differences in culture and incentive mechanisms (Metcalfe 1994).[3] And when innovations require labour with new skills, it may take a long time before the educational system is restructured and new appropriate courses are offered.

In addition to the fact that it is not self-evident that interactions occur between organisations, it is also unlikely that organisations are sufficiently flexible to implement innovations. In reality, almost by nature, organisations are not flexible and responsive, due to routines and path dependency (Nelson and Winter 1982). Thus, regions will reap the benefits from entrepreneurial activities of firms only when the actions of these key organisations are coordinated and form a system of innovation, and when regions have local organisations that respond quickly and smoothly to new developments. This is crucial for the long-term competitiveness of regions: some regions are more capable of making these connections and have more responsive organisations than other regions. This is a key systemic asset of a region that is almost impossible to copy by other regions.

Institutions play a crucial role in this respect (Nelson 1995). Apart from basic institutions like democracy and markets that support entrepreneurship and innovation, institutions also regulate and coordinate actions between organisations (Hodgson 1996). This task is fulfilled by formal institutions (such as laws) and informal institutions (like norms and habits) (Edquist and Johnson 1997). An example of a formal institution is a patent law system that protects inventors for some time while making information public. An example of an informal institution is a culture of shared trust, which is a local capability that supports inter-firm learning (Maskell 1999). Countries and regions accumulate different institutions over time, which is quite similar to the way the regional knowledge base accumulates. They are the outcome of a long history in a specific regional context that cannot easily be copied by public policy in other regions. Like the innovation process itself, institutions have a systemic dimension: they form systems that are territory-specific (Hall and Soskice 2001). At the international level, there are 'exit-based' and 'voice-based' institutional models (Ergas 1984), at the regional level, there are many more. Consequently, regions follow different institutional paths that yield comparable levels of economic development: there is more than one way regions can accomplish economic development.

Because institutions tend to be durable and resistant to change, they not only support but may also constrain new developments. When new institutions are formed and created alongside new economic activities, they fulfill a specific need (Murmann 2003), but once they are established, they may obstruct new developments. Powerful special-interest organisations may take over an economy, slowing down the capacity of regions to adopt new technologies and to reallocate resources to new activities (Olson 1982). What matters thus is whether institutions are flexible and responsive to change, in order to avoid regional lock-in (Freeman and Perez 1988): regions need a capacity to upgrade and transform institutions required for the development of new activities. This dynamic capability of organisations and institutions impacts on the long-term competitiveness of regions (Boschma 2004).

Conclusion

In a knowledge economy, regions depend on their ability to develop and apply new knowledge in their economies. Since knowledge tends to accumulate, new knowledge will not diffuse widely between firms and between regions. It requires absorptive capacity and institutions that bring agents together. Both of these intangible assets provide incentives and constraints within which the innovation process takes place. If region-specific, interregional variety may be a persistent feature of economies. Knowledge will spill over more intensively when regions are endowed with related industries that share a common knowledge base. Due to the systemic nature of innovation processes, regions also require a critical mass of organisations that meet the following conditions: (1) they have to be well connected, enabling flows of knowledge, capital, and labour; (2) however these ties should not be too strong, and not too focused on the region, avoiding problems of lock-in; and (3) local organisations and institutions must be flexible and responsive to new circumstances, overcoming inertial tendencies due to habits, routines, and path dependency (Boschma and Lambooy 1999).

11.3 SYSTEM FAILURES AS BASIS FOR REGIONAL INNOVATION POLICY

In this section, we sketch out some policy implications. While it is common to refer to market imperfections as basic underpinnings for public intervention, we will claim that system failures should be taken as the basic rationale for regional innovation policy.

A conventional market failure argument is that knowledge is a semi-public good. Because of the problem of non-appropriability, government should take action to overcome underinvestment in new knowledge through the provision of R&D subsidies or the establishment of property rights. Another standard argument is that knowledge is characterised by increasing returns to scale. For that reason, investments in public R&D, technology transfer and education are expected to foster economic growth (Hall 1994). As such, the government aims to encourage the dissemination of knowledge, through the public provision of infrastructure such as broadband Internet. This is especially relevant for lagging regions and small- and medium-sized firms that lack resources to invest in R&D.

By and large, the market failure argument suffers from two shortcomings. First of all, evolutionary economists argue that market imperfections are not necessarily a problem that needs to be corrected by public intervention. For instance, due to cognitive constraints, knowledge can be excluded from other firms to a greater or lesser extent. Market imperfections, such as knowledge spillovers, knowledge asymmetries, and monopolies can even be considered the real drivers of innovation and economic growth (Bryant 2001). Because knowledge asymmetries limit knowledge transfer, they provide a strong incentive to invest in knowledge creation. What is more, variety acts as a major source for exploration and economic renewal (see chapter 3). Therefore, variety is a key regional asset that needs to be cherished.[4] Consequently, public intervention aimed at tackling market imperfections could damage rather than benefit an economy. Secondly, the market failure argument is based too much on a linear model of innovation policy that focuses on R&D infrastructure and technology transfer, as if these automatically lead to innovation in regions. For instance, science and technology policy in the European Union is focused on enhancing R&D, and there is a strong belief that R&D policy will bring benefits to many regions. In reality, R&D-based policy favours only a few regions in Europe, that is, the ones that are already specialised in R&D (Morgan 1997; Simmie 2003). In addition, much of the newly created knowledge is not exploited economically in Europe but leaks away to countries like the US. This means European R&D policy is subsidising the exploitation of knowledge elsewhere. In other words, this linear model of innovation policy based on market failure will lead to poor results, if such policy does not account for the systemic nature of innovation and the importance of absorptive capacity and institutions for knowledge diffusion.

Therefore, we claim that system failures, rather than market failures, should be the starting point for policy intervention (Metcalfe 2003; Asheim et al. 2006). This is not to deny the relevance of market failures for underpinning regional innovation policy in some cases. On the contrary, poor access to information, for instance, should be tackled by policy intervention, but it requires additional policy actions to be effective. So, at best, policy based on market failures needs additional actions to be effective, at worst, it may seriously damage the driving forces of innovation in an economy.

There are three types of system failures that may result in poor innovative performance of firms in regions (Edquist 1997; Bryant 2001; Todtling and Trippl 2005). The first type is more quantitative in nature, instead of relational: there may be crucial parts of the innovation system that are underdeveloped. This so-called 'organisational thinness' refers to the fact that key organisations in innovation systems, such as research institutes, educational facilities, venture capitalists, and specialised suppliers, but also key regulations, are weakly developed. Such a situation of 'organisational thinness' is often found in peripheral regions, due to a lack of critical mass of local demand (Camagni 1995).

The second type of system failure is a purely relational one, of a more qualitative nature. As mentioned in section 2, relationships between organisations in innovation systems are not self-evident, but have to be constructed. As explained in chapter 7, inter-organisational collaboration, for instance, is often risky and frequently fails. When missing or badly managed, knowledge will not be exchanged, inter-firm learning will come to a halt, and investment opportunities will not be realized due to shortages of capital and skilled labour.

The third type of system failure is associated with processes of lock-in. A lack of flexibility in organisations and their relationships may lead to inertia, which undermine the ability of regions to adapt and to renew their economic base. As noticed before, local organisations may be too strongly oriented towards old routines and old specialisations, as the experience of mature industrial regions illustrates (Grabher 1993). Moreover, local organisations may have developed too strongly tied networks, which limit their access to new sources of information, and which makes it difficult to implement changes. It is crucial to underline that public organisations may be part of such a regional deadlock: public agents may contribute to the formation of closed and inward-looking systems through their policy programs and their direct participation in such networks (Hassink 2005).

11.4 HOW TO DESIGN REGIONAL INNOVATION POLICY?

When building regional innovation policy on system failures, a number of issues call out for clarification. Should policy select and target particular sectors and regions? Should one adopt a 'one-size-fits-all' policy approach? Can policymakers make regional economies develop in new directions, and if so, how? And to what extent should innovation policy be regionally-based? These issues are addressed below.

Targeting sectors and regions?

There is often a tendency in policy to select particular sectors and regions *a priori* as targets at the national level. Policymakers are inclined to support relatively new sectors such as biotech, nanotechnology, or gaming, because these sectors are expected to create jobs in the near future. In a similar way, some regions are identified as innovation hotspots or 'brain ports', because these are considered the drivers of national economic growth in the near future. However, one can question the usefulness and relevance of such a 'picking-the-winner' policy.

First of all, such policy overlooks the fact that it is impossible to predict which will be the new growth regions and sectors of the future. A 'picking-the-winners' policy at the national level is risky, as history shows, because one runs the risk of selecting the wrong regions and sectors. There is little understanding of how regions move into new directions or start up new growth paths (Iammarino and McCann 2006; Martin and Sunley 2006). What has been observed is that new industries are often the result of spontaneous processes, rather than the outcome of orchestrated policy interventions (Lambooy and Boschma 2001; Pack and Saggi 2006). This is not to deny, however, that governments often play a key role, as in Silicon Valley, where huge defensive expenditures by the US government gave the region an enormous boost.

Secondly, 'picking-the-winner' policy often results in picking the same winners by many countries and regions. When all regions are targeting the same sectors (like biotech), it is meanwhile likely that most of these sectors will cluster in only a small number of regions in the world;one can predict that the overwhelming majority of regions will fail to develop these industries, with huge losses of public resources (Boschma 2005b). An exception might be the public support of general purpose technologies (like the Internet): there is no doubt these will have long-term impacts, but it remains uncertain which parts of the economy will be most strongly affected in the next decades, and how.

Thirdly, 'picking-the-winner' policy at the national level denies the fact that, in principle, almost every region has growth potential in the knowledge economy. Growth or innovation potentials of regions can be measured in different ways. Indicators like R&D, creative workers, high-tech industries, and knowledge-intensive services identify different dimensions of the knowledge economy. Each of these indicators will reveal a different spatial pattern, as an empirical study of the Netherlands shows (Raspe et al. 2004). For example, R&D is located more often in the more peripheral parts of the Netherlands, while creative workers are concentrated in the central, urbanized part of the country. If all the maps of each indicator would be put on top of each other, it would be almost impossible to identify regions that lack innovation potential. That is, most of the Dutch regions participate in the knowledge economy in one way or another. Therefore, it would be wrong to exclude many regions from policy intervention from the very start, because it would leave regional potential untouched and unexploited.

Consequently, one should be cautious of focusing innovation policy too narrowly on R&D. As explained before, R&D is only one indicator to measure innovation potential, and it is grounded in a traditional linear model of innovation that simply equates innovation with R&D. In addition, innovation policy based on R&D potential has strong geographical implications. Since R&D activities are concentrated in a small number of affluent regions, R&D-based policy will benefit these leading regions even more (Oughton et al. 2002).

Regional innovation policy based on related variety will avoid the dangers of 'picking-the-winner' policy, because its objective is to broaden and diversify the regional economic base while building on region-specific resources and extra-regional connections. No particular regions need to be targeted. Each region can be made part of such a policy approach, no matter whether these regions are specialised or diversified, or whether these have a high or low degree of related variety. Nor do specific sectors (low or high tech, creative or not) have to be excluded from such a policy approach. As it aims to bring together activities with possible complementary pieces of knowledge, such policy leaves behind a narrow sectoral perspective. Having said that, there is no doubt that regional policy based on related variety needs focus to be effective: it needs to identify and target region-specific assets and extra-regional linkages that have obtained some critical mass in a region. However, the objective is not to make strong sectors even stronger, but the objective is to enhance interaction and exchange between different activities, in order to support new variety in the region.

No 'one-size-fits-all' policy?

To say that almost each region has innovation potential is not to say that all regions are equal. On the contrary, there is a strong need to account for a variety of innovation potentials between regions, because regions differ in terms of location, human capital, knowledge base, and institutional structure. Italy is a prime example: the north of Italy is strong in science-based organisations with a high R&D intensity, the Third Italy is characterised by industrial districts which consist of small- and medium-sized organisations that have formal and loosely structured relationships, and the south of Italy is characterised by a weak indigenous learning capability and weak networks of organisations due to poor institutional arrangements (Iammarino 2005). Because of such spatial variety, it would be wrong to apply a 'one-size-fits-all' policy, such as copying neo-liberal policies or a best practice like Silicon Valley, which is often the case in regional policy development (Todtling and Trippl 2005).

It would also be wrong to create regional policies from scratch. Effective policy making requires localised action embedded in, and attuned to available resources in regions. To a large extent, it is the regional history that determines available options and probable outcomes of policy (Lambooy and Boschma 2001). It means one should take the knowledge and institutional base in a region as starting point when broadening the region's sector base by stimulating new fields of application that give birth to new sectors. Accordingly, there is a need for differentiated, tailor-made policy strategies that are geared towards specific potentials, and that will focus on tackling specific bottlenecks in regions. In sum, regional policy needs to capitalise on region-specific assets, extending and renewing the economic base, rather than selecting from a portfolio of specific policy models and recipes that owe their success to different environments (Asheim et al. 2006).

'One-size-fits-all' regional policy models do not work in a highly fragmented economy (Cooke and Morgan, 1998). This implies that copying of best practices, as identified by benchmarking studies, are bound to fail, as regional policies aimed at imitating success stories such as those of Silicon Valley have demonstrated (Boschma 2004). Howells (2005) points out that 'best practice policies' are hard to adapt to local situations and difficult to understand and implement. Moreover, copying of success stories in practice often focuses on the success factors, rather than on the basis of a sound analysis of how public policy contributed to the success of that particular region. This is not to say that regions cannot learn from each other (Hassink and Lagendijk 2001). There may be advantages to best practice policies: they have proven their success elsewhere, they are more or less ready to use, and they may break down closed local networks that serve vested interests (Howells 2005). In other words, there may be disadvantages attached to region-specific policies: these are often unique, so it is not clear whether they will work, and local vested interests may dominate the design of new policy, excluding outsiders and newcomers (Fritsch and Stephan 2005).[5] This implies that region-specific policies should be designed in such a way that these potential problems are tackled. This means, for instance, that it might be a good thing to encourage public support of academic spinoffs everywhere, but for this to work it requires different public strategies that are adapted to regional circumstances (Degroof and Roberts 2004).

Policymakers are adapters

The rejection of 'one-size-fits-all' policy is in line with the view that policymakers are not fully informed and omnipotent. Therefore, policymakers have few degrees of freedom (Lambooy and Boschma 2001). As stated before, the trajectories that regions followed in the past, as accumulated in a particular knowledge base and a set of institutions, determine to a large extent the available options and probable outcomes of policy. Regional policy is likely to fail when local strategies deviate from their local context. The more the policy objectives are embedded in the surrounding environment, the larger the potential impact of policy. When adaptation to change is constrained by the spatial system, policy based on related variety may increase the probability of policy success, because it builds on existing structures, while its objective is to broaden or diversify the regional economy in new directions.

Such a policy approach takes a more contextualised view of how policy should intervene in a regional economy. It implies that the degree and nature of policy intervention should be different in different regions because their histories differ. As a consequence, the question of whether governments should intervene in a regional economy should be based on the institutional history of a region and on the type of intervention that better fits a region's situation, rather than being based on theoretical or ideological accounts (Fromhold-Eisebith and Eisebith 2005). In addition, there is general awareness that the state is only one of the actors in a region, although a key player (Kohler-Koch 1998). When taking a systemic innovation policy approach, we claim that governments do not only directly intervene (e.g., through regulations, public research, and education), but also take a role as broker and intermediary to an increasing extent, bringing together actors at the regional level (Cooke and Morgan 1998).

We should account for the fact that policymakers, just as firms, do not optimise, but they adapt (Metcalfe 1994). They operate in a world of uncertainty. In these circumstances, policy failures will occur, just as it is a rule that firms will eventually fail in markets (Ormerod 2005). Because policymakers cannot rely on 'one-size-fits-all' policies, regional innovation policy is necessarily based on trial-and-error, in which policymakers learn and adapt, based on experience (Schwerin and Werker 2003). To stimulate learning, the policy system should be open to newcomers and new ideas, leave room for policy experiments, and a system of constant policy evaluation should be put in place (Wegner 1997).

Regional dimension of regional innovation policy

Policymakers in many countries have embraced the view that innovation processes have a regional dimension, and have responded by adding a regional dimension to their innovation policy (Van Geenhuizen and Nijkamp 2006).[6] We stated earlier that diversity in regional innovation policy is something that should be encouraged. While it is essential to take the knowledge and institutional base of regions as a starting point, one should be cautious, however, not to overestimate the role of the region as a driver of innovation. This has implications for regional innovation policy.

First of all, knowledge relationships often cross regional boundaries. As noted in section 2, non-local linkages are often found to be crucial for learning and innovation, in order to avoid cognitive lock-in. Amin and Cohendet (1999) claim that non-local networks are crucial for path-breaking innovations, while local learning results in more incremental innovations. For firms, being connected may be as important, or even more so, than simply being co-located (Giuliani and Bell 2005). This means that policy intervention should not focus on the region alone, as if geographical proximity is sufficient for innovation (Boschma 2005a). On the contrary, it should encourage geographical exposure by means of cross-regional knowledge collaboration and inflows of human capital, in order to avoid regional lock-in.

Secondly, policy should take in consideration that knowledge transfer between local firms is not just accomplished by bringing them together. Knowledge does not spill over automatically between firms, even when they are located in the same region. Network analysis demonstrates that the position of firms in knowledge networks depends on their absorptive capacity (Giuliani and Bell 2005). The higher a firm's absorptive capacity, the more it is connected locally, the more central its position in the knowledge network, and the higher its innovative performance. Firms with a high absorptive capacity are also more connected to the world. They may even act as gatekeepers that bring new variety into the region. Whether this external knowledge will diffuse in the region depends on the absorptive capacity of all other firms in the region. This can be stimulated by policy, by enhancing the absorptive capacity of local firms through public research, and by education schemes.

Conclusion

There are good reasons to avoid a 'picking-the-winner' policy that targets only a few sectors or regions. The idea that it is possible to design 'one-size-fits-all' regional policies is no longer valid. The copying of best practices is almost impossible when it comes to intangible regional assets that are the result of long histories. Regions provide opportunities but also set limits to effective growth policies. When policy solutions are built on regional indigenous capacity, the probability of effective policy is likely to increase. To avoid regional lock-in, it is crucial that policy is open to newcomers, to new ideas, and to experimentation.

11.5 POLICY OPTIONS

Now, how can regional innovation policy tackle system failures? There are many policy options one could think of, too many to be mentioned here. Due to a lack of space, only selected policy options will be listed and discussed below.

The objective of regional innovation policy is to encourage and facilitate new ideas and innovation through the creation, diffusion, and exploitation (or commercialisation) of new knowledge. The government might directly intervene, through the supply of R&D, education, and capital that match the need of local firms, and which increase the absorptive capacity and innovative capability of firms. Public policy can also stimulate the effective transfer of knowledge through various mechanisms, such as spinoff dynamics, labour mobility, and collaborative networks.[7] Below, we briefly direct attention to these three mechanisms of knowledge transfer, because they tend to take place at the regional level, and they may provide inputs for regional innovation policy based on related variety.

Recent studies have shed light on the importance of spinoff dynamics for knowledge diffusion, entrepreneurship, and regional development (Helfat and Lieberman 2002). As noted before, spinoffs are new firms that are founded by entrepreneurs that have acquired relevant experience as far as market and/or technical knowledge is concerned. Empirical studies tend to show that spinoff companies often perform better than other types of entrants because they can build on relevant knowledge and experience acquired in parent organisations in the same or related industries (Klepper 2002). Since spinoffs tend to locate near their parents almost as a rule, they may provide a basis for regional innovation policy. Because the spinoff process has played a crucial role in the emergence of many new sectors, it may be seen as a mechanism that makes regional economics diversify into new sectors, while building on knowledge and competences available in existing sectors (Boschma and Wenting 2007). A policy option is to target potential entrepreneurs (not just supporting any new firm), by taking into consideration what kind of knowledge the founder brings into the new firm. Another policy option is to encourage spinoff policies at universities that may contribute to the growth potential of spinoff ventures (Lockett et al. 2005).

Regional innovation policy could also play a role in encouraging labour mobility, because it is a crucial mechanism through which skills and experience are transferred from one local company to the other (Camagni 1991). Since most labour mobility takes place at the regional level, policy promoting labour mobility may enhance knowledge transfer and innovation at the regional level. Since labour mobility may take away the incentive of firms to invest in their personnel, public policy should invest heavily in education and life-long learning at the same time. Aghion et al. (2006) argue that labour markets need to be more flexible in order to smooth the process of creative destruction and lower the costs of such adjustments. This again needs to be complemented by a policy of life-long-learning, because it increases the capability of individuals to confront changes and to move from one job to the other.

Another crucial policy measure is to encourage the immigration of skilled labour because it may bring new ideas and knowledge into the region. One way to achieve this is through international exchange programmes for students. Incoming students bring in new talents and skills from abroad, and combine these with new skills that are acquired in higher education institutes in the host country. If the host country is capable of maintaining this group of high-skilled students after graduation (policy can most certainly play a role here), they will contribute to the economy as skilled employees or as founders of new firms. Outgoing students will acquire new skills in research and education institutes abroad, and may return to their home region after a while, where they will exploit their newly acquired skills in an environment they are familiar with (Saxenian 2006). Policy could target those outgoing groups and provide incentives to return to their home region.

Another policy option is to stimulate networks as effective settings through which knowledge circulates and interactive learning takes place. As stated before, policymakers may act as intermediaries or knowledge brokers, or establish policy platforms that facilitate knowledge to spill over and diffuse from sectors to related ones. In doing so, policy should avoid that vested interests of established firms take over and dominate these networks, and deny access to small firms and newcomers. Some have conceived these vested interests as one of the weaknesses of the Innovation Platform, an initiative of the Dutch national government. In a similar vein, competition policy could aim at stimulating the establishment of network alliances between firms in related industries as a way of diversifying regional economies into new but complementary fields of activity.

This type of network policy should acknowledge that knowledge networks frequently cross regional boundaries (Gertler and Levitte 2005; Giuliani 2005). It is crucial that regional innovation policy stimulates extra-regional networks, because it brings new knowledge into the region. Besides new infrastructure development and international exchange programs, a way to accomplish this is to encourage foreign investments. In a study of small European countries, Dachs et al. (2007) found that foreign-owned companies in some European countries tend to show a higher innovation output and higher labour productivity, as compared to domestically-owned companies. What is more, affiliates of foreign multinationals were quite strongly embedded in the national innovation system, may of them even showing a higher propensity to cooperate with domestic partners, as compared to domestically-owned companies.

Universities may also play a crucial role in exploiting inter-regional linkages, because they are extremely well connected to international networks. After their graduation, students will exploit and diffuse this knowledge in the regional economy. Academic spinoff policy and other policy measures may be implemented to ensure that the knowledge of universities will be further exploited economically at the regional level (Feldman and Desrochers 2003). What would be risky though is that public policy specifies in detail which knowledge fields will be targeted (e.g., through the allocation of R&D subsidies). As outlined before, this would mean a 'picking-the-winners' policy that denies the crucial role of variety as a source of novelty.

11.6 CONCLUSIONS

We have built upon insights drawn from evolutionary economic geography to present some recommendations for effective regional innovation policy. Since knowledge tends to accumulate mainly at the firm level, variety is the rule, and the more diversified a regional economy is, the higher regional growth. However, knowledge may also diffuse between firms, having an additional impact on regional development. If knowledge externalities are geographically bounded, knowledge will also accumulate at the regional level, and the regional economy will benefit as a whole. In addition, knowledge will spill over more intensively when regions are endowed with related industries that share a common knowledge base. This makes regional economies diversify into new directions and start up new growth paths, which are crucial for long-term regional development.

However, knowledge creation and knowledge spillovers alone will not lead to innovation. Regions require a critical mass of organisations that provide necessary inputs to the innovation process, such as knowledge, skills, and capital. Besides a critical mass, these organisations need to connect and interact, to enable flows of knowledge, capital, and labour. In addition, organisations and institutions need to be flexible and responsive in order to implement change. In reality, almost by nature, organisations and institutions are not, because they suffer from lock-in, due to routines, sunk costs and path dependency.

We have used these insights as key inputs and underpinnings for effective regional innovation policy. Following system failure arguments, public policy has the task of establishing key organisations of innovation systems in regions where these are found to be missing, or public policy has to ensure that these missing inputs to the innovation process will flow into the region. Once available, public intervention should encourage key organisations to connect. For example, firms need to be linked with research institutes and capital suppliers. In addition, public policy can make organisations more flexible and innovative, for instance, by upgrading their routines through the supply of new knowledge and skills. Finally, regional innovation policy can stimulate the effective transfer of knowledge at the regional level by means of spinoff activity, labour mobility, and networks. Since these mechanisms of knowledge transfer are basically taking place at the regional level, and because they make regions move into new growth paths while building on existing assets, these policy actions put in practice the idea that related variety may contribute to long-term regional development.

To increase the probability of policy success, regional innovation policy needs to account for the region-specific context because it provides opportunities but also sets limits to what can be achieved by public policy. In doing so, it should neither apply 'one-size-fits-all' policies nor adopt 'picking-the-winner' policies. Instead of copying best practice models or selecting winners, policy should take the history of each region as a basic starting point and identify regional potentials and bottlenecks accordingly. To avoid regional lock-in, it is crucial that public policy is open to newcomers and new policy experiments.

NOTES

1 In the late nineteenth century, Marshall (1890) attributed the spatial clustering of industries to specialised labour markets, local access to specialised suppliers and large markets, and the presence of local knowledge spillovers.

2 The relevance of geographical proximity for knowledge exchange is associated with *tacit*, as opposed to *codified* knowledge (Brown and Duguid, 2000). Codified knowledge consists of information that can be written down and, consequently, can be exchanged over long distances. Tacit knowledge is more difficult to express in an explicit form. Tacit knowledge is acquired through experience, demonstration, and practice which require personal interactions. Therefore, geographical proximity is seen as facilitating the transfer of tacit knowledge (Johnson et al. 2002). There is increasing awareness, however, that the need for face-to-face contacts to exchange tacit knowledge does not automatically mean that individuals have to be located close to one another (Rallet and Torre, 2000). In many cases, face-to-face contacts can be arranged on a temporary basis, for example through business travels, conferences, or fairs (Gallaud and Torre, 2005).

3 Because collaboration between different types of organisations (e.g., between firms and universities) is not self-evident, it tends to take place at a lower spatial scale, as compared to collaborations between similar organisations (e.g., between firms) (Ponds et al. 2007). This suggests that geographical proximity may be helpful in overcoming cultural and other barriers between different types of organisations.

4 In a study on the Swedish economy, Neffke et al. (2007) found evidence that young industries (in contrast to mature industries) tend to benefit from Jacobs' externalities, that is, they perform best in regions with a high degree of sectoral variety. An additional finding was that the effect of Jacobs' externalities was especially strong and positive when the industry was in the exploration stage, in contrast to the exploitation phase.

5 There is real danger that the ideal circumstances for regional policy (a specialised region with a few strong players) cause a situation of institutional lock-in, with adverse impacts on regional development in the long run (Cheshire and Gordon, 1996).

6 It is important to realise that this regionalisation of innovation policy consists of many dimensions: it incorporates objectives, instruments and administration issues of policy, among other things (Fritsch and Stephan, 2005). We view regionalisation of innovation policy in a broad way, covering any policy action that accounts for region-specific features. So, it is not only about cluster promotion, which is now a popular policy objective, but which has also been subject to severe criticism (see Martin and Sunley, 2003).

7 By contrast, Dosi et al. (2006) claim that the European Union should shift its policy approach from a networking type (emphasis on interactions with local environment) to an actor type of approach (strengthening high quality basic research and the innovative capacity of corporate actors).

REFERENCES

Aghion, P., P.A. David and D. Foray (2006) 'Linking Policy Research and Practice in "STIG Systems": Many Obstacles, But Some Ways Forward', paper presented at SPRU conference, September 11-13, 2006.

Agarwal, A., I. Cockburn and J. McHale (2006) 'Gone but Not Forgotten: Knowledge Flows, Labor Mobility, and Enduring Social Relationships', *Journal of Economic Geography,* 6: 571-591.

Amin, A. and P. Cohendet (1999) 'Learning and Adaptation in Decentralised Business Networks', *Environment and Planning D Society and Space,* 17(1): 87-104.

Anselin, L., A. Varga and Z. Acs (2000) 'Geographic and Sectoral Characteristics of Academic Knowledge Externalities', Papers in Regional Science, 79: 435-443.

Arrow, K.J. (1962) 'The Economic Implications of Learning by Doing', *Review of Economic Studies,* 29: 155-173.

Asheim, B.T. (1996) 'Industrial Districts as 'Learning Regions': A Condition for Prosperity', *European Planning Studies,* 4(4): 379-400.

Asheim B.T. and A. Isaksen (2002) 'Regional Innovation Systems: The Integration of Local 'Sticky' and Global 'Ubiquitous' Knowledge', *Journal of Technology Transfer,* 27: 77-86.

Asheim B. et al. (2006) *Constructing Regional Advantage: Principles, Perspectives, Policies,* final report, DG Research, Brussels: European Commission.

Audretsch, D.B. and M. Feldman (1996) 'Spillovers and the Geography of Innovation and Production', *American Economic Review,* 86: 630-40.

Bathelt, H., A. Malmberg and P. Maskell (2004) 'Clusters and Knowledge: Local Buzz, Global Pipelines and the Process of Knowledge Creation', *Progress in Human Geography,* 28: 31-56.

Becattini, G. (ed.) (1987) *Mercato e forze locali. Il distretto industriale,* Bologna: Il Mulino.

Boschma, R.A. (1999) 'The Rise of Clusters of Innovative Industries in Belgium During the Industrial Epoch', *Research Policy,* 28: 853-871.

Boschma, R.A. (2004) 'Competitiveness of Regions From an Evolutionary Perspective', *Regional Studies,* 38(9): 1001-1014.

Boschma, R.A. (2005a) 'Proximity and Innovation: A Critical Assessment', *Regional Studies,* 39(1): 61-74.

Boschma, R.A. (2005b) 'Rethinking Regional Innovation Policy: The Making and Breaking of Regional History', in: G. Fuchs and P. Shapira (eds.) *Rethinking Regional Innovation and Change: Path Dependency or Regional Breakthrough?,* Dordrecht: Springer Verlag, 249-271.

Boschma, R., K. Frenken and J.G. Lambooy (2002) *Evolutionaire economie, een inleiding,* Bussum: Coutinho.

Boschma, R.A. and S. Iammarino (2007) 'Related Variety and Regional Growth in Italy', working paper series on Evolutionary Economic Geography, Utrecht: Utrecht University.

Boschma, R.A. and J.G. Lambooy (1999) 'Evolutionary Economics and Economic Geography', *Journal of Evolutionary Economics,* 9: 411-429.

Boschma, R.A. and R. Wenting (2007) 'The Spatial Evolution of the British Automobile Industry', *Industrial and Corporate Change*, 16(2): 213-238.

Breschi, S. and F. Lissoni (2003) 'Mobility and Social Networks: Localised Knowledge Spillovers Revisited', CESPRI Working Paper 142, available at http://www.cespri.unibocconi.it/.

Brown, J.S. and P. Duguid (2000) 'Mysteries of the Region: Knowledge Dynamics in Silicon Valley', in: C.M. Lee (ed.) *The Silicon Valley Edge: A Habitat for Innovation and Entrepreneurship*, Palo Alto: Stanford University Press, 16-39.

Bryant, K. (2001) 'Promoting Innovation: An Overview of the Application of Evolutionary Economics and Systems Approaches to Policy Issues', in: J. Foster and S. Metcalfe (eds.) *Frontiers of Evolutionary Economics: Competition, Self-organization and Innovation Policy*, Cheltenham: Edward Elgar, 361-383.

Camagni, R. (ed.) (1991) *Innovation Networks: Spatial Perspectives*, London/New York: Bellhaven Press.

Camagni, R. (1995) 'The Concept of Innovative Milieu and its Relevance of Public Policies in European Lagging Regions', *Papers in Regional Science*, 74(4): 317-340.

Carlsson, B. and R. Stankiewicz (1991) 'On the Nature, Function and Composition of Technological Systems', *Journal of Evolutionary Economics*, 1: 93-118.

Cheshire, P.C. and I.R. Gordon (1996) 'Territorial Competition and the Predictability of Collective (In)action', *International Journal of Urban and Regional Research*, 20: 383-399.

Cohen, W.M. and D.A. Levinthal (1990) 'Absorptive Capacity: A New Perspective on Learning and Innovation', *Administrative Science Quarterly*, 35: 128-153.

Cooke, P. (2001) 'Regional Innovation Systems, Clusters, and the Knowledge Economy', *Industrial and Corporate Change*, 10(4): 945-74.

Cooke P and K. Morgan (1998) *The Associational Economy: Firms, Regions, and Innovation*, Oxford: Oxford University Press.

Dachs, B., B. Ebersberger and H. Loof (2007) 'The Innovative Performance of Foreign-owned Enterprises in Small Open Economies', CESIS working paper 87.

Degroof, J. and E.B. Roberts (2004) 'Overcoming Weak Entrepreneurial Infrastructures for Academic Spin-off Ventures', *Journal of Technology Transfer*, 29: 327-352.

Dosi, G. (1997) 'Opportunities, Incentives and the Collective Patterns of Technological Change', *The Economic Journal*, 107(44): 1530-1547.

Dosi, G, C. Freeman, R. Nelson, G. Silverberg and L. Soete (eds.) (1988) *Technical Change and Economic Theory*, London: Pinter Publishers.

Dosi, G., P. Llerena and M. Sylos Labini (2006) 'The Relationships Between Science, Technologies and their Industrial Exploitation: An Illustration Through the Myths and Realities of the So-called "European paradox"', *Research Policy*, 35(10): 1450-1464

Edquist, C. (1997) 'Systems of Innovation Approaches: Their Emergence and Characteristics', in: Edquist (ed.) *Systems of Innovation: Technologies, Institutions and Organizations*, London/Washington: Pinter, 1-35.

Edquist, C. and B. Johnson (1997) 'Institutions and Organizations in Systems of Innovation', in: C. Edquist (ed.) *Systems of Innovation: Technologies, Institutions and Organizations*, London/Washington: Pinter, 41-63.

Ergas, H. (1984) *Why Do Some Countries Innovate More than Others?*, Brussels: CEP.

Essletzbichler, J. (2005) 'Diversity, Stability and Regional Growth in the U.S. (1975-2002)', *Papers in Evolutionary Economic Geography*, 5(13), Utrecht: Utrecht University.
Essletzbichler, J. and D.L. Rigby (2005) 'Competition, Variety and the Geography of Technology Evolution', *Tijdschrift voor Economische en Sociale Geografie*, 96(1): 48-62.
Feldman, M.P. and D.B. Audretsch (1999) 'Innovation in Cities: Science-based Diversity, Specialization and Localized Competition', *European Economic Review*, 43: 409-429.
Feldman, M.P. and P. Desrochers (2003) 'The Evolving Role of Research Universities in Technology Transfer: Lessons from the History of John Hopkins University', *Industry and Innovation*, 10: 5-24.
Freeman, C. (1987) *Technology Policy and Economic Performance: Lessons from Japan*, London: Pinter.
Freeman, C. and C. Perez (1988) 'Structural Crisis of Adjustment, Business Cycles and Investment Behaviour', in: G. Dosi, C. Freeman, R. Nelson, G. Silverberg and L. Soete (eds.) *Technical Change and Economic Theory*, London: Pinter, 38-66.
Frenken, K., F.G. van Oort and T. Verburg (2007) 'Related Variety, Unrelated Variety and Regional Economic Growth', *Regional Studies*, 41(5): 685-697.
Fritsch, M. and A. Stephan (2005) 'Regionalization of Innovation Policy: Introduction to the Special Issue', *Research Policy*, 34(8): 1123-1127.

Fromhold-Eisebith, M. and G. Eisebith (2005) 'How to Institutionalize Innovative Clusters? Comparing Explicit Top-down and Implicit Bottom-up Approaches', *Research Policy*, 34(8): 1250-1268.
Gallaud, D. and A. Torre (2005) 'Geographical Proximity and the Diffusion of Knowledge', in: G. Fuchs, P. Shapira and A. Koch (eds.) *Rethinking Regional Innovation and Change: Path Dependency or Regional Breakthrough?*, Dordrecht: Springer Verlag, 127-146.
Geenhuizen, M. van and P. Nijkamp (2006) 'Learning Regions in an Evolutionary Context: Policymaking for High Technology Firms', *International Journal of Entrepreneurship and Innovation Management*, 6(3): 265-282.
Gertler, M.S. (2003) 'Tacit Knowledge and the Economic Geography of Context or the Undefinable Tacitness of Being (There)', *Journal of Economic Geography*, 3: 75-99.
Gertler, M., and Y. Levitte (2005) 'Local Nodes in Global Networks: The Geography of Knowledge Flows in Biotechnology Innovation', *Industry and Innovation*, 13: 487-507.
Giuliani, E. (2005) 'The Structure of Cluster Knowledge Networks: Uneven and Selective, Not Pervasive and Collective', DRUID Working Paper 2005, 11.
Giuliani, E. and M. Bell (2005) 'The Micro-determinants of Meso-level Learning and Innovation: Evidence from a Chilean Wine Cluster', *Research Policy*, 34: 47-68.
Glückler, J. (2007) 'Economic Geography and the Evolution of Networks', *Journal of Economic Geography*, 7(5), forthcoming.
Grabher, G. (1993) 'The Weakness of Strong Ties: The Lock-in of Regional Development in the Ruhr Area', in: G. Grabher (ed.) *The Embedded Firm*, London: Routledge, 255-277.
Hall, P. (1994) *Innovation, Economics and Evolution: Theoretical Perspectives on Changing*

Technology in Economic Systems, Hemel Hempstead: Harvester Wheatsheaf.
Hall, P.G. and P. Preston (1988) *The Carrier Wave: New Information Technology and the Geography of Innovation 1846-2003*, London: Unwin Hyman.
Hall, P.A. and D. Soskice (2001) 'An Introduction to Varieties of Capitalism', in: P.A. Hall and D. Soskice (eds.) *Varieties of Capitalism: The Institutional Foundations of Comparative Ddvantage*, Oxford: Oxford University Press, 1-68.
Hassink, R. (2005) 'How to Unlock Regional Economies from Path Dependency? From Learning Region to Learning Cluster', *European Planning Studies*, 13(4): 521-535.
Hassink, R. and A. Lagendijk (2001) 'The Dilemmas for Interregional Institutional Learning', *Environment and Planning C. Government and Policy*, 19(1): 65-84.
Helfat, C.E. and M.B. Lieberman (2002) 'The Birth of Capabilities: Market Entry and the Importance of Pre-history', *Industrial and Corporate Change*, 11(4): 725-760.
Henderson, J.V., A. Kuncoro and M. Turner (1995) 'Industrial Development in Cities', *Journal of Political Economy*, 103: 1067-1085.
Hodgson, G. (1996) 'An Evolutionary Theory of Long-term Economic Growth', *International Studies Quarterly*, 40: 391-410.
Howells, J. (2005) 'Innovation and Regional Economic Development: A Matter of Perspective?', *Research Policy*, 34: 1220-1234.
Iammarino, S. (2005) 'An Evolutionary Integrated View of Regional Systems of Innovation: Concepts, Measures and Historical Perspectives', *European Planning Studies*, 13(4): 497-519.

Iammarino, S. and P. McCann (2006) 'The Structure and Evolution of Industrial Clusters: Transactions, Technology and Knowledge Spillovers', *Research Policy*, 35(7): 1018-1036.
Jacobs, J. (1969) *The Economy of Cities*, New York: Vintage Books.
Johnson B, E. Lorenz and B.A. Lundvall (2002) 'Why All This Fuss about Codified and Tacit Knowledge?' *Industrial and Corporate Change*, 11(2): 245-262.
Klepper, S. (2002) 'The Evolution of the US Automobile Industry and Detroit as its Capital', paper presented at 9th Congress of the International Schumpeter Society, Gainesville, Florida, March 27-30.
Klepper, S. and K.L. Simons (2000) 'Dominance by Birthright: Entry of Prior Radio Producers and Competitive Ramifications in the US Television Receiver Industry', *Strategic Management Journal*, 21: 997-1016.
Kohler-Koch, B. (1998) 'Europe and the Regions: The Issue of Multi-level Governance and Sovereignty', paper presented at conference on Democracy in Europe, Enschede: Twente University.
Lambooy, J.G. and R.A. Boschma (2001) 'Evolutionary Economics and Regional Policy', *Annals of Regional Science*, 35(1): 113-133.
Levinthal, D.A. (1998) 'The Slow Pace of Rapid Technological Change: Gradualism and Punctuation in Technological Change', *Industrial and Corporate Change*, 7(2): 217-247.
Lindgren, U. and R. Eriksson (2007) *Cluster Mobility: Impacts of Inter-firm Labour Mobility on Firm Performance*, Umea: Umea University.
Lockett, A., D. Siegel, M. Wright and M.D. Ensley (2005) 'The Creation of Spin-off Firms at Public Research Institutions: Managerial and Policy Implications', *Research Policy*, 34: 981-993.

Lorenzen, M. (2007) 'Internationalization vs. Globalization of the Film Industry', *Industry and Innovation,* 14(4): 349-357.
Marshall, A. (1890) *Principles of Economics*, London: MacMillan.
Martin, R. andP. Sunley (2003) 'Deconstructing Clusters: Chaotic Concept or Policy Panacea?', *Journal of Economic Geography,* 3(1): 5-35.
Martin, R. and P. Sunley (2006) 'Path Dependence and Regional Economic Evolution', *Journal of Economic Geography,* 6(4): 395-437.
Maskell, P. (1999) 'Social Capital, Innovation and Competitiveness', in: S. Baron, J. Field and T. Schuller (eds.) *Social Capital Collection*, Oxford: Oxford University Press, 1-17.
Maskell, P. (2001) 'Towards a Knowledge-based Theory of the Geographical Cluster', *Industrial and Corporate Change,* 10(4): 919-941.
Maskell, P. and A. Malmberg (1999) 'Localised Learning and Industrial Competitiveness', *Cambridge Journal of Economics,* 23(2): 167-186.
Metcalfe, S. (1994) 'The Economic Foundations of Technology Policy: Equilibrium and Evolutionary Perspectives', in: M. Dodgson and R. Rothwell (eds.) *The Handbook of Industrial Innovation*, Cheltenham: Edward Elgar, 409-512.
Metcalfe, J.S. (2003) 'Equilibrium and Evolutionary Foundations of Competition and Technology Policy: New Perspectives on the Division of Labour and the Innovation Process', in: P. Pelikan and G. Wegner (eds.) *The Evolutionary Analysis of Economic Policy*, Cheltenham: Edward Elgar, 162-190
Morgan, K. (1997) 'The Learning Region: Institutions, Innovation and Regional Renewal', *Regional Studies,* 31(5): 491-503.
Murmann, J.P. (2003) *Knowledge and Competitive Advantage: The Co-evolution of Firms, Technology, and National Institutions*, Cambridge: Cambridge University Press.
Neffke, F., M. Svensson Henning, R. Boschma, K. Lundquist and O. Olander (2007) 'Evolution of Agglomeration Externalities: How Industry Life Cycles and Innovation Stages Help Explain Variations in Agglomeration Externalities', Papers in Evolutionary Economic Geography, Utrecht: University of Utrecht.
Nelson, R.R. (1995) 'Co-evolution of Industry Structure, Technology and Supporting Institutions, and the Making of Comparative Advantage', *International Journal of the Economics of Business,* 2(2): 171-184.
Nelson, R.R. and S.G. Winter (1982) *An Evolutionary Theory of Economic Change*, Cambridge, MA/London: The Belknap Press.
Nooteboom, B. (2000) *Learning and Innovation in Organizations and Economies*, Oxford: Oxford University Press.
Nooteboom, B., W. van Haverbeke, G. Duysters, V. Gilsing and A. van den Oord (2007) 'Optimal Cognitive Distance and Absorptive Capacity', *Research Policy,* 36: 1016-1034.
Olson, M. (1982) *The Rise and Decline of Nations: Economic Growth, Stagflation and Social Rigidities*, New Haven: Yale University Press.
Ormerod, P. (2005) *Why Most Things Fail: Evolution, Extinction and Economics*, New York: Pantheon Books.
Oughton, C., M. Landabaso and K. Morgan (2002) 'The Regional Innovation Paradox: Innovation Policy and Industrial Policy', *Journal of Technology Transfer,* 27: 97-110.

Pack, H. and K. Saggi (2006) 'The Case for Industrial Policy: A Critical Survey', *World Bank Policy Research* 3839, February: 51.
Pasinetti, L.L. (1981) *Structural Change and Economic Growth*, Cambridge: Cambridge University Press.
Ponds, R., F.G. van Oort and K. Frenken (2007) 'Internationalization and Regional Embedding of Scientific Research in the Netherlands', in: A. Varga (ed.) *Universities and Regional Development*, Cheltenham/Northampton, MA.: Edward Elgar Press.
Porter, M. (1990) *The Competitive Advantage of Nations*, New York: Free Press.
Pred, A.R. (1966) *The Spatial Dynamics of Urban-industrial Growth 1800-1914: Interpretative and Theoretical Essays*, Cambridge, MA: MIT Press.
Rallet, A. and A. Torre (2000) 'Is Geographical Proximity Necessary in Innovation Networks in the Era of Global Economy?' *GeoJournal*, 49: 373-380.
Raspe, O., F.G. van Oort and P. de Bruijn (2004) *Spatial Patttern in the Dutch Knowledge Economy* (in Dutch), Rotterdam: Nai Publishers.
Rosenberg, N. (1982) *Inside the Black Box: Technology and Economics*, Cambridge: Cambridge University Press.
Saviotti, P.P. (1996) *Technological Evolution, Variety and the Economy*, Cheltenham: Edwrad Elgar.
Saviotti, P.P. and A. Pyka (2004) 'Economic Development by the Creation of New Sectors', *Journal of Evolutionary Economics*, 14: 1-35.
Saxenian, A.L. (2006) *The New Argonauts: Regional Advantage in a Global Economy*, Cambridge: Harvard University Press.
Schumpeter, J.A. (1942) *Capitalism, Socialism and Democracy*, New York: Harper and Row.
Schwerin, J. and C. Werker (2003) 'Learning Innovation Policy Based on Historical Experience', *Structural Change and Economic Dynamics*, 14: 385-404.
Shaw, G.K. (1992) 'Policy Implications of Endogenous Growth Theory', *The Economic Journal*, 102: 611-621.
Simmie, J. (2003) 'Innovation and Urban Regions as National and International Nodes for the Transfer and Sharing of Knowledge', *Regional Studies*, 37(6-7): 607-620.
Todtling, F. and M. Trippl (2005) 'One Size Fits All? Towards a Differentiated Regional Innovation Policy Approach', *Research Policy*, 34: 1203-1219.
Uzzi, B. (1997) 'Social Structure and Competition in Interfirm Networks: the Paradox of Embeddedness', *Administrative Science Quarterly*, 42: 35-67.
Wegner, G. (1997) 'Economic Policy from an Evolutionary Perspective: A New Approach', *Journal of Institutional and Theoretical Economics*, 153: 463-509.
Wegner, G. and P. Pelikan (2003) 'Introduction: Evolutionary Thinking on Economic Policy', in: P. Pelikan and G. Wegner (eds.) *The Evolutionary Analysis of Economic Policy*, Cheltenham: Edward Elgar, 1-14.

12 CONCLUSIONS FOR INNOVATION POLICY: OPENING IN FOURS

Bart Nooteboom and Erik Stam

12.1 INTRODUCTION

Innovation is needed to sustain economic growth, but the position of the Netherlands in terms of innovation is not a very good one. It was shown in chapter 2 that in recent years the EU has lagged behind the US in productivity growth, and within the EU the Netherlands takes up a middle position. The Netherlands performs below the EU average in six out of eight indicators of innovativeness. It was shown in chapter 5 that although in recent years the number of new firms has risen rather spectacularly, this can largely be attributed to employees continuing their activities in self-employment for fiscal and 'lifestyle' reasons. On average, small- and medium-sized enterprises (SMEs) in the Netherlands have become less rather than more innovative in the last decade, and the percentage of innovative SMEs is much lower than the EU average. Therefore, there is an urgent policy issue: how can innovation be increased?

We have argued that innovation is a system phenomenon, with multiple types of individual and collective agents, including firms, entrepreneurs, institutes for education and research, policymakers, regulatory agencies, and many types of services and intermediaries, interacting in a variety of ways. Actions and interactions are enabled by institutions and forms of organisation (of firms and between firms), and in turn affect those institutions. In this innovation system institutional logics and dynamics arise that are difficult to manage, and yield unexpected and often adverse effects.

As argued in chapters 1 and 2, for an adequate innovation policy we need an adequate understanding of the micro-level actions and interactions of agents, in competition and collaboration, in idea generation, implementation, and diffusion of innovations. Little is known of the micro-foundations and institutional conditions of innovation policy, and the purpose of this book is to contribute to the further development of that insight and corresponding policy.

In this book we have discussed theories of cognition, learning, and trust (chapters 3 and 7), and we have analysed the following elements of the innovation system: the generation and utilisation of ideas (chapter 4), entrepreneurship (chapter 5 and 6), and the internal (chapters 8 and 9), and the external organisation of innovation (chapters 10 and 11). In this final chapter we summarise the main lines of analysis and we present our conclusions. We begin with a summary of the conceptual and theoretical perspectives that we have used. We proceed with a critical discussion of targeted industrial policies, and a discussion of market, system, and government failures.

This is followed by policy implications regarding different parts of the innovation system: knowledge, entrepreneurship, organisation, networks, and regions.

12.2 PERSPECTIVES

As discussed in chapter 3, in this volume we take a perspective informed by an Austrian, in particular Hayekian, and an evolutionary view of markets, where diversity, local specificity, and idiosyncrasy of knowledge form sources of innovation, in an experimentalist society. This yields a plea to afford autonomy and to create conditions for lower-level actors to experiment with solutions of their own devising within broadly defined areas of public policy. For this, entrepreneurship is crucial, as discussed in chapter 5. We also take an institutionalist view of markets, in which markets and institutions co-evolve in the innovation system. Markets require institutions but also create institutions that may obstruct the emergence of new markets.

An evolutionary perspective yields an appreciation of radical uncertainty and the role of variety, for exploration, and the corresponding role of trial and error, the need for a selection environment of markets and institutions to select among them, and the transmission of success. We add a Hayekian awareness of the distributed nature of knowledge. Together, the two perspectives suggest modesty concerning the ability of governments, especially central government, to 'pick winners', choose the right parties for 'backing winners', and to design innovation trajectories. Yet, it does not leave policy empty-handed. In addition to the traditional market failure arguments for policy interventions in (innovations in) public sectors, one can identify 'system failures' in the processes of idea generation, innovation, and diffusion that governments should address, to enable and facilitate evolutionary processes, without claiming to be able to outguess innovation in the prediction of the outcomes of those processes. The unpredictability of innovation lead to the requirement of openness in the course of innovation processes and their outcomes, to allow for surprises and changes of direction along the way.

However, the evolutionary perspective is in danger of neglecting creativity and invention. The creation of variety, in new ideas, is not as blind as it is in biological evolution, and we need a theory for it. A central issue, analysed in chapter 3, but returning in several if not all other chapters, is how to combine exploration and exploitation, within and between organisations. In learning and innovation, on all levels, of people, teams within organisations, firms, and public policy, there is a tension as well as a mutual dependence between exploitation, or first-order learning, in the *application and improvement* of new ideas, principles, designs, or logics, and exploration, or second-order learning, in which they are *generated.* On the one hand the two build upon each other, or emerge from each other, but on the other hand they have different requirements, in terms of mentality, approach, modes of governance, and organisational conditions, and are difficult to combine at the same time and place. We find this tension in firms (between research and development on the one hand and operations and production on the other hand), in the tension between entrepreneurship and management (in the relation between university (exploration) and industry (exploitation)), and in the tension between the design and implementation of policy.

In economic policy, there has been a focus on exploitation to the neglect of exploration, which requires a dynamic perspective.

We see this tension also in privatization. An example is the concession of regional bus transport to private providers. Static efficiency requires a short concession period, as an incentive for efficiency in the short term, to qualify for the next concession period, but innovation requires a longer horizon. Efficiency of scale in the provision of transport, and efficiency of connections in travel require a large concession area, with a large volume of transport, but this excludes small contenders and limits variety for experimentation. Low cost is usually safeguarded in tight ex ante specifications of performance, but this excludes the room for experimentation needed for innovation.

The fundamental duality and tension between exploitation and exploration has to be faced in innovation policy, and its consequences have to be developed and translated into policy measures, in most if not all parts of the innovation system.

In chapter 3, we analysed a 'cycle of discovery', as a model for a 'knowledge ecology'. According to this model, exploration is stimulated by submitting established practice to new challenges, in new contexts of application, to gain fresh insight into its limits, to build motivation to change, and to find inspiration of possible elements and directions of change, as an avenue of discovery. The cyclical nature of learning, with exploration and exploitation building upon each other, yields the requirement of openness of entry for new players, outside entrepreneurs, and inventors. It also requires openness to the world, allowing for connections with outside communities, and for avenues of discovery. In policy making it requires openness to ideas, goals, and experiences of citizens.

As discussed in chapter 3, the Hayekian view of diverse and dispersed knowledge can be connected, going beyond Hayek, with an 'embodied' view of cognition, including perception, interpretation, and evaluation, as based on mental categories that constitute absorptive capacity and are constructed in interaction between people. Here, cognition is both individual and social: an individual construction based on interaction. This yields 'cognitive distance' between people to the extent that they have developed their cognition along different life paths. This yields both a problem, of imperfect understanding, and an opportunity, of new insights. Information needs to be absorbed to become knowledge. Knowledge is transmitted only imperfectly and with greater or lesser effort. This applies not only to 'tacit knowledge', as widely recognised, but also to codified knowledge, since that can also only be absorbed to the extent that it fits into absorptive capacity. Knowledge absorption, or assimilation into cognitive frameworks, entails transformation, to a greater or lesser extent. Thus learning in the sense of acquiring knowledge to some extent also entails the creation of knowledge. That, by the way, is also why it is difficult to completely separate the creation from the adoption of innovation (in diffusion), since adoption always entails transformation and in that sense is innovative. Absorptive capacity is subject to development: more knowledge enables the absorption of further knowledge. In other words, knowledge is cumulative. For innovation, diversity and cognitive distance are indispensable. The policy implications of this are that formal and informal institutions should be aimed not at eliminating that distance but at the ability to cross it. The benefits of cognitive distance yield an argument that for innovation actors (people, firms) should profit from outside collaboration with others, at optimal cognitive distance. On a fundamental level, that is what lies behind the currently fashionable notion of 'open innovation'.

On the governance side, a crucial theme is that of trust, discussed in chapter 7. As argued above, collaboration is especially necessary for innovation. We need collaboration to profit from cognitive distance, but especially in innovation governance of relationships by contractual and incentive control is problematic, due to the uncertainty of innovation and the novelty of emerging knowledge and competence, which limit contracting, monitoring and the evaluation of competence and performance. Especially in the earlier stages of innovation, i.e., exploration, one does not yet know what the goals, means, causalities between them, and requisite resources will be. In other words: problems are 'untamed'. As a result, relevant stakeholders have to iterate in their collaboration, in the hope of converging on a basis for exploitation. This is difficult to govern by means of contracts, monitoring, and control. But what is trust, how does it work, and what are its limits? Crucial for trust is openness of communication ('voice'), in which one voices concerns, reports one's weaknesses and mistakes, and responds constructively to such openness of partners. To further innovation, but more generally to provide scope for professionalism, we should get away from current tendencies towards excessive control and monitoring of work, and master the art of voice in the building of trust. chapter 7 provided indications on how to do that.

Our key message is that innovation policy should create and maintain openness, in four dimensions: openness to uncertainty in the innovation process; openness for collaboration with others at a fruitful 'cognitive distance'; openness for new entrants; and openness to the world outside. This fourfold openness for innovation stands in contrast to much theory, policy, and practice of innovation, which in many cases locks up innovation in pre-conceived targets, established players, national programmes, isolated activities, and relationships at arms length that lack collaboration and openness of communication. The different chapters of this book serve to make these dimensions of openness more concrete, and to specify ways to achieve them. Those are summarised in the following sections. First, we proceed with a critical discussion of targeted industrial policies.

12.3 TARGETED INDUSTRIAL POLICIES

In chapter 1 we noted that one should expect radical, path-breaking, paradigm-switching innovations to be rare, particularly in a small country like the Netherlands. Most innovation will be exploitative, incremental, or imitative. There, in a wide range of application across the economy, invention achieves its greatest economic impact. Clearly, then, a priority for policy is to enable and stimulate entrepreneurship to conduct this wide range of exploitative innovation. That is why, in a later section, we indicate how entrepreneurship may be enhanced, based on the analysis in chapter 5.

The question next is in what areas 'we', as a nation, should choose to be leaders, accepting that we are mostly followers. Here, a temptation arises for policymakers to make a choice in targeted industrial policies, on the basis of expected unique strengths in the future ('picking winners'), or to enhance strengths that have been proven in the market ('backing winners'). The former approach was taken, in the Netherlands, in the 1980s, with a choice of 'arrowhead' sectors, and the latter policy is being pursued now, with the choice of 'key areas'. Picking winners is now recognised as going against the unpredictability of future success. One cannot outguess the evolutionary process, and as much as possible one should leave the errors of choice up to private business, in a wide range of experimentation in diverse directions that occasionally, somewhere will yield unpredictable success. The argument for backing winners is that in proven success errors and misfits have been weeded out. Another argument that is used for backing winners is in fact an old argument of comparative advantage, now applied to areas of knowledge or competence. If we are successful in certain areas this reveals comparative advantage, upon which we should capitalise. However, who says that present success guarantees future success? Again, we cannot outguess evolution. Also, success is often the result of strength developed from challenge and struggle, and may be weakened rather than further strengthened by support. Innovation when cuddled may only survive in an artificial way.

An argument that has also been used for backing winners is an argument of concentration: we cannot afford to spread money and attention around everywhere. For the sake of efficiency we have to concentrate it where it is best spent, and that is where quality is highest. In chapter 1 we acknowledged that economies of scale do occur, sometimes, for example in transaction costs, but there are also diseconomies of scale, and moreover any advantages of scale have to be traded off against loss of variety. Spreading experimentation over a variety of many small-scale trials increases the chance of breakthroughs while limiting the losses that arise from failure, compared to fewer bets on larger projects.

Furthermore, the question with backing winners is why they should be backed if they are winners? If they have been proven in the market, they are generating profits for their expansion. Such policy carries the risk of confirming what exists, in established ideas, technologies, and players, rather than stimulating the emergence of novelty generated by outsiders. Backing winners may confirm or even raise obstacles to entry for new challengers. In our view, the priority should lie in keeping the system open to outsiders. This entails prioritising space and incentives for entrepreneurship.

Upon closer inspection and debate it turns out that what is in fact intended by the Dutch policy of backing winners, is better labelled as 'mobilising unrealized potential', i.e., eliminating obstacles for the full utilisation of proven strengths. Of course, government certainly has a task in seeing that potential is realised, for example by breaking down barriers to innovations. Barriers may arise from the lack of appropriate infrastructure, skills, education, technical standards, vested interests, problems of high costs due to initial small scale and lack of learning by doing, and systemic complexities. There may be market or system failures, and then there is an argument, in principle, for government intervention, but those barriers then have to be specified. We consider such barriers in a subsequent section on failures of governments, markets, and systems.

On what basis would government make choices? Part of the problem is that in the design of programmes, networks, clusters, and subsidies, one needs insights from industry, and these tend to be offered and taken from established large business (small firms hardly have the staff to spare, for example to participate in such committees). Naturally, their suggestions are coloured by their experience and interests – hen they slide into positions where they can protect their interests, it would be naive to assume that they would not. In particular, in innovation large incumbent firms are tempted to engage in 'pre-emptive participation'. Here, they do invest in the R&D needed for innovation, if only to have a basis for appropriating innovation if it does break through, but hold back on implementation and breakthrough as long as the innovation would cannibalise existing, sunk investments, when those innovations are incompatible with it. This is not a matter of 'evil plots' but of institutional logic. In the Netherlands, this phenomenon is intensified by the country's peculiar industrial structure, with a hydrocephalus ('waterhead') of a relatively large number of very large firms, in combination with a small number of medium-sized firms, whereby the large firms are likely to exercise a disproportionately large influence on public policy.

Can one expect the established oil and chemical industry to play along with degradable bio-plastics that do not fit into the existing assets geared to petrochemicals, or with hydrogen fuel that does not fit into the assets geared to gasoline (refineries, pipelines, gas stations)? And is government able to resist these forces of conservatism? On the other hand, one should allow corporations to conduct early research in new technologies or fields of activity, and to pull out when they conclude that it does not fit their strategic portfolio of activities. So, how do we know when they are engaging in pre-emptive participation and when in reasonable pull-out? In the latter case, one might expect them to be liberal in licensing the results of their research to others who do see a worthwhile opportunity in it. This may also take the form of supporting spin-offs from the company into such ventures.

There are also policy bandwagon effects. If one targets a particular industry or technology in innovation policy, everyone in that area wants to take part, and on what basis would government refuse them? It is not only a matter of whom to admit but also of whom to exclude. On what basis would government determine that exclusion is justified or unjustified? Moreover, other areas worth their salt would lobby to also be recognised as a key area, or at least a potential one. Which area would not consider itself a potential key area? In this way backing winners might lead to picking winners (or: winners picking governments) again.

This is what we observed in the Netherlands of the 1980s, when government, upon the advice of the 'Wagner committee', instituted a similar policy of 'arrowhead' sectors. It started out with seven, increased to 14 and ended up with 21. Similarly, while in present policy there are now four key areas (flowers and food, high-tech systems and materials, water, and creative industry), four more sectors have been recognised as potential new ones (chemistry, life sciences and health, logistics, and durable energy), and two more have laid a claim (the finance sector, and the Hague as 'residence of peace and justice').

Finally, there are obstacles of bureaucracy, in government failures. We will discuss those in more detail later, but here the point is that participation of the government in the funding, facilitation, and monitoring of consortia for innovation entails bureaucratic entanglement, with consequent delays, irritations, and frustrations among business partners. It is naïve to see this as a simple matter of lack of competence, will, or commitment on the part of civil service. It is due, again, to an institutional logic that follows from the need for public spending to be publicly justifiable and accountable, with a division of competencies and authority across different sections and levels of a government department, and in innovation often also across departments, to accommodate novel combinations.

Clearly also, there are sectors of activity where, for classic arguments concerning public goods and externalities, government needs to take the lead, such as health, education, infrastructure, mobility, energy, environment, security, and, increasingly, the provision and control of water. In matters of societal interest with a public goods nature, government should not just back winners but also pick them, even in spite of the inevitable errors in trying to outguess evolution, because private firms do not take the lead. If private firms cannot be tempted to make choices that are liable to mistake, then government must make them.

Next, there is the issue of openness to the world. In particular, do targeted industrial policies satisfy our condition of openness to the world? Or are they in danger of locking innovation up in a country while the priority may lie precisely in linking national activities into globally fragmented value chains, as argued in chapter 1? The latter danger is not unrealistic: in a survey of measures of Dutch innovation policy, in chapter 10, it was shown that of 21 measures only one was explicitly directed at international collaboration (the Innovation Subsidy for Collaborative Projects). It is true that implementation of policies for chosen key areas does occasionally include measures to make international connections. Typically, this occurs *ex post*. However, in the basic choice and design of key areas there is the simple point of institutional logic that an EU nation is not inclined to involve organisations from other EU countries in its national innovation policy, so that the basic design is necessarily nationally oriented, while the choice and design should be taken in an international perspective from the very beginning. In other words, the choice, *ex ante*, of what to do, and of whom to involve, should be internationally oriented, and not only the choice of how to implement national choices. An example of how national boundaries constrain innovation projects is given below.

"The University of Hasselt, in Belgium, cooperated with the University of Maastricht, in the Netherlands, in the development of an instrument to measure a plant's vitality by the state of its chlorophyll, and they needed a development subsidy. Dutch authorities did not want to subsidize cross-border collaboration, and, likewise, Belgian authorities required subsidies to be spent only on Belgian firms and institutes" (*Financieele Dagblad* 2007).

Also, if we want to be leaders rather than followers in some areas, then in those areas we should not take a parochial view of excellence and be satisfied with the best that we have at home, but instead we should take a global view, where we build what is best in the world. Innovation policy should beware of policy measures that are nationally oriented and may have the adverse effect of locking innovation up in a country when instead innovation policy should enable international connections.

In sum, utmost caution should be taken in any policy of backing winners. The emphasis should lie on enabling challengers. There may be arguments for lowering obstacles in the realisation of proven innovative potential, but these obstacles should be made explicit as a basis for policy and its design. But what, then, is our answer to the question of where to be leaders and where to be followers in innovation? We maintain that one should exercise restraint in backing winners, except in cases of market or system failures, since in principle winners in markets should reap their own funds for expansion.

12.4 FAILURES OF GOVERNMENTS, MARKETS, AND SYSTEMS

12.4.1 GOVERNMENT FAILURES

Before undertaking any government policy for innovation, the customary question, which is still valid, is why government should or should not act? Why do markets fail, in respect of innovation, what other failures may there be in the innovation system, and what are potential government failures? Usually, the analysis starts with market failures and ends with government failures. Let us do it the other way around.

The main government failure, in line with Hayek's ideas, is that government does not have an adequate grasp of the varied – to a greater or lesser extent idiosyncratic and context-specific – local knowledge and insights that are distributed across society. In view of diverse and distributed knowledge, government should be modest about its grasp of what can be done, and should tap into local knowledge rather than impose its own.

Government is vulnerable to being taken hostage by established interests. It needs the knowledge and support of industry in the design of subsidies and other schemes, but design and implementation then inevitably carry the imprint of their interests, as we discussed above. Also, in the discussion of the foundations and limits of trust, in chapter 7, we noted that government is more vulnerable to loss of reputation, and has to be more careful with commitments and openness of information than industrial firms. Since government is more vulnerable to opportunism than business, it should beware of confirming established interests by getting involved in close relationships with industry.

Earlier, we noted that bureaucratic entanglement is inevitable when government becomes involved in programmes or projects for innovation, as a result of the imperative of public accountability. The question is how this can be fulfilled without generating the delays, irritations, and frustrations that result for business partners. Here, the principle comes to mind of an 'account manager', which is already familiar in business: there should be a single point of contact. This account manager carries the responsibility, and is given the authority, of coordinating across departments and across segments and levels within departments. We see this as a basis not only for resolving coordination problems, but also for policy learning. The need for an account manager to coordinate across departments or segments within departments exerts a pressure to take cognisance of inconsistencies or incoherence between regulations, rules, or processes, as a basis for simplifying or revising them. This pressure is greater than when it is left to the outside partners to deal with the problems of bureaucratic entanglement. In this way, the public internalisation of transaction costs may best stimulate their reduction by streamlining and integration. The problem of bureaucratic entanglement is exacerbated by the condition, argued in chapter 3, that in innovation, particularly when exploration plays a large role, there should be openness concerning the course of a project, to allow for the surprises and changes of direction that are inherent in innovation. This is at odds with any *ex ante* specification of 'deliverables' that must be achieved for payment of the subsidy. An alternative then is to attach an official to the process, to monitor progress, discuss and authorise changes of direction, and to assess the value of outcomes that deviate from expectations. This entails a fundamental switch from the authorisation of deliverables that need to be specified up front, to the authorisation of the process as it unfolds, and/or the value of outcomes, *ex post*. Thus, a public account manager has the task not only of coordinating public contacts, but also of authorising progress and legitimating outcomes.

Earlier we noted the risk of government becoming hostage to private interests, when playing an inside role in collaborative programmes or projects. Now the additional problem is that of temptations of favoritism or outright corruption. Can this problem be resolved? We suggest a solution. Public account management and process authorisation should be accompanied by safeguards of accountability. This probably requires a supervisory board, to specify boundary conditions and for accountability *ex post*. Account managers submit verifiable reports according to standards specified by the board, which are scrutinised, integrally or by sampling, *ex post*. This entails correction of the conduct of account managers, and a basis for learning, in an adjustment of rules and procedures. This increases transaction costs, but it is not clear *a priori* that total transaction costs, for government and business, increase. They may well decrease, even substantially. That is an issue for further investigation. Crucial here is that the arrangements allow for more openness of innovation.

12.4.2 MARKET FAILURES

Knowledge does not suffer as much from the market failure of non-excludability as has been assumed in earlier economic theory, since it cannot always spill over easily, due to cognitive distance and limited absorptive capacity. As discussed in chapter 4, a policy question is to what extent protection of intellectual property (IP) is still needed. We suggest that the degree of IP protection should be weakened – by shortening the period of protection, by raising the originality bar, and by making compulsory licensing and parallel imports easier (Chang 2007: 143).

Where patent protection is still needed, for small, independent innovators there are transaction costs with effects of scale in the acquisition, monitoring, and protection of intellectual property rights. Acquiring a patent is costly, especially for smaller firms and especially the first time when the procedure is unfamiliar, as is the cost of determining if a certain patent already exists, and also the monitoring and fighting of patent infringement. The procedure is also too slow, especially for small firms, who often have to move fast and lack the resources to wait long. There should be a single EU patent, not complicated by language issues, available through a faster process, and a lower price for searching, filing, and renewing a patent, as well as support for small firms in the identification and redress of infringement. This is on the agenda of present innovation policy, but its urgency should again be highlighted.

As indicated above, market failure in the appropriation of rewards from knowledge is not as strong as previously thought, due to cognitive distance and limitations in absorptive capacity. However, the reverse side of that coin is that there are serious market failures in diffusion of innovation. Knowledge 'transfer' is a highly misleading term. Firms with limited absorptive capacity cannot easily capture innovative opportunities. Often this creates a problem particularly for smaller and more traditional firms, who for reasons of scale lack specialised expertise. Also, particularly in more traditional smaller firms knowledge is often highly tacit, transmitted via 'learning by doing' rather than formal learning. Tacit knowledge is harder to assess by outsiders than codified knowledge, and tends to be self-evident to the bearer of the knowledge, which makes criticism of existing knowledge and practice, and hence adoption of innovations, more problematic. Lack of absorptive capacity and a high degree of tacit knowledge yield what one might call 'cognitive' transaction costs, and these are relatively high for many small firms (Nooteboom 1993). Therefore, there is a good rationale for a subsidised knowledge transfer service to small firms, as offered by the Dutch 'Syntens' organisation.

Consider the recent policy measure in the Netherlands of 'innovation vouchers', aimed at triggering the 'market' for knowledge 'transfer', with which small firms can obtain free advice from universities, polytechnics and technical institutes. There is a potential problem here. Knowledge intensive firms will mostly find their own way to knowledge institutes, and hardly require a subsidy. The more traditional, less knowledge intensive firms that form the main target group often do not have the absorptive capacity to deal with especially universities, to whom cognitive distance is often large. There is a risk that more time will be needed than expected to build mutual understanding and trust, and to properly frame the research question, before knowledge can be shared. This may far exceed the worth of the voucher, and in the end the result may even be counter-productive, with the small firm feeling that with the voucher it was lured into a waste of time and effort. A recent evaluation of the scheme showed that firms that applied for the voucher were indeed larger and more experienced in innovation than average firms (see CPB 2007).

As argued in chapter 3, exploration requires time and slack resources to deal with uncertainties of goals, means, causal relations between them, and resources needed, in iterations with relevant stakeholders, and this slack is likely to be eliminated by extreme price competition. This constitutes a newly recognised market failure for innovation. Here, the failure is not that the market does not work, but that it works too well, in the sense that price competition is so intense as to squeeze out resources for exploration. We suspect that this is one of the reasons why industry has cut down, or even abolished, its more fundamental 'blue sky' research, and is now falling back on universities for such research.

For collaboration for innovation we argued that, as elaborated in chapter 7, a minimal duration of relationships is needed to make and recoup relation-specific investments in mutual understanding and trust, and that we should strive for optimum, not maximum, flexibility of relationships. This has implications for competition policy. Competition policy should allow for dedicated collaborative relationships, with specific investments needed to utilise cognitive distance, to have an adequate duration, even if this means temporary exclusivity and hence limitation of competition.

Mobility of labour helps innovation, in both the diffusion of new practice and exploration. In particular, innovation is stimulated by movement between exploitation and exploration. However, for labour also the logic applies that a certain stability of a job is needed to elicit the specific investments in mutual understanding and trust that is needed to cross the cognitive distance that is conducive to innovation. The argument extends to the ownership of firms, in issues of corporate governance. There also, ownership should not be so volatile as to discourage specific investment needed to build networks, mutual understanding, commitment, and corresponding trust. This means that in relations of work and ownership of firms, we need a corrective to the present discourse of maximum flexibility, aiming for optimal, not maximum flexibility.

12.4.3 SYSTEM FAILURES

System failures arise from mismatches between elements of the innovation system. These are many. In the previous section we discussed the institutional logic of 'pre-emptive participation'. Some system failures go beyond the scope of this book. One of those is lack of adequate education and training to match new ideas or practices. Other system failures have been discussed in various chapters in this book, and conclusions will be presented below. One of those is an apparent lack of connection between the generation of new knowledge, at universities, and its application, in industry, as discussed in chapter 4. A second is lack of finance for entrepreneurship, discussed in chapter 5. A third is lack of adequate management and organisation to deal with the need to combine exploration and exploitation, discussed in chapter 8, and to provide conditions for the functioning of creative teams, discussed in chapter 9. These failures, and implications for policy, are discussed in subsequent sections. Another system failure is lack of trust and ability to collaborate, in alliances for innovation, discussed in chapter 7. Another system failure is the reverse, in a sense: lock-in into established interests and positions, discussed partly in chapter 7 and partly in chapter 11.

In our critical discussion of the policy of 'backing winners' we granted that there may be problems of coordination in the collaboration needed to utilize opportunities of novelty, and that this might yield an argument for government intervention. Here we will focus on that issue. Regularly, stakeholders get stuck in stalemates of manoeuvring for position, in choosing which option to take, which standard to establish, and how to divide influence, costs, risks, and revenues.

In chapter 6 we gave several examples. One was the development of the 'whisper' bus , reducing noise, pollution, and energy consumption. This was stalled due to lack of coordination between the municipality of the city of Apeldoorn that wanted to promote this bus, the Ministry of the Environment that was prepared to subsidise the scheme, the public transport companies that would need to introduce the busses, and the bus manufacturers who would produce them. There was no alignment of interests and everybody was waiting for everybody to take the first step. Another example was the development of an electronic patient dossier, which stalled due to lack of alignment between a wide variety of physicians and medical specialists, psychotherapists, physiotherapists, nurses, hospitals, home care providers, pharmacies, and the ministry of health.

In terms of game theory, players can get caught in equilibria of non-cooperation ('prisoner's dilemmas'). Often, players need each other but are also rivals. Examples are technical standards or market structuring. Some solutions are closer to established practice and knowledge of a given player than others. It may be difficult to bring manoeuvring for position to an end. Once equilibrium is found it may be very difficult to find new equilibrium when conditions change. An outside party, like the government, may be required to take steps to solve the dilemma. We have heard this argument for government intervention several times, also from industry. We are sceptical about this. There are conditions where intervention is indeed needed to unlock such lock-ins. This may be a task for government, but there may be alternative intermediaries. And if, for want of adequate intermediaries, there is indeed a need for government to intervene, it may have to be only a short-term 'nudge', after which government can retire, rather than maintain ongoing involvement. As indicated earlier, government can be taken hostage in tugs of influence.

The real issue, it seems to us, is that rather than running to government for help as soon as deadlocks arise, industry should face up to the challenge of developing the capabilities needed for collaboration and network formation, in the new networked knowledge economy. Also, as argued in chapter 7, there are interesting entrepreneurial opportunities in a new branch of business services to provide the roles of go-betweens to facilitate collaboration for innovation. There is considerable knowledge concerning the issues involved and ways to deal with them, which can be exploited commercially (Nooteboom 2004).

Chapter 6 showed how specific the problems can be, in unique configurations of positions and interests of different stakeholders that are difficult to translate into general regulations. Solutions then need to be case-specific, to be effective and to forestall further accumulation of regulations. There may be a role for a 'deblocking brigade'[1] that is expert in managing complex relationships and has the authority to cut across different areas of departmental and jurisdictional authority, and can be called in when unyielding systemic failures occur. We propose this deblocking brigade partly as an alternative to the targeted industrial policies that we criticised earlier.[2] In sum, government should exercise utmost restraint in participating in the configuration, design, planning, and progress of collaboration between private actors. It should stimulate private actors to take their own responsibilities, and to develop their own capabilities in collaboration, and it might stimulate the development of business services that facilitate collaboration. Government should intervene only when obstacles arise that are inveterate to the point that they cannot otherwise be resolved.

Earlier, we discussed government failures, and the de-blocking brigade should be consistent with that analysis. First, clearly officials at the brigade should be masters in the management of collaborative relationships and networks for innovation. Second, the process requires an 'account manager' with sufficient authority. This 'account manager' should provide a single point for integrated and consistent access to the public stakeholders involved. Third, when the issue at hand involves a prolonged process of innovation, this official should have the authority to authorise changes of direction in order to maintain openness to innovation. The official should be able, or have the resources, to establish a valid judgement about the quality of the outcome of the process, when it does deviate from expectations. Fourth, to cover the risk of favouritism or even corruption that such a position of power and trust could elicit, the brigade should be subject to a supervisory board that establishes boundary conditions, and to which project managers report for possible correction, *ex post*. This board would be accountable to parliament, through a minister that has this in her portfolio. The reporting procedure should provide a basis for policy learning.

12.5 GENERATION AND UTILISATION OF KNOWLEDGE

12.5.1 VALORISATION

In chapter 4, we considered the system failure of the supposed 'knowledge paradox', according to which in the EU in general and the Netherlands in particular scientific performance is fine, but fails to be adequately carried forward into innovation. Presently, in the Netherlands, this problem of utilisation, or 'valorisation' as it is called, carries high priority in innovation policy.

In chapter 4 we noted a perverse effect of forcing universities to ask for a fee for any knowledge they supply to industry. Pricing this knowledge provision is not efficient in view of the transaction costs involved. Furthermore, even from a commercial perspective it is better to allow for an ample threshold of free advice that forms the basis for judging the merits and feasibility of a contract for a more substantial research project.

In the preparation of this book we developed some ideas on how the interaction between university and industry in exploration could possibly be improved, and we tested those ideas in a series of interviews with four universities, a polytechnic, two technical institutes allied with universities, two medium-sized firms (MOTEK and OTB), and three large corporations (Unilever, Shell, and Philips).

The main result is that among our respondents the problem of valorisation is not perceived to be quite as large or urgent as it is among policymakers in 'Dutch government. Many forms of collaboration have already been developed. Lessons can be learned from this, improvements can be made, and it may be useful to collect and diffuse the resulting principles, summarised below. The principles that emerged from our own analysis, in chapter 3, plus feedback from the informants, are as follows.

A key principle, in our view, is that the relation between university and industry is not only a one-way process of putting university research to use, in exploitation, but a two-way process in which inspiration is also provided from practice for more fundamental university research, in exploration. Theoretically, this follows from the analysis of the 'cycle of discovery', in chapter 3, in which exploitation and exploration build upon each other. Most respondents acknowledged the validity of this principle.

Another key principle, to ensure exploration, is that meetings between people generate ideas and projects, around a broad theme, rather than that projects are specified beforehand for participants to subscribe to. This is related to the Hayekian principle that innovation projects should tap into the diversity of knowledge and ideas spread around different communities, and the principle that projects should be sufficiently open to allow for surprise and change of direction.

It proved useful also for other reasons to employ the distinction between exploitation and exploration. Exploitation emerges in more or less straightforward contract research, in which industry can specify and contract for desired outcomes. Exploration arises in collaboration between universities and industry to literally explore uncertain but possible developments, in new technologies, areas of application or problems to be investigated. In line with our earlier analysis, in chapter 7, which yielded a recommendation for optimal rather than maximum flexibility, it was recognised that in such collaboration for exploration each actor should take enough time to get to know and understand each other ('speak the same language') and build trust, since in such settings contractual control, for example of property rights, is difficult, cumbersome, unworkable, or even counter-productive. On the other hand, there should be enough turnover of people to maintain variety of ideas and sufficient cognitive distance.

Concerning the issue of intellectual property and 'spillover', respondents noted a growing awareness that a relaxation was in order: to get knowledge one should offer and risk knowledge, in many cases exploration was sufficiently 'pre-competitive' not to yield a direct threat, contractual control would not work in early stages of exploration (there isn't yet anything well defined to appropriate), and people simply could not make progress without each others' knowledge. This does not mean that issues of intellectual property have disappeared, but that in early exploration they are less urgent or even do not arise. As one respondent said: patents from early exploration are likely to have elapsed before exploitation is reached.

For some respondents it was fine to conduct exploration more or less ad hoc, with different groups of partners on different occasions. The advantage of this is flexibility. Others were aiming at more durable, 'strategic' relations, as a basis for building understanding and trust. An element of such relations was also the provision of temporary staff and exchange of personnel, for the duration of a project. Some respondents expressed apprehension at too close and extensive an involvement of firms in universities. That may interfere with the independence of universities and their task, generally seen as legitimate and important, to do independent fundamental research, and it may be seen by others (for example, politicians and the public) as appropriation of public institutes by business.

A solution here may be the concept of a 'third space', between university and industry, in the form of an actual institute, with residential facilities, dedicated to their interaction, financed jointly by business and the government, with the express task of enabling and facilitating meetings and joint projects to utilise cognitive distance for the sake of exploration. Such separation from both university and industry may serve to protect the integrity of the university and ensure that the work does not fall back into exploitation induced by commercial pressures. A condition for this to work is that participation by academics is seen as legitimate, and is given a place in the system of performance evaluation. For industry, a condition for this to work is that their staff views participation as a good career move within the firm. A third space could perhaps also function as a platform for entrepreneurial spin-offs. In that case the set-up may also involve venture capitalists. Mostly, the respondents from university and technical institutes claimed that they practised what we suggested, and that something like 'third spaces' already were a reality. We acknowledge that, but not all potential participants seem to be aware of all relevant options, and we believe that the people involved can still learn, mainly from each other.

Valorisation may be less of a problem than is perceived by policymakers. The actors involved, in industry and academia, have developed forms of interaction for both exploitation and exploration, oriented at both application and inspiration of university research. However, options and forms can still be further developed and improved. Government has only a limited role here – to ensure that adequate incentives are in place for university staff to participate in interaction with industry and to disseminate experience in modes of collaboration.

These arrangements are appropriate for large firms and for small firms with adequate absorptive capacity, such as high-tech firms or spin-offs from universities or large firms. However, the limited absorptive capacity of most SMEs makes Dutch institutes of higher vocational education a more suitable collaboration partner for them than universities. To play this role, institutes of higher vocational education should be given the resources to conduct applied research.

12.5.2 FUNDING UNIVERSITY RESEARCH

An issue in relation to university research, is to what extent such research should be programmed and rewarded according to societal needs. We do acknowledge the legitimacy to indicate broad areas of public priorities, such as energy, environment, water, health care, and aging. However, in line with our Hayekian perspective, we are wary of bureaucrats in central offices, at ministries, and national science foundations, specifying programmes or projects for which scientists or businesses may then submit tenders. Why should they know better than the totality of differentiated and dispersed knowledge? There is a danger also of fashionable 'hypes' directing research. Ideas should be allowed to arise from below, from the wellsprings of variety. This would also mean that within universities ideas for Ph.D. projects should come more often from Ph.D. candidates themselves than from their supervisors, as is the rule in the US but occurs in the Netherlands only at some universities. It must be possible to allow for this and still achieve sufficient coherence and connection between projects.

On the basis of a traditional logic of public goods, we propose to back winners in scientific achievement, with awards that yield funds for expansion to proven researchers and research leaders. The merit of such a policy is that it avoids the problem of trying to predict successful new ideas, and provides the means for success to expand where it cannot generate its own profits to do so. As discussed in chapter 4, in the Netherlands we have such awards, notably the 'Spinoza prize'. Note that the award does not yield a higher income to the researcher, but funds to expand his patently excellent line of research. What successful firms get out of profits researchers should get out of awards. The logic of this principle is that in order to avoid the paradox of outguessing evolution, and as a means of extending the scope of success, one gives a reward afterwards for proven success if the research cannot get such a reward from a market. We applaud awards for proven personal excellence, allocated in second-stream funding. In addition to that, however, funding on the basis of prior assessment of the merits of a research proposal is also needed, to allow for newcomers and outsiders, but this is more problematic, in view of the difficulty to predict the value of exploratory research. As argued in chapter 4, here we have a preference for the first stream, in view of the greater variety and richness of local knowledge and perspectives for generating and evaluating proposals.

A further issue is inter-disciplinarity of university research, as discussed in chapter 4. While there are good reasons for disciplinarity, both radical novelty and application often arise across the frontiers of disciplines. However, universities are mostly oriented towards, and organised according to, disciplines. This was noted above as one of the complications in university-industry collaboration. Indeed, one of the reasons for university-industry collaboration, possibly in third spaces, is to stimulate interdisciplinary exploration. In this respect, measures are needed to facilitate inter-disciplinary research, for example with special programmes at the Science Foundation, and measures to recognise and reward interdisciplinary research at universities, in formal assessment systems and Ph.D. programmes.

12.6 ENTREPRENEURSHIP

In view of our evolutionary argument for variety generation, and our plea for openness to surprise and to challengers of the status quo, entrepreneurship is of central importance for innovation. In chapter 5, we distinguished between Schumpeterian entrepreneurship, in breakthrough innovations, and Kirznerian entrepreneurship, in more incremental innovation and 'arbitrage' in filling gaps between supply and demand. The former are needed to produce innovations and the latter to diffuse and fully profit from them. While the second type does not and should not form the focus of innovation policy, it is important for economic policy more widely, in order to diffuse innovation into all relevant parts of of society and to profit fully from innovation in economic growth. Self-employment may also yield an escape from unemployment. Beyond economics, self-employment has an important societal value as a stabilising political factor, for example as an avenue for emancipation, an escape from discrimination in labour markets, and for social acceptance.

The majority of self-employed people are of the second, Kirznerian type, if innovative at all. That is not surprising: as argued in chapter 1, one should expect only few breakthrough innovations relative to incremental innovation and diffusion. For innovation policy we focus on the more dynamic, Schumpeterian entrepreneurs. Research tries to identify these in new technology-based firms, corporate venturing, spin-offs, and high-growth start-ups.

On the basis of recent research it was shown in chapter 5 that notwithstanding a recent increase in the number of self-employed people in the Netherlands, on average small- and medium-sized enterprises (SMEs) have become less and not more innovative in the last decade (1999-2007), and the percentage of innovative SMEs is much lower than the EU average. The Netherlands is lagging behind internationally with respect to entrepreneurial activities in general and ambitious entrepreneurship in particular. In view of the importance we attach to entrepreneurship for an open system of innovation this is a serious matter.

Another striking finding is that, counter to received wisdom, there is a relatively low 'fear of failure' in the Netherlands. Since this does not explain lack of innovative entrepreneurship, then what does? It may be lack of ambition. If it is, such a cultural feature is hard to change, certainly in the short term. So, we turn to markets and institutions. In particular, while venture capital has developed in the Netherlands, there is still a perennial problem in the Netherlands, as elsewhere, in the provision of early stage finance ('seed capital'). Here we make a link to the need for major efforts for innovation in public sector activities such as those related to energy, the environment, climate change and water management, ageing and health care, and congestion in transport. Since government will have to make a major investment in research and development in these areas, this might also be used to stimulate and mobilise entrepreneurship. In view of institutional logic that tends to favour larger firms, it would be a good idea to systematically channel public funds for R&D to smaller firms, in imitation of the SBIR programme in the US.

Corporate spin-offs are a much more important phenomenon than university spin-offs, both in number and in impact. In chapter 5 we discussed the conditions for corporate spin-offs. The advantage of such spin-offs, from the perspective of combining exploration and exploitation, is that they carry relevant experience from business into new ventures (See chapter 3). Chapter 11 also offered evidence of the importance of spin-offs for innovation. Corporate spin-offs might be stimulated by lowering barriers such as anti-competition clauses. Innovation policy and innovation strategies of large firms should stimulate corporate venturing to a larger extent, for example by showcasing best practices.

In our analysis of interactions between university and industry, in chapters 3 and 4, we noted the danger of too many market pressures within universities. That provided a basis for considering 'third spaces' between university and industry. This notion of a 'third space' also has implications for the development of 'business incubators' at universities. Facilitating the growth of promising university spin-offs by privatised incubators both provides these incubators with more competent staff, and protects the university in its role as producer of public knowledge.

The fragmentation of present innovation policy in the Netherlands, and of schemes and regulations more widely, yields high transaction costs, which weigh more heavily for small firms due to effects of scale in transaction costs (Nooteboom 1993). Often, these costs are in the nature of a fixed effort of search, contracting and monitoring compliance, which weigh more heavily on small volumes of business than on large ones. In addition to such familiar types of transaction costs, we noted cognitive transaction costs, related to absorptive capacity and tacit knowledge, which also exhibit an effect of scale. The case studies in chapter 6 illustrated barriers to innovation, in established interests, standards, procedures, coordination failures, and perverse institutional logics, which can obstruct innovation generally but are especially difficult to overcome for small firms.

12.7 ORGANISATION

An important part of organisation for innovation lies in the collaboration between firms and other organisations, in open innovation. This has been discussed above, and it reappears in the discussion of networks. Here the focus is more on internal organisation. Public policy has at best an indirect role to play here, for example in the diffusion of best practices in organising innovation, and hence this section is modest in scope. In chapters 8 and 9 we analysed organisational conditions for innovation, and perspectives for organisational innovation, concluding that there are enormous as yet unexploited opportunities. Perhaps this is the cause of the Netherland's apparent lag in total factor productivity relative to the US, which appears to be ahead in this regard.

There are many opportunities to improve the efficiency of exploitation, but the big challenge is how to combine, within an organisation, or to connect, between organisations, exploration and exploitation, in 'ambidextrous organisation'. In chapter 8 we found ways to do this. In chapter 9 we identified conditions for the functioning of creative teams within organisations. The problem, however, is that the more short-term oriented and familiar conditions of exploitation tend to prevail and to squeeze out exploration, which is also more vulnerable to pressures of price competition. This appears to lead to a conservatism in management and organisation that the Netherlands can ill afford.

In chapter 9, we noted that a number of organisational conditions for innovative teams are consistent with dimensions of the quality of labour: intrinsic motivation, autonomy, openness, psychological safety, and inspirational leadership. There appears to be an opportunity here for a new alliance between management and labour, to both stimulate innovation and improve the quality of labour.

Related to the combination of exploration and exploitation, but going beyond that, profiting from new opportunities of ICT, there are several new forms of platform organisations, in which users can to a greater or lesser extent configure their product, and even contribute to its innovation, and open source communities, where users are the producers. Particularly for government services there are obstacles in present authority relations and in rules of accountability, there are opportunities to make mistakes, and there is low status for 'front-office' workers. There are opportunities to improve innovation, in both quality and speed, not least in government services, by employing the concept of platform organisations, and even, perhaps, of open source communities. For government services, this will require a change of perspective concerning the role and the authority of the 'front office', in suggesting and leading innovation. We have only touched lightly upon the wide issue of innovation in government. The further exploration of opportunities and their exploitation in the innovation of organisation in government services by itself merits further study.

12.8 NETWORKS AND REGIONS

We also see conservatism and insufficient learning in the limited ability of public and private organisations to collaborate with other organisations, in open innovation. Too often, organisations engage in the power play of mergers and acquisitions, to maintain hierarchical authority, rather than engage in the more difficult game of collaboration in mutual dependence, where that would be better for innovation. Inter-organisational collaboration yields more flexibility of (re)configuring activities and greater variety and cognitive distance, conducive to innovation. Also, as argued in chapter 7, formal hierarchical authority and control are less appropriate especially for exploration. There is increasing insight into how to manage such relationships, in mastering the art of trust, as discussed in an earlier section.

As discussed in chapter 5, there is complementarity between small and large firms, which should also be taken into account in regional policy. Small firms provide the variety of trial and error needed for experimentalism, and large firms provide a platform for spin-offs of new entrepreneurial firms, a basis for large-scale exploitation (production, distribution) of innovations, and a home for deep specialisation of labour (with a corresponding advanced demand for labour). Also, as argued earlier, large firms are often more interesting partners for universities than small firms, because they generally offer higher levels of knowledge and specialisation, lower transaction costs, and deeper pockets, to engage in collaboration in exploration, possibly in 'third spaces'. For most small firms polytechnics are more natural partners, in view of a smaller cognitive distance.

In chapter 10, a survey was given of features of networks of collaboration for innovation, their effects on innovative performance, and opportunities to gear policy to favourable network variables, such as variety of potential participants, centrality, density, and small world structure, which was also discussed in chapter 7. For example, policies to further entrepreneurship contribute to the number and variety of potential participants in networks. As noted in chapter 7, governments have a task in maintaining institutions that support reliability and trustworthiness of actors.

As discussed in chapters 7 and 10, structure and strength of ties in networks have effects on both competence and governance in networks for innovation. On the one hand, dense and strong ties are needed, especially in exploration, to ensure accessibility to contacts under the volatility of networks, to pool absorptive capacity for understanding sources of knowledge, triangulation for accuracy, and reputation mechanisms, coalition formation and trust building as an alternative to contractual control. On the other hand, strong and dense ties can limit variety and cognitive distance, and can yield cognitive and relational lock-in, thus reducing the variety and flexibility of configuration needed for innovation. A solution to this dilemma lies in 'small world' structures, where local communities with dense and strong ties are complemented with less dense and weaker outside ties to other, similarly structured communities.

The cycle of discovery, discussed in chapter 3, provided the foundations for a call for openness to new contexts, including foreign countries. In chapter 7 we noted that an alternative to external weak and sparse ties is frequent entry and exit of players, to maintain related variety and cognitive distance.

As discussed in chapter 11, in geography there is a notion of 'related variety', concerning activities that on the one hand differ but on the other hand are still similar. The claim is that such related variety contributes most to regional employment. Both more variety and more specialisation yield less growth. Note the correspondence between the notions of cognitive distance and related variety, where both plea for difference that is enough for potential novelty but not too much to utilise that potential. However, while related variety contributes to innovation, unrelated variety contributes to the spread of risk.

Regionally embedded, diverse, but also related, activities, that are upgraded with processes of localised learning, are less footloose, and do not dissipate as easily to emerging economies (such as China and India) as isolated activities or technologies do. However, there is a danger of locking activities into regional or local 'clusters' that are 'over-embedded', with too high strength and density of ties. The concept of small worlds suggests that there must also be 'channels' that connect an agglomeration with comparable, competing, and complementary agglomerations elsewhere. We call the result an 'open agglomeration'. Here, we wonder whether perhaps universities can provide the connecting nodes of small worlds, connecting regional innovation systems to similar systems elsewhere in the world (cf. Benneworth et al. 2006, Kitagawa 2005). Universities are geared to such access to internationally dispersed communities.

This potential role for universities is to be connected with the idea of 'third spaces' between universities and industry, discussed above. Here, third spaces connect universities with industry, in regional innovation systems, while the universities provide the necessary channels to the world. Local collaboration is probably more geared to exploiting the potential of novel combinations, on the basis of experience and experiments from local related variety, with an interdisciplinary structure and orientation, while the outside, global connections of universities, organised along disciplinary lines, provide a deepening and renovation in specialised areas (Ponds and Van Oort 2006). Universities as connecting nodes, via a third space that connects them to local industry, may also help firms to gain access to new markets of inputs and outputs. That role of providing outside connections is also played by large firms.

The situation in the Dutch region of Twente appears to confirm the danger of too much local embedding without sufficient outside connections in the world. In spite of an exceptional involvement of the University of Twente in local business, development of the region has been disappointing (Benneworth et al. 2006). Perhaps here local involvement has become too strong relative to international linkages. Or is the local 'buzz' in Twente too limited?

In present policy there is an inclination to identify 'best practice' in regional systems, in the form of the configuration of activities and their organisation and governance (the renowned Silicon Valley and Italian industrial districts), and transplant them to one's own environment. This is illusory. As argued in chapter 11, local and regional systems are the historical outcome of the confluence of locally specific and highly path-dependent conditions. The context specificities of these systems are essential for structure and performance and cannot be universalised. Their success must be unraveled in underlying logics, such as those we are trying to set out here. As argued in chapter 11, there is spatial division of labour, with some regions being strong in research, others in innovation, and yet others in production. Different industries tend to concentrate in different regions and within an industry firms may look different in different places. As a result, different regions have potential in different types of activity. It is myopic to focus policy on regions that are strong in R&D, for example, while neglecting potential in application and production. Different indicators, such as R&D, creative workers, and high-tech industries reveal different strengths in different regions, and in the Netherlands it is hardly possible to find a region that is not strong on one or more of these indicators (Raspe et al. 2004).

The argument for regionally embedded policy gains further force when we take into account the dynamics of clusters. As discussed in chapter 7, in the early stage of exploration one would expect a relatively high need for local embedding, in strong and dense ties. Later, one would expect a certain amount of disembedding, to utilise the potential of emerging innovations in distant markets, and to achieve access to novel sources of novelty to replenish local variety and restore cognitive distance. This raises considerable complications for a policy for regional clusters and innovation systems. Are policymakers able to correctly identify the stage of development that a local cluster is in, and are they able to implement policy in time, before development has reached the next stage, where the policy may be counter-productive? One may wind up furthering local embedding by the time that disembedding is needed. Regional government seems better informed and better able to act quickly than national government. However, regional/local government may not be able to break up local structures if they become counterproductive under the creative destruction of radical innovation, illustrated in chapter 11.

In view of local specificity of knowledge, conditions, successes and failures, and the need to choose and implement policies in time to fit the needs of different stages of development, central government should be modest in its ambitions to design regional development, and leave it as much as possible to 'bottom-up' regional initiatives, in the locally specific configuration of relevant variables. Such variables are: related variety, complementarity between large and small firms, collaboration between universities and large firms, and between polytechnics and small firms, spin-off formation, features of network structure, education and training, labour mobility, linkages outside the region ('small worlds'), and inward and outward mobility of firms and people.

National government could then focus on the linkages between regional clusters and other clusters, at home and abroad, the possible role of universities in establishing and maintaining linkages abroad, entry and exit of organisations, establishing safeguards against local or regional clientism and possible corruption, and providing pressure and support to break up regional structures when they become counter-productive under the creative destruction of radical innovation.

There should be a division of labour between central and regional government. This we connect with the notion of experimentalist governance or directly-deliberative polyarchy, discussed in chapter 3, where:

"... lower level actors ... experiment with solutions of their own devising within broadly defined areas of public policy. In return they furnish central or higher-level units with their rich information regarding their goals as well as the progress they are making..., and agree to respect in their actions framework rights of democratic procedure and substance... With periodic pooling of results... (that) reveals the defects of parochial solutions, and allows the elaboration of standards for comparing local achievements, exposing poor performers to criticism from within and without, and making of good ones (temporary) models for emulation" (Gerstenberg & Sabel 2002).

And finally, we wonder whether for an overall policy concept we could think of the Netherlands as a 'hub of buzz'; an open knowledge economy or ecology, where as an extension of its traditional function as a portal to Europe, as a hub of streams of goods, it could also function as a hub for meetings of explorers and exploiters of a variety of knowledge. As discussed in chapter 7, this would entail meetings between scientists, producers and users of technology, traders and businesspeople, designers and artists, politicians, diplomats, lawyers, security and police officials, publishers, marketers, and distributors. To support such a hub, we would need a variety of supporting services, in law, finance, transport and distribution, conferencing, education, communication, languages and publishing, accommodation and housing, with attractive spatial, recreational and cultural environments. For all this, we would need a renaissance of traditional openness to other cultures. Hopefully, the Netherlands might be a place where people meet at a fruitful cognitive distance and where trust is built. A place where identity matters little and processes of identification take place (WRR 2007). Perhaps the Dutch can again, and even more widely than in the past, assume the role of 'go-betweens' to help other people cross their cognitive distances.

NOTES

1 We were inspired partly by the establishment, by the Dutch foundation 'Nederland Kennisland', of a 'Kafka brigade' to help people who are caught up in institutional tangles.

2 The Dutch 'Innovation Platform' was advertised in terms that suggest that it should have acted as a 'de-blocking brigade', but in fact it did not quite get around to carrying out that function.

REFERENCES

Benneworth, P., G.J. Hospers, and B.W.A.Jongbloed (2006) 'New Economic Impulses in Old Industrial Regions: The Case of the University of Twente', in: A. Prinz, A.E. Steenge and J. Schmidt (eds.), *Innovation: Technical, Economic and Institutional Aspects*, Münster: LIT, 1-24.

Chang, H.-J. (2007) *Bad Samaritans: Rich Nations, Poor Policies and the Threat to the Developing World*, London: Random House.

CPB (2007) *The Effectiveness of the Innovation Voucher 2004 and 2005* (in Dutch), The Hague: Central Planning Bureau.

Kitagawa, F. (2005) 'The Fukuoka Silicon Sea-belt Project – An East Asian Experiment in Developing Transnational Networks', *European Planning Studies*, 13(5): 793-799.

Gerstenberg, O. and C.F. Sabel (2000) 'Directly Deliberative Polyarchy: An Institutional Ideal for Europe?' in C. Joerges andR. Dehousse (eds.) *Good Governance in Europe's Integrated Market*, Oxford: Oxford University Press, 289-341.

Nooteboom, B. (1993) 'Firm Size Effects on Transaction Costs,' *Small Business Economics*, 5: 283-295.

Nooteboom, B. (2004) *Inter-firm Collaboration, Learning and Networks: An Integrated Approach*, London: Routledge.

OECD (2005) *Innovation in the Business Sector*, OECD Economics Departmentworking papers no. 4, Paris: OECD.

Raspe, O., F.G. Van Oort, and P. De Bruijn (2004) *Kennis op de kaart. Ruimtelijke patronen in de kenniseconomie*, Rotterdam: NAi Uitgevers & Ruimtelijke Planbureau.

Ponds, R. and F. Van Oort (2006) *Kennishubs in Nederland. Ruimtelijke Patronen van onderzoekssamenwerking*, Rotterdam/The Hague: NAi Uitgevers/Ruimtelijk Planbureau.

WRR (2007) *Identificatie met Nederland*, Amsterdam: Amsterdam University Press.

For Product Safety Concerns and Information please contact our EU
representative GPSR@taylorandfrancis.com
Taylor & Francis Verlag GmbH, Kaufingerstraße 24, 80331 München, Germany

www.ingramcontent.com/pod-product-compliance
Ingram Content Group UK Ltd.
Pitfield, Milton Keynes, MK11 3LW, UK
UKHW050234200926
475702UK00020B/3025

* 9 7 8 9 0 5 3 5 6 5 8 2 7 *